Lecture Notes in Computer Science 6499

Commenced Publication in 1973
Founding and Former Series Editors:
Gerhard Goos, Juris Hartmanis, and Jan van Leeuwen

James F. Peters Andrzej Skowron
Chien-Chung Chan Jerzy W. Grzymala-Busse
Wojciech P. Ziarko (Eds.)

Transactions on Rough Sets XIII

 Springer

Editors-in-Chief

James F. Peters
University of Manitoba, Winnipeg, Manitoba, Canada
E-mail: jfpeters@ee.umanitoba.ca

Andrzej Skowron
Warsaw University, Warsaw, Poland
E-mail: skowron@mimuw.edu.pl

Guest Editors

Chien-Chung Chan
The University of Akron, OH, USA
E-mail: chan@uakron.edu

Jerzy W. Grzymala-Busse
The University of Kansas, Lawrence, KS, USA
E-mail: jerzy@eecs.ku.edu

Wojciech P. Ziarko
University of Regina, SK, Canada
E-mail: ziarko@cs.uregina.ca

ISSN 0302-9743 (LNCS)   e-ISSN 1611-3349 (LNCS)
ISSN 1861-2059 (TRS)   e-ISSN 1861-2067 (TRS)
ISBN 978-3-642-18301-0   e-ISBN 978-3-642-18302-7
DOI 10.1007/978-3-642-18302-7
Springer Heidelberg Dordrecht London New York

Library of Congress Control Number: 2010942518

CR Subject Classification (1998): I.2, H.3, H.2.8, I.4, F.2, G.2

Typesetting: Camera-ready by author, data conversion by Scientific Publishing Services, Chennai, India

Printed on acid-free paper

Springer is part of Springer Science+Business Media (www.springer.com)

Preface

Volume XIII of the *Transactions on Rough Sets* (TRS) consists of extended versions of selected papers presented at the Rough Sets and Current Trends in Computing Conference (RSCTC 2008) organized in Akron, OH, USA, in October 2008 that are part of *Lecture Notes in Artificial Intelligence* volume 5306 edited by Chien-Chung Chan, Jerzy W. Grzymala-Busse and Wojciech P. Ziarko; it also includes some regular papers. The selection of papers accepted for RSCTC 2008 was made by the editors and the authors were invited to submit extended versions to this issue. The 13 submissions received on this invitation went through two rounds of reviews, and 10 papers were accepted for publication.

The editors of the special issue are particularly grateful to all the authors of submitted papers, and to the following reviewers: Salvatore Greco, Masahiro Inuiguchi, Wojciech Jaworski, Richard Jensen, Łukasz Kobyliński, Tianrui Li, Tsau Young Lin, Pawan Lingras, Dun Liu, Dariusz Maylyszko, Georg Peters, Sheela Ramanna, Jerzy Stefanowski, Marcin Szczuka, Guoyin Wang, Wei-Zhi Wu and JingTao Yao. Their laudable efforts made possible a careful selection and revision of submitted manuscripts.

The articles of this special issue on *Rough Sets and Current Trends in Computing* introduce a number of new advances in both the foundations and the applications of rough sets. The advances in the foundations concern mathematical structures of generalized rough sets in infinite universes, approximations of arbitrary binary relations, and attribute reduction in decision-theoretic rough sets. Methodological advances introduce rough set-based and hybrid methodologies for learning theory, attribution reduction, decision analysis, risk assessment, and data-mining tasks such as classification and clustering. In addition, this volume contains regular articles on mining temporal software metrics data, C-GAME discretization method, perceptual tolerance intersection as an example of a near set operation and compression of spatial data with quadtree structures.

The editors and authors of this volume extend their gratitude to Alfred Hofmann, Anna Kramer, Ursula Barth, Christine Reiss and the LNCS staff at Springer for their support in making this volume of the TRS possible.

The Editors-in-Chief have been supported by the State Committee for Scientific Research of the Republic of Poland (KBN) research grants N N516 368334, N N516 077837, the Natural Sciences and Engineering Research Council of Canada (NSERC) research grant 185986, Canadian Network of Excellence (NCE), and a Canadian Arthritis Network (CAN) grant SRI-BIO-05.

September 2010

Chien-Chung Chan
Jerzy W. Grzymala-Busse
Wojciech P. Ziarko
James F. Peters
Andrzej Skowron

LNCS Transactions on Rough Sets

The *Transactions on Rough Sets* series has as its principal aim the fostering of professional exchanges between scientists and practitioners who are interested in the foundations and applications of rough sets. Topics include foundations and applications of rough sets as well as foundations and applications of hybrid methods combining rough sets with other approaches important for the development of intelligent systems. The journal includes high-quality research articles accepted for publication on the basis of thorough peer reviews. Dissertations and monographs up to 250 pages that include new research results can also be considered as regular papers. Extended and revised versions of selected papers from conferences can also be included in regular or special issues of the journal.

Table of Contents

Bit-Vector Representation of Dominance-Based Approximation Space

Chien-Chung Chan[1] and Gwo-Hshiung Tzeng[2,3]

[1] Department of Computer Science, University of Akron
Akron, OH, 44325-4003, USA
chan@uakron.edu

[2] Department of Business and Entrepreneurial Administration, Kainan University
No. 1 Kainan Road, Luchu, Taoyuan County 338, Taiwan
ghtzeng@mail.knu.edu.tw

[3] Institute of Management of Technology, National Chiao Tung University
1001 Ta-Hsueh Road, Hsinchu 300, Taiwan
ghtzeng@cc.nctu.edu.tw

Abstract. Dominance-based Rough Set Approach (DRSA) introduced by Greco et al. is an extension of Pawlak's classical rough set theory by using dominance relations in place of equivalence relations for approximating sets of preference ordered decision classes. The elementary granules in DRSA are P-dominating and P-dominated sets. Recently, Chan and Tzeng introduced the concept of indexed blocks for representing dominance-based approximation space with generalized dominance relations on evaluations of objects. This paper shows how to derive indexed blocks from P-dominating and P-dominated sets in DRSA. Approximations are generalized to any family of decision classes in terms of indexed blocks formulated as binary neighborhood systems. We present algorithms for generating indexed blocks from multi-criteria decision tables and for encoding indexed blocks as bit-vectors to facilitate the computation of approximations and rule generation. A new form of representing decision rules by using interval and set-difference operators is introduced, and we give a procedure of how to generate this type of rules that can be implemented as SQL queries.

Keywords: Rough sets, Dominance-based rough sets, Multiple criteria decision analysis (MCDA), Neighborhood systems, Granular computing.

1 Introduction

The Dominance-Based Rough Set Approach (DRSA) to multiple criteria decision analysis was introduced by Greco, Matarazzo and Slowinski [1, 2, 3] as an extension of Pawlak's classical rough sets (CRS) [8, 9, 10]. In DRSA, attributes with preference-ordered domains are called criteria, and the indiscernibility relation in CRS is generalized to a dominance relation that is reflexive and transitive. It is also assumed that decision classes are ordered by some preference ordering. The fundamental principle in DRSA is that the assignment of decision classes

J.F. Peters et al. (Eds.): Transactions on Rough Sets XIII, LNCS 6499, pp. 1–16, 2011.
© Springer-Verlag Berlin Heidelberg 2011

to objects based on evaluation values of given criteria follows the monotonic dominance principle. The dominance principle requires that if criteria values of object x are better than those of object y, then x should be assigned to a class not worse than y. In DRSA, an approximation space of decision classes is defined by dominating and dominated sets of objects. Let x be an object and P be a set of evaluation criteria, then the P-dominating set of x consists of those objects with evaluation values at least as good as x's with respect to P, and objects with evaluation values at most as good as x's belong to the P-dominated set of x. In DRSA, decision classes considered for approximations are upward and downward unions of decision classes. It has been shown to be an effective tool for multiple criteria decision analysis problems [12] and has been applied to solve multiple criteria sorting problems [4, 5].

The representation of dominance-based approximation space by a family of indexed blocks was introduced in [13]. Indexed blocks are sets of objects indexed by pairs of decision values. The basic idea is to use a binary relation on decision values as indices for grouping objects in light of dominance principle where inconsistency is defined as a result of violating the dominance principle. A set of ordered pairs is derived from objects violating dominance principle involving a pair of decision values, which is used as the index for the set of ordered pairs. For example, objects that are consistently assigned to decision class i form a set of ordered pairs with decision index (i, i). A set of ordered pairs with decision index (i, j), $i \neq j$, corresponds to objects that violate dominance principle with respect to decision values i and j. Each indexed set of ordered pairs induces a set of objects, called a block, which is indexed by the same pair of decision values. These blocks are called indexed blocks, which are elementary sets (granules) of a dominance-based approximation space. In this approach, approximations of any union of decision classes are computed by neighborhoods of indexed blocks.

In this paper, indexed blocks are formulated in terms of binary neighborhood systems [15]. We show how to derive indexed blocks from P-dominating and P-dominated sets. Basically, an object y in P-dominating of x or P-dominated set of x is in a inconsistent indexed block $B(i, j)$, if the decision class assignment to x and y violates the dominance principle. Strategies for generating indexed blocks from multi-criteria decision tables were introduced in [13]. Here, we introduce algorithms to compute indexed blocks from inconsistent ordered pairs. To facilitate the computation of approximations and the generation of decision rules, bit-vector encodings for indexed blocks is introduced. In our approach, conditional elements of decision rules are formulated as conjuncts of intervals, which may be modified with EXCEPT clauses corresponding to set-difference operators. This type of rules is suitable for SQL (Standard Query Language)implementations. We introduce a procedure to generate these conjuncts from approximations of any unions of decision classes by using bit-vector representation of indexed blocks.

The remainder of this paper is organized as follows. Related concepts of rough sets, dominance-based rough sets, and indexed blocks are given in Section 2. The relationship between indexed blocks and dominating and dominated sets and definitions of approximations by neighborhoods of indexed blocks are given in this section. In Section 3, we introduce algorithms for computing indexed blocks from multi-criteria decision tables. In Section 4, we consider how to generate decision rules from indexed blocks. The bit-vector encoding of indexed blocks is introduced, and a strategy is presented for deriving decision rules formulated by using interval and set-difference operators. Finally, conclusions are given in Section 5.

2   Related Concepts

2.1   Information Systems and Rough Sets

In rough set theory, information of objects in a domain is represented by an information system $IS = (U, A, V, f)$, where U is a finite set of objects, A is a finite set of attributes, $V = \cup_{q \in A} V_q$ and V_q is the domain of attribute q, and $f : U \times A \to V$ is a total information function such that $f(x, q) \in V_q$ for every $q \in A$ and $x \in U$. In many applications, we use a special case of information systems called *decision tables* to represent data sets. In a decision table $(U, C \cup D = \{d\})$, there is a designated attribute d called *decision attribute* and attributes in C are called *condition attributes*. Each attribute q in $C \cup D$ is associated with an equivalence relation R_q on the set U of objects such that for each x and $y \in U$, $x R_q y$ means $f(x, q) = f(y, q)$. For each x and $y \in U$, we say that x and y are *indiscernible* on attributes $P \subseteq C$ if and only if $x R_q y$ for all $q \in P$.

2.2   Dominance-Based Rough Sets

Dominance-based rough set was introduced by Greco et al., where attributes with totally ordered domains are called *criteria* which are associated with outranking relations [8]. Let q be a criterion in C and S_q be an outranking relation on the set U of objects such that for each x and $y \in U$, $x S_q y$ means $f(x, q) \geqslant f(y, q)$. For each x and $y \in U$, it is said that x *dominates* y with respect to $P \subseteq C$, denoted by $x D_P y$, if $x S_q y$ for all $q \in P$. The dominance relation is reflexive and transitive. The equivalence relation in Pawlak's rough sets is replaced by the dominance relation D_P. Furthermore, assuming that the set U of objects is partitioned into a finite number of classes by the decision attribute d, let $Cl = \{Cl_t | t \in T\}, T = \{1, 2, ..., n\}$, be a set of decision classes such that for each $x \in U$, x belongs to one and only one $Cl_t \in Cl$ and for all r, s in T, if $r > s$, the decision from Cl_r is preferred to the decision from Cl_s. The fundamental assumption of applying Dominance-base Rough Set Approach (DRSA) to multi-criteria decision analysis is that the process of decision making follows the "dominance principle", namely, the assignment of decision classes to objects based on evaluations of their condition criteria should be monotonic. More precisely, the assignment of decision classes to objects $x, y \in U$ obeys the dominance principle based on a set of criteria $P \subseteq C$ when (1) $x D_P y$ if and only if $f(x, q) \geqslant f(y, q)$ for all $q \in P$ and (2) if $x D_P y$, then $f(x, d) \geqslant f(y, d)$.

In addition, it is useful to consider decision classes with upward and downward unions, which are defined as:

$$Cl_t^{\geq} = \cup_{s \geq t} Cl_s, \quad Cl_t^{\leq} = \cup_{s \leq t} Cl_s, \quad t = 1, 2, ..., n.$$

An object x is in $Cl_t^{\geq}$ means that x at least belongs to class Cl_t, and x is in $Cl_t^{\leq}$ means that x at most belongs to class Cl_t.

Given $P \subseteq C$ and $x \in U$, the elementary sets in dominance-based rough sets are $D_P^+(x)$ and $D_P^-(x)$, which are called P-*dominating set* of x and P-*dominated set* of x, respectively, and they are defined as:

$$D_P^+(x) = \{y \in U: yD_Px\} \text{ and } D_P^-(x) = \{y \in U: xD_Py\}.$$

The P-*lower* and P-*upper approximation* of upward union $Cl_t^{\geq}$ are defined by using the P-*dominating set* of x as:

$$\underline{P}(Cl_t^{\geq}) = \{x \in U : D_P^+(x) \subseteq Cl_t^{\geq}\}, \overline{P}(Cl_t^{\geq}) = \bigcup_{x \in Cl_t^{\geq}} D_P^+(x), \text{ for } t = 1, ..., n.$$

The P-*lower* and P-*upper approximation* of downward union $Cl_t^{\leq}$ are defined by using the P-*dominated set* of x as:

$$\underline{P}(Cl_t^{\leq}) = \{x \in U : D_P^-(x) \subseteq Cl_t^{\leq}\}, \overline{P}(Cl_t^{\leq}) = \bigcup_{x \in Cl_t^{\leq}} D_P^-(x), \text{ for } t = 1, ..., n.$$

2.3   Indexed Blocks as Granules

The concept of indexed blocks was introduced in [13]. Indexed blocks are subsets of U indexed by pairs of decision values. They are computed by using a generalized dominance relation, and the assignment of decision classes to objects $x, y \in U$ obeys the dominance principle with respect to a set of criteria $P \subseteq C$ when (1) xD_Py if and only if $f(x, q) \geq f(y, q)$ for some $q \in P$ and (2) if xD_Py, then $f(x, d) \geq f(y, d)$. Consequently, blocks indexed by decision value pairs $(i, j), i \neq j$, denote objects with inconsistent class assignment related to decision values i and j. Objects in blocks $B(i, i)$ are considered satisfying the dominance principle, even though they may be incompatible. The family of indexed blocks derived from a Multi-Criteria Decision Table (MCDT) is a binary neighborhood system [14]. The use of neighborhood systems to study approximations of database and knowledge-based systems was first introduced by Lin [15]. Let V and U be two sets. A binary neighborhood system is a mapping $B : V \rightarrow 2^U$. Each binary relation of $V \times U$ corresponds to one binary neighborhood system [16].

Let $(U, C \cup D = \{d\})$ be a multi-criteria decision table with condition criteria C and decision attribute d. Decision class labeled by decision value i is denoted as D_i. Let B be a mapping: $V_D \times V_D \rightarrow 2^U$ defined as follows.

$B(i, i) = D_i - \bigcup_{j \neq i} B(i, j)$, and
$B(i, j) = \{x \in D_i : \exists y \in D_j, i \neq j, \text{ such that } \forall q \in C, f(x, q) \geq f(y, q)$
  and $i < j$, or $f(x, q) \leq f(y, q)$ and $i > j\}$.

A block $B(i, j)$ indexed by the decision value pair (i, j) is called *inconsistent indexed block*, and it contains objects violating the dominance principle when evaluating all criteria in C with respect to the decision pair $(i, j), i \neq j$. It is clear that indexed blocks can be derived from the P-dominating and P-dominated sets as follows.

$$B(i,j) = \{x \in D_i : \exists y \in D_j, i \neq j, \text{ such that } y \in D_P^+(x) \text{ and } i > j, \text{ or }$$
$$y \in D_P^-(x) \text{ and } i < j\}.$$

The mapping B or equivalently the family of indexed blocks $\{B(i,j)\}$ derived from the mapping is a binary neighborhood system. The indexed blocks are called *elementary neighborhoods* or *elementary granules*. Let $P \subseteq C$ and let B_P denote the restriction of the mapping B to P, then B_P (equivalently, $\{B_P(i,j)\}$) is also a binary neighborhood system.

Furthermore, a second level binary neighborhood system on $\{B(i,j)\}$ is a mapping $BNS : \{B(i,j)\} \to \wp(2^U)$ defined as, for all $i, j \in D$,

$$BNS(B(i,j)) = \{B(k,l) : \{k,l\} \cap \{i,j\} \neq \varnothing, \forall k, l \in D\}.$$

The exclusive neighborhood of an indexed block $B(i,j)$ is defined as $BNS(B(i,j)) - B(i,j)$, for all i and j.

2.4 Approximations by Neighborhoods of Indexed Blocks

By a singleton MCDT we mean a multi-criteria decision table where criteria are assigned with singleton value. In the following, we will consider definability of decision classes in a singleton MCDT represented by a binary neighborhood system of indexed blocks [14].

Let $S = (U, C \cup D = \{d\})$ be a MCDT. Let $K \subseteq V_D$ and $\neg K = V_D - K$. Let $D_K = \bigcup_{i \in K} D_i$ and $D_{\neg K} = \bigcup_{i \in \neg K} D_i$. Then, S is a singleton MCDT table if and only if for $i \in V_D$, $D_{\{i\}} \cap D_{\neg\{i\}} = \varnothing$ and $D_{\{i\}} \cup D_{\neg\{i\}} = U$.

Let $B = \{B(i,j)\}$ be the binary neighborhood system derived from a MCDT table, and let $K \subseteq V_D$.

The upper approximation of D_K by B is defined as

$$\overline{B}D_K = \bigcup\{B(i,j) : B(i,j) \cap D_K \neq \varnothing\} = \bigcup\{B(i,j) : \{i,j\} \cap K \neq \varnothing\}.$$

Similarly, the upper approximation of $D_{\neg K}$ by B is defined as

$$\overline{B}D_{\neg K} = \bigcup\{B(i,j) : B(i,j) \cap D_{\neg K} \neq \varnothing\}$$
$$= \bigcup\{B(i,j) : \{i,j\} \cap \neg K \neq \varnothing\}.$$

The boundary set of D_K by B is defined as

$$BN_B(D_K) = \overline{B}D_K \cap \overline{B}D_{\neg K} = BN_B(D_{\neg K}).$$

The lower approximation of D_K by B is defined as

$$\underline{B}D_K = \bigcup\{B(i,j) : B(i,j) \subseteq D_K\} - BN_B(D_K)$$
$$= D_K - BN_B(D_K),$$

and the lower approximation of $D_{\neg K}$ by B is defined as

$$\underline{B}D_{\neg K} = \bigcup \{B(i,j) : B(i,j) \subseteq D_{\neg K}\} - BN_B(D_{\neg K})$$
$$= D_{\neg K} - BN_B(D_{\neg K}).$$

From the above definitions, we say that decision class D_K is B-*definable* if and only if $\underline{B}D_K = \overline{B}D_K$ if and only if $\underline{B}D_{\neg K} = \overline{B}D_{\neg K}$.

Example 1. Consider the following multi-criteria decision table, where U is the universe of objects, q_1 and q_2 are condition criteria and d is the decision with preference ordering $3 > 2 > 1$. Table 2 shows the indexed blocks $\{B(i,j)\}$ derived from Table 1.

Table 1. Example of a multi-criteria decision table

U	q_1	q_2	d
x_1	1	2	2
x_2	1.5	1	1
x_3	2	2	1
x_4	1	1.5	1
x_5	2.5	3	2
x_6	3	2.5	3
x_7	2	2	3
x_8	3	3	3

Table 2. Indexed blocks $\{B(i,j)\}$

$D \times D$	1	2	3
1	$\{2, 4\}$	$\{1, 3\}$	$\{3, 7\}$
2		$\varnothing$	$\{5, 7\}$
3			$\{6, 8\}$

From Table 2, we have the following neighborhoods of indexed blocks:

$$BNS(B(1,1)) = \{B(1,1), B(1,2), B(1,3)\},$$

$$BNS(B(1,2)) = \{B(1,1), B(1,2), B(1,3), B(2,3)\},$$

$$BNS(B(1,3)) = \{B(1,1), B(1,2), B(1,3), B(2,3), B(3,3)\},$$

$$BNS(B(2,3)) = \{B(1,2), B(1,3), B(2,3), B(3,3)\}, and$$

$$BNS(B(3,3)) = \{B(1,3), B(2,3), B(3,3)\}.$$

Example 2. From the MCDT given in Table 1, the dominating and dominated sets for each object $x \in U$ are shown in Table 3. Table 4 shows objects with inconsistent decision values based on Table 3. The column labeled with $I^+(x)$ denotes objects in $D_P^+(x)$ that are assigned with an inconsistent decision value.

Table 3. Dominating and dominated sets derived from Table 1

U	$D_P^+(x)$	$D_P^-(x)$	d
1	$\{1, 3, 5, 6, 7, 8\}$	$\{1, 4\}$	2
2	$\{2, 3, 5, 6, 7, 8\}$	$\{2\}$	1
3	$\{3, 5, 6, 7, 8\}$	$\{1, 2, 3, 4, 7\}$	1
4	$\{1, 3, 4, 5, 6, 7, 8\}$	$\{4\}$	1
5	$\{5, 8\}$	$\{1, 2, 3, 4, 5, 7\}$	2
6	$\{6, 8\}$	$\{1, 2, 3, 4, 5, 7\}$	3
7	$\{3, 5, 6, 7, 8\}$	$\{1, 2, 3, 4, 7\}$	3
8	$\{8\}$	U	3

Table 4. Objects violating dominance principle based on Table 3

U	$I^+(x)$	$I^-(x)$	d
1	$\{3\}$	$\varnothing$	2
2	$\varnothing$	$\varnothing$	1
3	$\varnothing$	$\{1, 7\}$	1
4	$\varnothing$	$\varnothing$	1
5	$\varnothing$	$\{7\}$	2
6	$\varnothing$	$\varnothing$	3
7	$\{3, 5\}$	$\varnothing$	3
8	$\varnothing$	$\varnothing$	3

Similarly, $I^-(x)$ column denotes those objects in $D_P^-(x)$ with inconsistent decision class assignment. It is clear that we can derive the indexed blocks shown in Table 2 from Table 4.

Example 3. Consider the singleton MCDT shown in Table 1 and its indexed blocks in Table 2. The decision classes are $D_1 = \{x_2, x_3, x_4\}$, $D_2 = \{x_1, x_5\}$, and $D_3 = \{x_6, x_7, x_8\}$. Let $K = \{1, 2\}$, then $\neg K = \{3\}$. We have $D_K = \{x_1, x_2, x_3, x_4, x_5\}$ and $D_{\neg K} = \{x_6, x_7, x_8\}$. The upper approximations are

$$\overline{B}D_K = \overline{B}D_{\{1,2\}} = B(1,1) \cup B(1,2) \cup B(1,3) \cup B(2,3) = \{x_1, x_2, x_3, x_4, x_5, x_7\},$$

$$\overline{B}D_{\neg K} = \overline{B}D_{\{3\}} = B(1,3) \cup B(2,3) \cup B(3,3) = (x_3, x_5, x_6, x_7, x_8\}.$$

The boundary sets are $BN_B(D_K) = \overline{B}D_K \cap \overline{B}D_{\neg K} = BN_B(D_{\neg K}) = \{x_3, x_5, x_7\}$, and the lower approximations are

$$\underline{B}D_K = \bigcup\{B(i,j) : B(i,j) \subseteq D_K\} - BN_B(D_K)$$
$$= D_K - BN_B(D_K) = \{x_1, x_2, x_4\}, and$$

$$\underline{B}D_{\neg K} = \bigcup\{B(i,j) : B(i,j) \subseteq D_{\neg K}\} - BN_B(D_{\neg K})$$
$$= D_{\neg K} - BN_B(D_{\neg K}) = \{x_6, x_8\}.$$

3   Computing Indexed Blocks from MCDT

In this section, we show how to compute indexed blocks from a MCDT. We will first consider using single criterion, and then, we will consider the case of multiple criteria.

3.1   Based on Single Criterion

Let q be a criterion in C, and let $I_q(i)$ denote the set of values of criterion q for objects in decision class D_i. Let $min_q(i)$ denote the minimum value of $I_q(i)$, and let $max_q(i)$ denote the maximum value, then $I_q(i) = [min_q(i), max_q(i)]$ is the interval with $min_q(i)$ and $max_q(i)$ as its lower and upper boundary values. From $I_q(i)$, intervals related to pairs of decision values is defined by the mapping $I_q(i,j) : D \times D \to \wp(V_q)$ as:

$$I_q(i,j) = I_q(i) \cap I_q(j) \text{ and } I_q(i,i) = I_q(i) - \bigcup_{j \neq i} I_q(i,j)$$

where $D = V_D$ and $\wp(V_q)$ denotes the power set of V_q.

Intuitively, the set $I_q(i,i)$ denotes the values of criterion q where objects can be consistently labeled with decision value i. For $i \neq j$, values in $I_q(i,j) = I_q(j,i)$ are conflicting or inconsistent in the sense that objects with higher values of criterion q are assigned to a lower decision class or vice versa, namely, the dominance principle is violated.

For each $I_q(i,j)$ and $i \neq j$, the corresponding set of *ordered pairs* $[I_q(i,j)]$: $D \times D \to \wp(U \times U)$ is defined as

$$[I_q(i,j)] = \{(x,y) \in U \times U : f(x,d) = i, f(y,d) = j \text{ such that}$$
$$f(x,q) \geqslant f(y,q) \text{ for } f(x,q), f(y,q) \in I_q(i,j)\}.$$

For simplicity, the set $[I_q(i,i)]$ is taken to be reflexive. For each set $[I_q(i,j)]$ of ordered pairs, the restrictions of $[I_q(i,j)]$ to i and j are defined as:

$$[I_q(i,j)]_i = \{x \in U : \text{there exists } y \in U \text{ such that } (x, y) \in [I_q(i,j)]\} \text{ and}$$
$$[I_q(i,j)]_j = \{y \in U : \text{there exists } x \in U \text{ such that } (x, y) \in [I_q(i,j)]\}.$$

The corresponding indexed block $B_q(i,j) \subseteq U$ of $[I_q(i,j)]$ can be derived by

$$B_q(i,j) = [I_q(i,j)]_i \cup [I_q(i,j)]_j.$$

3.2   Based on Multiple Criteria

Three rules for updating indexed blocks by combining criteria incrementally were introduced in [13]. Here, we propose an update procedure consisting of two steps based on the idea of neighborhood systems defined in Subsection 2.3. To combine two criteria q_1 and q_2, the first step is to update sets of ordered pairs for inconsistent decision value pairs as $[I_{\{q_1,q_2\}}(i,j)] = [I_{q_1}(i,j)] \cap [I_{q_2}(i,j)]$. From the ordered pair set $[I_{\{q_1,q_2\}}(i,j)]$, we can obtain the indexed block $B_{\{q_1,q_2\}}(i,j)$ by $[I_{\{q_1,q_2\}}(i,j)]_i \cup [I_{\{q_1,q_2\}}(i,j)]_j$. The second step is to update indexed block $B_{\{q_1,q_2\}}(i,i)$ by $D_i - ENB(B_{\{q_1,q_2\}}(i,i))$ where ENB is the exclusive neighborhood of $B_{\{q_1,q_2\}}(i,i)$.

Based on the incremental update procedure for indexed blocks, we have the following procedure for generating an indexed block table from a MCDT.

procedure Generate_Indexed_Block_Table
inputs: a MCDT with criteria $q_1, ..., q_m$ and decision criterion d with a finite number K of decision classes;
outputs: an indexed block table IBT for the MCDT;
begin
for $q_i = 1$ to m **do**
  begin
    Sort the MCDT in terms of d, followed by q_i;
    for $i = 1$ to K **do**
    for $j = i$ to K **do**
      Compute inconsistent intervals $I_{q_i}(i, j)$;
  end; //for
for $q_i = 1$ to m **do**
  for $i = 1$ to K **do**
    for $j = i$ to K **do**
    Generate sets of ordered pairs $[I_{q_i}(i, j)]$;
//Combine ordered pair sets by multiple criteria
for $q_i = 2$ to m **do**
  for $i = 1$ to K **do**
    for $j = i$ to K **do**
      $[I_C(i, j)] = [I_{q_1}(i, j)] \cap [I_{q_i}(i, j)]$;
for $i = 1$ to K **do**
  for $j = i + 1$ to K **do**
    Generate inconsistent blocks $B(i, j)$ from $[I_C(i, j)]$;
for $d_i = 1$ to K **do**
    // Generate indexed blocks from its exclusive neighborhood by
    $B(i, i) = D_i - ENB(B_C(i, i))$;
end; //Generate_Indexed_Block_Table

Example 4. For convenience, we use the multi-criteria decision table taken from [6], which is shown in Table 5. The result of applying sorting to the table on decision attribute d followed by sorting on q_1 is shown in Table 6. The corresponding inconsistent intervals for criterion q_1 are shown in Table 7. Similarly, inconsistent intervals for criteria q_3 and q_3 can be obtained, and they are shown in Tables 8 and 9.

From the tables, we have the sets of ordered pairs for criterion q_1 as:

$[I_{q1}(1, 1)] = [I_{q1}(2, 2)] = \varnothing$,
$[I_{q1}(1, 2)] = \{(4, 12), (3, 12), (7, 12), (7, 6), (9, 12), (9, 6), (14, 12), (14, 6), (14, 13), (14, 1), (14, 2), (14, 10), (14, 15)\}$,
$[I_{q1}(1, 3)] = \varnothing$,
$[I_{q1}(2, 3)] = \{(11, 8), (11, 16)\}$,
$[I_{q1}(3, 3)] = \{(17, 17), (5, 5)\}$.

The sets of ordered pairs for criterion q_2 are:

$[I_{q2}(1, 1)] = \{(4, 4)\}$,
$[I_{q2}(2, 2)] = \varnothing$,

$[I_{q2}(1, 2)] = \{(7, 13), (3, 13), (3, 16), (14, 13), (14, 6), (9, 13), (9, 6), (9, 10),$
$(9, 1), (9, 12)\}$,
$[I_{q2}(1, 3)] = \varnothing$,
$[I_{q2}(2, 3)] = \{(11, 8), (11, 16), (15, 8), (15, 16), (15, 5), (2, 8), (2, 16), (2, 5)\}$,
$[I_{q2}(3, 3)] = \{(17, 17)\}$.

Table 5. A multi-criteria decision table

U	q_1	q_2	q_3	d
1	1.5	3	12	2
2	1.7	5	9.5	2
3	0.5	2	2.5	1
4	0.7	0.5	1.5	1
5	3	4.3	9	3
6	1	2	4.5	2
7	1	1.2	8	1
8	2.3	3.3	9	3
9	1	3	5	1
10	1.7	2.8	3.5	2
11	2.5	4	11	2
12	0.5	3	6	2
13	1.2	1	7	2
14	2	2.4	6	1
15	1.9	4.3	14	2
16	2.3	4	13	3
17	2.7	5.5	15	3

Table 6. Result after sorting in terms of d followed by q_1

U	q_1	d
3	0.5	1
4	0.7	1
7	1	1
9	1	1
14	2	1
12	0.5	2
6	1	2
13	1.2	2
1	1.5	2
2	1.7	2
10	1.7	2
15	1.9	2
11	2.5	2
8	2.3	3
16	2.3	3
17	2.7	3
5	3	3

Table 7. Inconsistent intervals $I_{q_1}(i, j)$

$D \times D$	*1*	*2*	*3*
1	[ ]	[0.5, 2]	[ ]
2		[ ]	[2.3, 2.5]
3			[2.7, 3.0]

Table 8. Inconsistent intervals $I_{q_2}(i, j)$

$D \times D$	*1*	*2*	*3*
1	[0.5, 0.5]	[1, 3]	[ ]
2		[ ]	[3.3, 5]
3			[5.5, 5.5]

Table 9. Inconsistent intervals $I_{q_3}(i, j)$

$D \times D$	*1*	*2*	*3*
1	[1.5, 2.5]	[3.5, 8]	[ ]
2		[ ]	[9, 14]
3			[15, 15]

Table 10. Inconsistent indexed blocks $B_{\{q_1,q_2,q_3\}}(i, j)$

$D \times D$	**1**	**2**	**3**
1		{6, 9, 14}	∅
2			{8, 11}
3			

And, the sets of ordered pairs for criterion q_3 are:

$[I_{q3}(1, 1)] = \{(4, 4), (3, 3)\}$,
$[I_{q3}(2, 2)] = \emptyset$,
$[I_{q3}(1, 2)] = \{(9, 1), (9, 6), (14, 10), (14, 6), (14, 12), (7, 10), (7, 6), (7, 12), (7, 13)\}$,
$[I_{q3}(1, 3)] = \emptyset$,
$[I_{q3}(2, 3)] = \{(2, 5), (2, 8), (11, 5), (11, 8), (1, 5), (1, 8), (15, 5), (15, 8), (15, 16)\}$,
$[I_{q3}(3, 3)] = \{(17, 17)\}$.

From the above sets of ordered pairs, we can derive their corresponding ordered pair sets by combining criteria q_1, q_2, and q_3, and we have

$[I_{\{q_1,q_2,q_3\}}(1, 2)] = \{(14, 6), (9, 6)\}$,
$[I_{\{q_1,q_2,q_3\}}(1, 3)] = \emptyset$,
$[I_{\{q_1,q_2,q_3\}}(2, 3)] = \{(11, 8)\}$.

Table 11. Indexed blocks $B_{\{q_1,q_2,q_3\}}(i,j)$

$D \times D$	1	2	3
1	{3, 4, 7}	{6, 9, 14}	∅
2		{1, 2, 10, 12, 13, 15}	{8, 11}
3			{5, 16, 17}

Then, we can derive inconsistent blocks from the above combined ordered pair sets. The result is shown in Table 10. From Table 10, consistent indexed blocks can be computed by using corresponding exclusive neighborhoods. Finally, the set of indexed blocks after combining q_1, q_2, and q_3 is shown in Table 11. It represents the dominance-based approximation space generated by the criteria $\{q_1, q_2, q_3\}$ from the multi-criteria decision table shown in Table 5.

4   Generate Decision Rules from Indexed Blocks

In rough set approach, approximations are used to generate rules from decision tables. In the following, we consider representing indexed blocks as bit-vectors to facilitate generation of decision rules by SQL queries.

4.1   Bit-Vector Encodings of Indexed Blocks

For a multi-criteria decision table with N objects and K decision values, each object is encoded as a bit-vector of K bits. The encoding of a decision table can be represented by an $N \times K$ Boolean matrix generated from the table of indexed blocks as follows. Each indexed block is encoded by a vector of N bits where 1 means an object is in the block. The encoding for decision class i is computed by taking a logical OR of indexed blocks in the i-th row and i-th column of the indexed block table.

Let each decision class be encoded as a vector v of K bits where the i-th bit is set to 1 for a decision class with decision value i. A vector of all zero bits is called a *NULL* vector, denoted by **0**. Two bit-vectors v_1 and v_2 are *compatible*, if the logical AND of v_1 and v_2 is not *NULL*, i.e., $v_1 \wedge v_2 \neq \mathbf{0}$. Two compatible vectors v_1 and v_2 are *equal* if they have exactly the same bit patterns, i.e., $v_1 \wedge v_2 = v_1 = v_2$. A bit-vector v_1 is a *subvector* of v_2, $v_1 \subseteq v_2$, if $v_1[i] = 1$ then $v_2[i] = 1$ for all bits $i = 1, \ldots, K$.

Let $(U, C \cup D = \{d\})$ be a multi-criteria decision table. Let $E(x)$ denote the decision bit-vector encoding of object $x \in U$. Then, the lower approximation $\underline{C}v$ and upper approximation $\overline{C}v$ for an encoded decision class v are defined as

$\underline{C}v = \{x | E(x) \subseteq v\}$ and
$\overline{C}v = \{x | E(x) \wedge v \neq \mathbf{0}\}$.

The boundary set of v is $BN(v) = \overline{C}v - \underline{C}v$.

Example 5. Consider the indexed block table shown in Table 11. The set of objects in the indexed block (1, 1) is {3, 4, 7}. The encoding for the block is shown in the following table:

1	2	3	4	5	6	7	8	9	10	11	12	13	14	15	16	17
0	0	1	1	0	0	1	0	0	0	0	0	0	0	0	0	0

Similarly, the encoding for the indexed block (1, 2) = {6, 9, 14} is

1	2	3	4	5	6	7	8	9	10	11	12	13	14	15	16	17
0	0	0	0	0	1	0	0	1	0	0	0	0	1	0	0	0

The result of taking a logical OR of the above two vectors is

1	2	3	4	5	6	7	8	9	10	11	12	13	14	15	16	17
0	0	1	1	0	1	1	0	1	0	0	0	0	1	0	0	0

Applying the procedure to all indexed blocks of Table 11, the resulted encoding for all three decision values {1, 2, 3} is shown in Table 12.

Example 6. Consider the encodings in Table 12. The bit-vector of decision class 1 is the vector $v_{\{1\}} = (1, 0, 0)$. The following is the lower approximation, upper approximation, and boundary set of the decision class:

$\underline{C}v_{\{1\}} = \{3, 4, 7\}$,
$\overline{C}v_{\{1\}} = \{3, 4, 6, 7, 9, 14\}$, and
$BN(v_{\{1\}}) = \{6, 9, 14\}$.

The bit-vector of decision class 2 is the vector $v_{\{2\}} = (0, 1, 0)$. Then,

$\underline{C}v_{\{2\}} = \{1, 2, 10, 12, 13, 15\}$,
$\overline{C}v_{\{2\}} = \{1, 2, 6, 8, 9, 10, 11, 12, 13, 14, 15\}$, and
$BN(v_{\{2\}}) = \{6, 8, 9, 11, 14\}$.

The bit-vector of decision classes {1, 2} is the vector $v_{\{1,2\}} = (1, 1, 0)$. Then,

$\underline{C}v_{\{1,2\}} = \{1, 2, 3, 4, 6, 7, 9, 10, 12, 13, 14, 15\}$,
$\overline{C}v_{\{1,2\}} = \{1, 2, 3, 4, 6, 7, 8, 9, 10, 11, 12, 13, 14, 15\}$, and
$BN(v_{\{1,2\}}) = \{8, 11\}$.

4.2   Decision Rules

Typically, decision rules are derived from description of objects as logical conjuncts of attribute-value pairs. We introduce a form of decision rules by using interval and set operators similar to those constructs available in the SQL query language for relational database systems. The conditional elements of a decision rule is a *conjunct of terms*, and each *term* is an interval represented as (q BETWEEN $min(q)$ and $max(q)$) where q is a condition criterion, $min(q)$ is the lower boundary value, and $max(q)$ is the upper boundary value. In addition, a decision rule may be qualified by exceptional conditions, which correspond to

Table 12. Bit-vector encoding for indexed blocks of Table 11

U	d1	d2	d3
1	0	1	0
2	0	1	0
3	1	0	0
4	1	0	0
5	0	0	1
6	1	1	0
7	1	0	0
8	0	1	1
9	1	1	0
10	0	1	0
11	0	1	1
12	0	1	0
13	0	1	0
14	1	1	0
15	0	1	0
16	0	0	1
17	0	0	1

the set-difference operation of set theory. More precisely, a rule for determining if object x belongs to decision class D_K is of the form:

if $(f(x, q_1)$ BETWEEN $min(q_1)$ and $max(q_1))$ and $(f(x, q_2)$ BETWEEN $min(q_2)$ and $max(q_2))$ and ... $(f(x, q_m)$ BETWEEN $min(q_m)$ and $max(q_m))$
then x is in decision class D_K
EXCEPT when x is in the conjunct $\wedge t_i$, where t_i's are terms denoting exceptional intervals.

The main task of generating rules with the proposed format is to determine lower and upper boundary values for condition criteria such that the query based on conditional element of the rule will return a subset of the decision class D_K. The following is a strategy for identifying those boundary values from lower and upper approximations.

procedure CONJUNCT
inputs: a MCDT with V_D decision classes, $K \subseteq V_D$,
        G = lower approximation or upper approximation of D_K, and
        a bit vector table derived from the MCDT.
outputs: Conditional elements of decision rules for G
begin
   $I_{q_i}(G)$ = the set of min and max boundary values from objects in G;
   $[I_{q_i}(G)]$ = retrieve objects from MCDT using $I_{q_i}(G)$;
   if $([I_{q_i}(G)] - G)$ is empty
   then return $I_{q_i}(G)$ as the conjunct
   else {apply CONJUNCT to the set $[I_{q_i}(G)] - G$ for exceptional conditions;
       return EXCEPT qualified conjunct; }
end;

Example 7. Let us consider the MCDT in Table 5 with bit-vector encodings for its indexed blocks shown in Table 12. Let $K = \{2\}$, then the bit-vector encoding for decision class $D_K = (0, 1, 0)$. From Table 12, the lower approximation of $\underline{B}(D_K) = \{1, 2, 10, 12, 13, 15\}$. Applying the CONJUNCT procedure to target set $G = \underline{B}(D_K)$, we can identify the following set of intervals:

$I_{q_i}(G) = \{(q_1 \text{ BETWEEN } 0.5 \text{ and } 1.9), (q_2 \text{ BETWEEN } 1 \text{ and } 5), (q_3 \text{ BETWEEN } 3.5 \text{ and } 14)\}$.

The objects retrieved by applying a query with the intervals to the MCDT is

$[I_{q_i}(G)] = \{1, 2, 6, 7, 9, 10, 12, 13, 15\}$.

Because $[I_{q_i}(G)] - G = \{6, 7, 9\} \neq \emptyset$, the CONJUNCT procedure is applied to the set $G' = \{6, 7, 9\}$ to find an exceptional conjunct with the following intervals:

$I_{q_i}(G') = \{(q_1 \text{ BETWEEN } 1.0 \text{ and } 1.0), (q_2 \text{ BETWEEN } 1.2 \text{ and } 3), (q_3 \text{ BETWEEN } 4.5 \text{ and } 8)\}$.

Therefore, we have a certain rule for decision class 2 as

Object x is in D_2,
if (q_1 BETWEEN 0.5 and 1.9) and (q_2 BETWEEN 1 and 5) and (q_3 BETWEEN 3.5 and 14)
EXCEPT (q_1 BETWEEN 1.0 and 1.0) and (q_2 BETWEEN 1.2 and 3) and (q_3 BETWEEN 4.5 and 8).

5   Conclusions

In this paper, we presented the formulation of indexed blocks as binary neighborhood systems and showed how to derive indexed blocks from dominating and dominated granules in DRSA. A bit-vector encodings of indexed blocks is introduced to facilitate the generation of decision rules in the form of interval conjuncts with except modifiers. Procedure is introduced for rule generation that can be implemented as SQL queries, which will be implemented and optimized as future work.

References

1. Greco, S., Matarazzo, B., Slowinski, R.: Rough approximation of a preference relation by dominance relations. ICS Research Report 16/96, Warsaw University of Technology, Warsaw (1996); European Journal of Operational Research 117(1), 63–83 (1999)
2. Greco, S., Matarazzo, B., Slowinski, R.: A new rough set approach to evaluation of bankruptcy risk. In: Zopounidis, C. (ed.) Operational Tools in the Management of Financial Risks, pp. 121–136. Kluwer Academic Publishers, Dordrecht (1998)

3. Greco, S., Matarazzo, B., Slowinski, R.: The use of rough sets and fuzzy sets in MCDM. In: Gal, T., Stewart, T., Hanne, T. (eds.) Advances in Multiple Criteria Decisions Making, ch. 14, pp. 14.1–14.59. Kluwer Academic Publishers, Dordrecht (1999)
4. Greco, S., Matarazzo, B., Slowinski, R.: Rough sets theory for multicriteria decision analysis. European Journal of Operational Research 129(1), 1–47 (2001)
5. Greco, S., Matarazzo, B., Slowinski, R.: Rough sets methodology for sorting problems in presence of multiple attributes and criteria. European Journal of Operational Research 138(2), 247–259 (2002)
6. Greco, S., Matarazzo, B., Slowinski, R., Stefanowski, J.: An algorithm for induction of decision rules consistent with the dominance principle. In: Ziarko, W.P., Yao, Y. (eds.) RSCTC 2000. LNCS (LNAI), vol. 2005, pp. 304–313. Springer, Heidelberg (2001)
7. Greco, S., Matarazzo, B., Slowinski, R., Stefanowski, J., Zurawski, M.: Incremental versus non-incremental rule induction for multicriteria classification. In: Peters, J.F., Skowron, A., Dubois, D., Grzymała-Busse, J.W., Inuiguchi, M., Polkowski, L. (eds.) Transactions on Rough Sets II. LNCS, vol. 3135, pp. 33–53. Springer, Heidelberg (2004)
8. Pawlak, Z.: Rough sets: basic notion. International Journal of Computer and Information Science 11(15), 344–356 (1982)
9. Pawlak, Z.: Rough sets and decision tables. In: Skowron, A. (ed.) SCT 1984. LNCS, vol. 208, pp. 186–196. Springer, Heidelberg (1985)
10. Pawlak, Z., Grzymala-Busse, J., Slowinski, R., Ziarko, W.: Rough sets. Communication of ACM 38(11), 89–95 (1995)
11. Roy, B.: Methodologie Multicritere d'Aide a la Decision. Economica, Paris (1985)
12. Fan, T.F., Liu, D.R., Tzeng, G.H.: Rough set-based logics for multicriteria decision analysis. European Journal of Operational Research 182(1), 340–355 (2007)
13. Chan, C.-C., Tzeng, G.H.: Dominance-based rough sets using indexed blocks as granules. Fundamenta Informaticae 94(2), 133–146 (2009)
14. Chan, C.-C., Chang, F.M.: On definability of sets in dominance-based approximation space. In: Proceedings of IEEE SMC Conference, San Antonio, TX, October 2009, pp. 3032–3037 (2009)
15. Lin, T.Y.: Neighborhood systems and approximation in database and knowledge base systems. In: Proceedings of the Fourth International Symposium on Methodologies of Intelligent Systems, Poster Session, October 12-15, pp. 75–86 (1989)
16. Lin, T.Y.: Granular computing on binary relations I: data mining and neighborhood systems. In: Polkowski, L., Skowron, A. (eds.) Rough Sets in Knowledge Discovery, pp. 107–120. Physica-Verlag, Heidelberg (1998)

Approximations of Arbitrary Binary Relations by Partial Orders: Classical and Rough Set Models

Ryszard Janicki[*]

Department of Computing and Software,
McMaster University,
Hamilton, ON, L8S 4K1 Canada
janicki@mcmaster.ca

Abstract. The problem of approximating an arbitrary binary relation by a partial order is formally defined and analysed. Five different partial order approximations of an arbitrary binary relation are provided and their relationships analysed. Both the classical relational algebra model and a model based on the Rough Set paradigm are discussed.

1   Introduction

Consider the following problem: we have a set of data that have been obtained in an empirical manner. From the nature of the problem we know that the set should be partially ordered, but because the data are empirical it is not. In a general case, this relation may be arbitrary. What is the "best" partially ordered approximation of an arbitrary relation and how this approximation can be computed?

It appears that the concept of approximation has two different intuitions in mathematics and science. The first one stems from the fact that all empirical numerical data have some errors, so in reality we do not have an exact value x, but always an interval $(x - \varepsilon, x + \varepsilon)$, i.e. the upper approximation and the lower approximation. Rough Sets [16,17] and various concepts of closure [2,18] exploit this idea for more general data structures.

The second intuition can be illustrated by the *least square approximation* of points in two dimensional plane (c.f. [23]). Here we know or assume that the points should be on a straight line and we are trying to find the line that fits the data best. In this case the data have a structure (points in two dimensional plane, i.e. a binaryrelation) and should satisfied a desired property (be linear). This is even more apparent when the *segmented least square* approximation is considered (c.f. [14]). Note that even if we replace a solution $f(x) = ax + b$ by two lines $f_1(x) = ax + b + \delta$ and $f_2(x) = ax + b - \delta$, where δ is a standard error (c.f. [23]), there is no guarantee that any point resides between $f_1(x)$ and $f_2(x)$. Hence this is not the case of an upper, or lower approximation in any sense. However this approach assumes that there is a well defined concept of a *metric* which allows us to minimize the distance, and this concept is not obvious, and often not even possible for non-numerical objects (see for instance [9]).

---

[*] Partially supported by NSERC grant of Canada.

J.F. Peters et al. (Eds.): Transactions on Rough Sets XIII, LNCS 6499, pp. 17–38, 2011.

The approach presented in this paper is a mixture of both intuitions, there is no metric, but the concept of "minimal distance" is simulated by a sequence of property-driven lower and/or upper approximations.

The first part of this paper is an extended and revised version of [10]. It presents a solution based on the classical theory of relations [15,18]. The question of how such a solution relates to Rough Set Theory was posed in [10] as a research problem. In the second part of this paper we provide some solutions to this problem. Our solution is derived from ideas presented initially in [24,25].

The paper is structured as follows. In the next section we will discuss the problem that provided both the intuition and the major motivation for our solution, namely, the problem of Pairwise Comparisons Non-numerical Ranking. Sections 3 and 4 provide mathematical bases (sets and relations), slightly structured to satisfy our needs. The formal definition and the theory of approximation of arbitrary binary relations by partial orders, within the standard theory of relations, are given in Sections 5 and 6.

Section 7 recalls the classical Rough Set approach to the approximation of relations. In Section 8, probably the most novel in the entire paper, the concept of Property-Driven Rough Approximation is introduced and discussed. In the following two sections, Sections 9 and 10, we use the ideas of Section 8 to approximate an arbitrary binary relation by a partial order. Section 10 contains final comments.

2   Intuition and Motivation: Pairwise Comparisons Non-numerical Ranking

While ranking the importance of *several* objects is often problematic (as the "perfect ranking" often does not exist [1]), it is often much easier when to do restricted to *two* objects. The problem is then reduced to constructing a global ranking from the set of partially ordered pairs. The method could be traced to the Marquis de Condorcet's 1795 paper [13]. At present the numerical version of pairwise comparisons based ranking is practically identified with the controversial Saaty's Analytic Hierarchy Process (AHP, [19]). On one hand AHP has respected practical applications, on the other hand it is still considered by many (see [4]) as a flawed procedure that produces arbitrary rankings. We believe that most of the problems with AHP stem mainly from the following two sources:

1. The final outcome is always expected to be totally ordered (i.e. for all a, b, either $a < b$ or $b > a$),
2. Numbers are used to calculate the final outcome.

An alternative, non-numerical method was proposed in [11] and refined in [8,9]. It is based on the concept of partial order and the concept of *partial order approximation of an arbitrary binary relation*. In [9] the non-numerical approach has been formalised as follows.

A *ranking* is just a partial order $Rank = (X, <^{rank})$, where X is the set of objects to be ranked and $<^{rank}$ is a ranking relation. We assume that $<^{rank}$ is a weak or total order. The ranking relation $<^{rank}$ is unknown and the *ranking problem* is to construct $<^{rank}$ on the basis of *ranking data*.

A *pairwise comparisons ranking data* is a tuple $PCRD = (X, R_0, R_1, ..., R_k)$, where $k \geq 1$, and R_i's are relations satisfying $R_0 \cup R_1 \cup ... \cup R_k = X \times X$ and $R_k \subseteq R_{k-1} \subseteq ... \subseteq R_1$. The relation R_0, interpreted as *indifference*, is symmetric and reflexive, the relations $R_1, ..., R_k$, interpreted as *preferences*, are asymmetric and irreflexive. In [8], the case $PCRD = (X, \approx, \sqsubset, \subset, <, \prec)$, with the following interpretation $a \approx b$: a and b are *indifferent*, $a \sqsubset b$: *slightly in favour* of b, $a \subset b$: *in favour* of b, $a < b$: b is *strongly better*, $a \prec b$: b is *extremely better*, was considered in some detail. The list $\sqsubset, \subset, <, \prec$ may be shorter or longer, but not empty and not much longer (due to limitations of the human mind [3]).

We may now state the *ranking problem* more precisely: "*derive the ranking relation $<^{rank}$ from a given pairwise comparison ranking data PCRD*". Note that in general, *none* of the relations R_i, $i = 1, ..., k$, could be even a partial order. The problem is that X is believed to be partially or weakly ordered by the ranking relation $<^{rank}$ but the data acquisition process may be so influenced by informational noise, imprecision, randomness, or expert ignorance that the collected data $R_1, R_2, ..., R_k$ are only some relations on X. We may say that they give a fuzzy picture of ranking, and to focus it, we must do some pruning and/or extending.

The methods of finding $<^{rank}$ presented in [9,11] are in principle based on the following three concepts

- the partial order approximation $<_R$ of an arbitrary relation R,
- the partial order approximation $<_{(R, \lhd)}$ of a pair of relations R and $\lhd$, where R is an arbitrary relation, $\lhd$ is a partial order included in R, and $\lhd \subseteq <_{(R, \lhd)}$ (the relation $\lhd$ represents the part of R that already is a precise ranking),
- the weak order approximation of a given partial order.

While weak order approximations of partial orders are *upper approximations* (as they are supersets of given partial orders), the least weak order approximation usually does not exist [5]; the partial orders that can intuitively be considered as approximations of an arbitrary relation are usually neither upper nor lower approximations (neither supersets nor subsets of a given relation). It appears that for partial order approximations of arbitrary relations the concepts "upper" and "lower" approximations alone are of limited use. However one may use a sequence of "lower" and/or "upper" approximations as a kind of a "metric". In this way we may obtain several different approximations, and different partial orders could be considered as "the best" approximations in various circumstances.

The approximation of R proposed in [9,11], denoted $(R^+)^\bullet$ in this paper, can be described as follows: "first compute the transitive closure of R, and then remove all cycles from it". The technique could be traced to E. Schröder's 1895 paper [20] and is often considered as the only one solution. It seems to work nicely in many cases [8,9,11], but *not always*.

Consider the following example. Suppose we have four objects a, b, c, d, each of them is characterised by a vector of real numbers $(x_1, ..., x_4)$, so $a = (x_1^a, x_2^a, x_3^a, x_4^a)$, etc. Suppose that the measurements have errors so each x_i is only an estimation. We define the relation $<_{(1)}$ on real numbers as follows $x <_{(1)} y \iff y - x \geq 1$. The relation $<_{(1)}$

is a partial order, in fact it is a semi-order [5] (semi-orders are often used to model cases when errors of data are taken into account). We now define:

$$(x_1, x_2, x_3, x_4) \leftarrow (y_1, y_2, y_3, y_4) \iff (\exists i.\ x_i <_{(1)} y_i) \land (\forall i.\ \neg(y_i <_{(1)} x_i)).$$

In other words, if either $x_i <_{(1)} y_i$ or x_i and y_i are incomparable w.r.t. $<_{(1)}$ and at least for one j, $x_j <_{(1)} y_j$. This looks like a reasonable way of comparing *two* objects. Let

$$a = (1.0, 0.5, 0.5, 0.1),\ b = (0, 1.0, 0.5, 0.5),\ c = (0.5, 0, 1.0, 0.5),\ d = (0.9, 0.5, 0, 0.5).$$

We now have: $d \leftarrow c \leftarrow b \leftarrow a$, but the relation $\leftarrow$ *is not* transitive, as we have $\neg(c \leftarrow a)$, $\neg(d \leftarrow a)$, $\neg(d \leftarrow b)$. Using the technique of [8,11] we obtain the following totally ordered ranking: $d <^{rank} c <^{rank} b <^{rank} a$. However, since all numerical values are only estimates and we can say that one is bigger than another only if the difference between them is greater or equal 1, the rank $<^{rank} = \emptyset$, i.e. a, b, c and d are *incomparable*, is what we would intuitively expect! In this paper we will propose a solution to this problem[1]. Note that for $d = (0.5, 0.5, 0, 1.1)$ and the same a, b, c, we have $a \leftarrow d \leftarrow c \leftarrow b \leftarrow a$, so the technique of [8,11] produces $<^{rank} = \emptyset$, as expected.

In this paper we will introduce and analyse five different kinds of partial order approximations denoted R^{C}, $R^{\mathsf{C} \wedge \bullet}$, $(R^\bullet)^{\mathsf{C}}$, $(R^\bullet)^+$, $(R^+)^\bullet$, respectively. First we derive them using some desired properties formulated in the standard theory of relations. Next we apply the Rough Sets paradigm.

3   Relations and Partial Orders

In this section we recall some fairly known concepts and results that will be used in the following sections [5,18].

Let X be a finite set, fixed for the rest of this paper. For every relation $R \subseteq X \times X$, let $R^+ = \bigcup_{i=1}^{\infty} R^i$, denote the *transitive closure* of R, $id = \{(x,x) \mid x \in X\}$ denote the identity relation, and let $R^\circ = R \cup id$ denote the *reflexive closure* of R (see [18] for details).

For each relation R and each $a \in X$ we define:

$$Ra = \{x \mid xRa\} \qquad aR = \{x \mid aRx\}.$$

A relation $< \in X \times X$ is a *(sharp) partial order* if it is irreflexive and transitive, i.e. if $\neg(a < a)$ and $a < b < c \implies a < c$, for all $a, b, c \in X$.

We write $a \sim_< b$ if $\neg(a < b) \land \neg(b < a)$, that is if a and b are either *distinctly incomparable* (w.r.t. $<$) or *identical* elements. We also write

$$a \equiv_< b \iff (\{x \mid a < x\} = \{x \mid b < x\} \land \{x \mid x < a\} = \{x \mid x < b\}).$$

The relation $\equiv_<$ is an *equivalence relation* (i.e. it is reflexive, symmetric and transitive) and it is called *the equivalence with respect to* $<$, since if $a \equiv_< b$, there is nothing in $<$ that can distinguish between a and b (see [5] for details). We always have $a \equiv_< b \implies a \sim_< b$.

---

[1] This example also illustrates that the generalisation from two to the whole space is often more problematic than usually anticipated when the pairwise comparisons paradigm is used. However this problem will not be discussed in this paper.

A partial order is

- *total* or *linear*, if $\sim_<$ is empty, i.e., for all $a, b \in X$. $a \neq b \implies (a < b \lor b < a)$.
- *weak* or *stratified*, if $a \sim_< b \sim_< c \implies a \sim_< c$, i.e. if $\sim_<$ is an equivalence relation.

If a partial order $<$ is weak than $a \equiv_< b \iff a \sim_< b$ (see [5]).

The sets $R^\circ a$ and aR° allow some characterisation of relations in terms of set theory inclusion. We have two folklore results.

Lemma 1. *For every relation R:*

1. $bR^\circ \subset aR^\circ \implies aRb$,
2. $R^\circ a \subset R^\circ b \implies aRb$.

Proof. (1) Let $bR^\circ \subset aR^\circ$. Since $b \in bR^\circ$, then $b \in aR^\circ$, i.e. $aRb \lor a = b$. But $a = b$ implies $bR^\circ = aR^\circ$, so aRb.

(2) Dually to (1). □

Lemma 2. *If R is a partial order then the following three statements are equivalent:*

1. aRb,
2. $bR^\circ \subset aR^\circ$,
3. $R^\circ a \subset R^\circ b$.

Proof. (2) $\implies$ (1) and (3) $\implies$ (1) follow from Lemma 1.

(1) $\implies$ (2): Let aRb and $x \in bR^\circ$. If $x = b$ then aRb implies $b \in aR^\circ$. If $x \neq b$ then aRb and bRx, which implies aRx, i.e. $x \in aR^\circ$. Hence $bR^\circ \subseteq aR^\circ$. But $aRb \implies a \neq b \land \neg bRa$, so $a \notin bR^\circ$, which means $bR^\circ \subset aR^\circ$.

(1) $\implies$ (3): Similarly to (1) $\implies$ (2). □

Lemma 2 simply says that "*a is smaller than b if and only the set of all elements smaller than a is included in the set of all elements smaller than b, and if and only if the set of all element bigger than b is included in the set of all elements bigger than a*".

We will call properties (2) and (3) of Lemma 2 *inclusion properties*, and say that the relation R has *inclusion properties* if it satisfies $bR^\circ \subset aR^\circ \land R^\circ a \subset R^\circ b$. Lemma 2 just says that R has inclusion properties if and only if it is a partial order.

A relation R is *acyclic* if and only if $\neg xR^+x$ for all $x \in X$.

For every relation R, define the relations R^{cyc}, R_{id}^{cyc} and $R^\bullet$ as

- $aR^{cyc}b \iff aR^+b \land bR^+a$,
- $aR_{id}^{cyc}b \iff aR^{cyc}b \lor a = b$,
- $aR^\bullet b \iff aRb \land \neg(aR^{cyc}b)$,

We will call $R^\bullet$ an *acyclic refinement of R*. If $aR^{cyc}b$ we will say that a and b belong to some cycle(s).

Corollary 1

1. R_{id}^{cyc} is an equivalence relation and R is acyclic if and only if $R^{cyc} = \emptyset$,
2. $R^{\bullet} \subseteq R$, $R^{\bullet}$ is acyclic (i.e. also irreflexive), and $aR^{\bullet}b \iff aRb \wedge \neg(bR^{+}a)$,
3. if R is a partial order then $R = R^{+} = R^{\bullet}$.   □

In this paper expressions like $(R^{\bullet})^{+}$ are interpreted as $(R^{\bullet})^{+} = Q^{+}$ where $Q = R^{\bullet}$. Also for each equivalence relation $E \subseteq X \times X$, $[x]_E$ will denote the equivalence class of E containing x and X/E will donote the set of all equivalence classes of E.

Lemma 3 (Schröder [20]). *For every relation* $R \subseteq X \times X$, *let* $\prec_R \subseteq (X/R_{id}^{cyc}) \times (X/R_{id}^{cyc})$ *be the following relation:*

$$[x]_{R_{id}^{cyc}} \prec_R [y]_{R_{id}^{cyc}} \iff xR^{+}y \wedge \neg yR^{+}x.$$

The relation $\prec_R$ *is a partial order on* X/R_{id}^{cyc}.   □

4   Inclusion Property and Equivalence w.r.t. a Given Relation

In this section two concepts initially introduced for partial orders will be extended to arbitrary relations. The first one is *inclusion property*.
   For every relation R, define the relation R^C as follows :

- $aR^C b \iff bR^{\circ} \subset aR^{\circ} \wedge R^{\circ}a \subset R^{\circ}b.$

We will call R^C the *inclusion property kernel* of R.

Corollary 2.   *1.* $R^C \subseteq R$ *and* R^C *is a partial order.*
   2. If R *is a partial order then* $R = R^C$.   □

The relation R^C is a partial order, but *can it be considered as a property-driven approximation of* R? It may happen that $aR^C b \wedge aR^{cyc}b$ for some a and b, see Figure 1, so in the preliminary version of this paper [10] we said 'No'. The major motivation was that if R is considered an approximation of a ranking then $aR^{cyc}b$ is often interpreted that a and b are indifferent. In concurrency theory if R were the union of all individual observations, it would mean no casual relationship between a and b [12]. However after a more detailed analysis of empirical data collected to confirm or refute some claims from [8], we have noticed that in some cases (although not often) we had $aR^{cyc}b$ for empirical R, while it was $a <^{rank} b$ in reality. If R represents arbitrary empirical data with arbitrary interpretation, some links in $aR^{cyc}b$ could just be the results of errors or lack of precision. In concurrency theory it might depend on how we define "simultaneity" [12]. Hence in this paper we will allow R^C to be considered as a property driven partial order approximation of R. We will later call it "a weak approximation" becasue of the fact that $aR^{cyc}b$ might happen.
   The second concept is the relation $\equiv_<$ which can easily be extended to an arbitrary relation R.

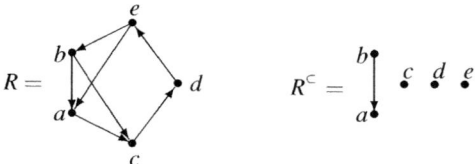

Fig. 1. An example of R such that $xR^C y \wedge xR^{cyc}y$ for some x and y. We have $R^{\circ}a = \{a,c\} \subset \{a,b,c\} = R^{\circ}b$ and $bR^{\circ} = \{b,c\} \subset \{a,b,c\} = aR^{\circ}$, so $aR^C b$ and $aRcRdReRbRa$ so $aR^{cyc}b$.

For every relation R, define the relation $\equiv_R$ as follows :

- $a \equiv_R b \iff aR = bR \wedge Ra = Rb$.

The relation $\equiv_R$ is an *equivalence relation* (i.e. it ir reflexive, symmetric and transitive) and it is called *the equivalence with respect to R*, since if $a \equiv_R b$, there is nothing in R that can distinguish between a and b. *This means that any reasonable partial order approximation of R should preserve the relation $\equiv_R$.*

Note also that:
$$a \equiv_R b \iff \forall x. (xRa \iff xRb) \wedge (aRx \iff bRx).$$

Lemma 4. *For every two relations R and Q:* $a \equiv_R b \wedge a \equiv_Q b \implies a \equiv_{R \cap Q} b$.

Proof. $a \equiv_R b \wedge a \equiv_Q b \implies aR = bR \wedge Ra = Rb \wedge aQ = bQ \wedge Qa = Qb \implies a(R \cap Q) = b(R \cap Q) \wedge (R \cap Q)a = (R \cap Q)b \iff a \equiv_{R \cap Q} b$. $\qquad\square$

It turns out that such operations as transitive closure, acyclic refinement and inclusion property kernel, preserve the equivalence with respect to R.

Lemma 5. *For every relation R we have:*

1. $a \equiv_R b \implies a \equiv_{R^+} b$,
2. $a \equiv_R b \implies a \equiv_{R^{\bullet}} b$,
3. $a \equiv_R b \implies a \equiv_{R^C} b$.

Proof. (1) $xR^+a \iff xRx_1R...Rx_nRa$. But $a \equiv_R b \implies (x_nRa \Leftrightarrow x_nRb)$, so $xR^+a \iff xRx_1R...Rx_nRb \iff xR^+b$. Similarly we show $aR^+x \iff bR^+x$, hence $a \equiv_{R^+} b$.

(2) Since $xR^{\bullet}a \iff xRa \wedge \neg aR^+x$ and $xRa \iff xRb$, then $xR^{\bullet}a \implies xRb$. Suppose bR^+x, i.e. $bRx_1R...x_kRx$. But $bRx_1 \iff aRx_1$, so $bR^+x \iff aR^+x$, a contradiction as $xR^{\bullet}a \implies \neg aR^+x$. Hence $xR^{\bullet}a \implies xR^{\bullet}b$. By replacing a with b, we immediately get $xR^{\bullet}b \implies xR^{\bullet}a$, i.e. $xR^{\bullet}a \iff xR^{\bullet}b$. In an almost identical manner we show $aR^{\bullet}x \iff bR^{\bullet}x$, so $a \equiv_{R^{\bullet}} b$.

(3) Note that if $a = b$ then clearly $a \equiv_{R^C} b$, so assume $a \neq b$.
First we show that $a \equiv_R b$ implies $\forall x. R^{\circ}x \subset R^{\circ}a \iff R^{\circ}x \subset R^{\circ}b$. Suppose $R^{\circ}x \subset R^{\circ}a$, i.e. $Rx \cup \{x\} \subset Ra \cup \{a\}$. Since $Ra = Rb$, then $R^{\circ}x = Rx \cup \{x\} \subseteq Rb \cup \{a\}$.
We now have to consider two cases:

Case 1: $a \in Rb$. Since $Ra = Rb$ then $a \in Ra$, so we have $R^{\circ}x \subset R^{\circ}a \cup \{a\} = Ra = Rb \subseteq Rb \cup \{b\} = R^{\circ}b$, so $R^{\circ}x \subset R^{\circ}b$.
Case 2: $a \notin Rb$. First we show that $a \in Rx \implies a \in Rb$. We have $a \in Rx \iff aRx \iff bRx \iff b \in Rx$ and $b \in Rx \subseteq R^{\circ}x \subset Ra \cup \{a\} \implies bRa \vee a = b$. Since $a \neq b$ then

bRa. Because $a \equiv_R b$ the we have $Ra = Rb$ and $aR = bR$, so $bRa \wedge Ra = Rb \implies bRb$, while $bRb \wedge aR = bR \implies aRb$, i.e. $a \in Rb$. This means $a \notin Rb$ implies $a \notin Rb \wedge a \notin Rx$. Hence we have: $R°a = R°a \setminus \{a\} \subset (Rb \cup \{a\}) \setminus \{a\} = Rb \subseteq R°b$, so $R°x \subset R°b$. In this way we have proved $\forall x.\ R°x \subset R°a \implies R°x \subset R°b$. Similarly we prove that $a \equiv_R b$ implies $\forall x.\ xR° \subset aR° \implies xR° \subset bR°$, which means that $a \equiv_R b$ implies

$$\forall x.\ (R°x \subset R°a \wedge xR° \subset aR°) \implies (R°x \subset R°b \wedge xR° \subset bR°).$$

By replacing a with b we get an inverse inclusion, so in fact we proved:

$$\forall x.\ (R°x \subset R°a \wedge xR° \subset aR°) \iff (R°x \subset R°b \wedge xR° \subset bR°),$$

i.e. $\forall x.\ (xR^C a \iff xR^C b)$. In almost identical way we can prove $\forall x.\ (aR^C x \iff bR^C x)$. Hence $a \equiv_{R^C} b$. $\square$

5 Approximating Relations by Partial Orders

We will start with a formal definition of a (*property-driven*) *partial order approximations* of a relation R. *Four* definitions will be discussed.

Definition 1. *A partial order* $< \subseteq X \times X$ *is a* **partial order approximation** *of a relation* $R \subseteq X \times X$ *if it satisfies the following three conditions:*

1. $a < b \implies aR^+ b$,
2. $a < b \implies \neg aR^{cyc} b$ *(or, equivalently* $a < b \implies \neg bR^+ a$*)*,
3. $aR^C b \wedge aR^\bullet b \implies a < b$,
4. $a \equiv_R b \implies a \equiv_< b$. $\square$

Since R^+ is the smallest transitive relation containing R (see [18]), and due to informational noise, imprecision, randomness, etc., some parts of R might be missing (in any interpretation of R), it is reasonable to assume that R^+ is the upper bound of $<$.

Condition (2) was already discussed in the previous section. It says that if $aR^{cyc}b$ then usually a and b are incomparable. If R is interpreted as an estimation of a ranking, then in most cases $aR^{cyc}b$ is interpreted that a and b are indifferent. Similar interpretations exist in concurrency theory. When $a < b \implies aR^+b$, then $\neg aR^{cyc}b$ can be replaced by $\neg bR^+a$.

Condition (3) defines the lower bound. Note that the greatest partial order included in R often does not exist, however if R is interpreted as an estimation of a ranking, it is reasonable to assume that the inclusion property refinement in included in the ranking order. However, as Figure 1 shows, it may happen that $aR^C b$ and $aR^{cyc}b$ so we need to add $aR^\bullet b$ to avoid a contradiction.

Condition (4) ensures preservation of the equivalence with respect to R.

Note that R^C is not a partial order approximation as it does not satusfy (2). However in some cases, including ranking, the condition (2) is too strong as some links in $aR^{cyc}b$ could be just due to errors. The version without the condition (2) is the following.

Definition 2. *A partial order* $< \subseteq X \times X$ *is a* **weak partial order approximation** *of a relation* $R \subseteq X \times X$ *if it satisfies the following three conditions:*

1. $a < b \implies aR^+ b$,
2. $aR^C b \wedge aR^\bullet b \implies a < b$,
3. $a \equiv_R b \implies a \equiv_< b$. $\square$

Clearly every partial order approximation is also a weak partial order approximation.

If R is an empirical ranking data (see [10]) constructed on the basis of pairwise comparisons paradigm, it may happen that aRb makes sense only locally, when the domain is restricted to $\{a,b\}$, and it needs to be pruned in global setting (as the relation $\leftarrow$ from Section 1). The situation that aRb makes sense only locally may appear in other applications as well. In such cases we may require $a <^{rank} b \implies aRb$, which leads to the following definition.

Definition 3. *A partial order* $< \subseteq X \times X$ *is an* **inner partial order approximation** *(***inner weak partial order approximation***) of a relation* $R \subseteq X \times X$, *if it is a partial order approximation (weak partial order approximation) of* R, *and satisfies:*

$$a < b \implies aRb. \qquad \square$$

Every partial order is transitive, acyclic and equal to its inclusion property kernel. An arbitrary relation R may not have these properties but we may try to refine R using transitive closure, acylcic refinement and finding the inclusion property kernel, in various orders or simultaneously (i.e. using set theory intersection). We will show that there are exactly four partial order approximations that can be derived in this way.

Let us first define the relation $R^{c \wedge \bullet}$ as follows:

$$aR^{c \wedge \bullet}b \iff aR^c b \wedge aR^{\bullet}b.$$

We can now formulate the main results of the first part of this paper.

Theorem 1

1. *The relations* $R^{c \wedge \bullet}$, $(R^{\bullet})^c$, $(R^{\bullet})^+$, $(R^+)^{\bullet}$ *are partial order approximations of* R.
2. *The relations* $R^{c \wedge \bullet}$ *and* $(R^{\bullet})^c$ *are inner partial order approximations of* R.
3. $R^{c \wedge \bullet} \subseteq (R^{\bullet})^c \subseteq (R^{\bullet})^+ \subseteq (R^+)^{\bullet}$ *and* $R^{c \wedge \bullet} \subseteq R^c$.
4. *If* R *is transitive, i.e.* $R = R^+$, *then* $R^{c \wedge \bullet} = (R^{\bullet})^c = (R^{\bullet})^+ = (R^+)^{\bullet}$.
5. *If* R *is a partial order, then* $R = R^c = R^{c \wedge \bullet} = (R^{\bullet})^c = (R^{\bullet})^+ = (R^+)^{\bullet}$.
6. *If* R *is acyclic, i.e.* $R = R^{\bullet}$, *then* $R^c = R^{c \wedge \bullet} = (R^{\bullet})^c$ *and* $(R^{\bullet})^+ = (R^+)^{\bullet}$.
7. *If a partial order* $<$ *is a partial order approximation of* R, *then*
   $$aR^{c \wedge \bullet}b \implies a < b \implies a(R^+)^{\bullet}b.$$
8. $aR^{cyc}b \implies a \equiv_{(R^+)^{\bullet}} b$.
9. *The realtions* $R^{c \wedge \bullet}$, $(R^{\bullet})^c$, $(R^{\bullet})^+$, $(R^+)^{\bullet}$ *are the only partial order approximations of* R *that can be derived from* R *by using operations* '$\cap$', '^{c}', '$^+$' *and* '$^{\bullet}$'.
10. R^c *is an inner weak partial order approximation of* R.
11. *The realtions* R^c, $R^{c \wedge \bullet}$, $(R^{\bullet})^c$, $(R^{\bullet})^+$, $(R^+)^{\bullet}$ *are the only weak partial order approximations of* R *that can be derived from* R *by using operations* '$\cap$', '^{c}', '$^+$' *and* '$^{\bullet}$'. $\qquad \square$

With the exception of (8), the above theorem is practically self-explanatory. Assertion (8) says that if a and b belong to a cycle in R then they are equivalent with respect to $(R^+)^{\bullet}$. This indicates that if we have a reason to believe that all cycles result from errors, informational noise, etc., and all elements of a cycle should be interpreted as indifferent, then $(R^+)^{\bullet}$ is most likely the best partial order approximation of R.

Proof of Theorem 1. First we show that the relations $R^{c\wedge\bullet}$, $(R^\bullet)^c$, $(R^\bullet)^+$, $(R^+)^\bullet$ are partial orders. Consider $R^{c\wedge\bullet}$. Clearly $aR^{c\wedge\bullet}b \iff aR^c b \wedge aR^\bullet b \iff aR^c b \wedge \neg bR^+ a$. By Corollary 1(2) the relation $R^{c\wedge\bullet}$ is irreflexive so we need only to prove its transitivity. Suppose that $aR^{c\wedge\bullet}b$ and $bR^{c\wedge\bullet}c$. This means $aR^c b$, $bR^c c$, $\neg bR^+ a$ and $\neg cR^+ b$. By Corollary 1(3), R^c is transitive, so $aR^c c$, and by Lemma 1, aRb, bRc and aRc. Hence we only need to show that $\neg cR^+ a$. Suppose $cR^+ a$. Then $cR^+ a$ and aRb implies $cR^+ b$, a contradition as $aR^{c\wedge\bullet}c$ implies $\neg cR^+ b$. Therefore $R^{c\wedge\bullet}$ is a partial order.

Consider $(R^\bullet)^c$. From Corollary 1(3) it immediately follows that the relation $(R^\bullet)^c$ is a partial order.

Consider $(R^\bullet)^+$. By Corollary 1(2), we have $aR^\bullet b \iff aRb \wedge \neg(bR^+ a)$. The relation $(R^\bullet)^+$ is clearly transitive, we need only to show $\neg(a(R^\bullet)^+ a)$ for all $a \in X$. Since $aRb \wedge \neg(bR^+ a) \implies a \neq b$, then $\neg aR^\bullet a$. Suppose $a(R^\bullet)^+ a$. Since $\neg aR^\bullet a$, this means $aR^\bullet b(R^\bullet)^+ a$, for some $b \neq a$. But $aR^\bullet b \implies aRb$ and $b(R^\bullet)^+ a \implies bR^+ a$, so we have $aRb \wedge bR^+ a$, contradicting $aR^\bullet b$. Hence $\neg(a(R^\bullet)^+ a)$, i.e. $(R^\bullet)^+$ is a partial order.

Consider $(R^+)^\bullet$. Notice that $a(R^+)^\bullet b \iff aR^+ b \wedge \neg bR^+ a \iff [x]_{R_{id}^{cyc}} \prec_R [y]_{R_{id}^{cyc}}$, where $\prec_R$ is the relation from Lemma 3. Hence, by Lemma 3, the relation $(R^+)^\bullet$ is a partial order.

We will now prove the first part of (3), i.e. $R^{c\wedge\bullet} \subseteq (R^\bullet)^c \subseteq (R^\bullet)^+ \subseteq (R^+)^\bullet$. Suppose $aR^{c\wedge\bullet}b$, i.e. $aR^c b \wedge \neg bR^+ a$. Then aRb and $\neg bR^+ a$, so $a \in (R^\bullet)^\circ a \cap (R^\bullet)^\circ b$. Assume that $x \in R^\bullet a$ and $x \notin R^\bullet b$. Since $aR^{c\wedge\bullet}b \implies aR^c b$, then we have $Ra \subset Rb$. But $R^\bullet a \subseteq Ra$, so $x \in Rb$. We now have $x \in Rb$ and $x \notin R^\bullet b$, i.e. $bR^+ x$. Since $x \in R^\bullet a$ means xRa, then $bR^+ ax$ and xRa give us $bR^+ a$, a contradiction as, $aR^{c\wedge\bullet}b \implies \neg bR^+ a$. Hence $R^\bullet a \subseteq R^\bullet b$. Since $a \neq b$ then $R^\bullet a \neq R^\bullet b$, so $(R^\bullet)^\circ a \subset (R^\bullet)^\circ b$. Similarly we show $b(R^\bullet)^\circ \subset a(R^\bullet)^\circ$, hence $a(R^\bullet)^c b$. Therefore $R^{c\wedge\bullet} \subseteq (R^\bullet)^c$.

By Lemma 1 we have $(R^\bullet)^c \subseteq R^\bullet$, and clearly $R^\bullet \subseteq (R^\bullet)^+$, hence $(R^\bullet)^c \subseteq (R^\bullet)^+$. Suppose $a(R^\bullet)^+ b$. Recall that $x(R^+)^\bullet y \iff xR^+ y \wedge \neg yR^+ x$. By Corollary 1(2), we have $R^\bullet \subset R$. Hence $a(R^\bullet)^+ + b \implies aR^+ b$. Suppose $bR^+ a$. Then $aR^{cyc}b$, i.e. $\neg aR^\bullet b$, a contradiction. Hence $a(R^+)^\bullet b$, i.e. $(R^\bullet)^+ \subseteq (R^+)^\bullet$. Therefore we have proved the first part (3). The second part of (3), i.e. $R^{c\wedge\bullet} \subseteq R^c$, follows trivially from appropriate definitions. Hence the assertion (3) holds.

Note that (3) together with the fact that all $R^{c\wedge\bullet}$, $(R^\bullet)^c$, $(R^\bullet)^+$, $(R^+)^\bullet$ are partial orders imply that $R^{c\wedge\bullet}$, $(R^\bullet)^c$, $(R^\bullet)^+$, $(R^+)^\bullet$ satisfy (1),(2) and (3) of Definition 1. By Lemma 5, $(R^\bullet)^+$ and $(R^+)^\bullet$ satisfy (4) of Definition 1; and by Lemmas 5 and 4, $R^{c\wedge\bullet}$ and $(R^\bullet)^c$ satisfy satisfy (4) of Definition 1. Therefore assertion (1) of the above theorem does hold.

Assertion (1) and Corollary 2(2) yield the assertion (2).

Hence (1), (2) and (3) hold. We will now prove (4). It suffices to show that if $R = R^+$ then $(R^+)^\bullet \subseteq R^{c\wedge\bullet}$. Note that in this case $a(R^+)^\bullet b \iff aRb \wedge \neg bRa$. If $R = R^+$ then $(R^+)^\bullet = R^\bullet$, so we only need to show $(R^+)^\bullet \subseteq R^c$. Let $a(R^+)^\bullet b$. This means $a \neq b$ and $\neg bRa$. Furthermore $\neg bRa$ implies $a \notin bR \wedge b \notin Ra$. Assume $x \in bR^\circ$. If $x = b$ then aRb implies $b \in aR$, i.e. $x \in aR^\circ$. If $x \neq b$ then $x \in bR^\circ \implies bRx$. Since R is transitive $aRb \wedge bRx \implies aRx \implies x \in Ra \implies x \in R^\circ a$. Hence $bR^\circ \subseteq aR^\circ$. Since $a \neq b$ and $a \notin bR$, then $a \notin bR^\circ$, which means $bR^\circ \subset aR^\circ$. Dually we show $R^\circ a \subset R^\circ b$, i.e. $aR^c b$, so we have proved (4).

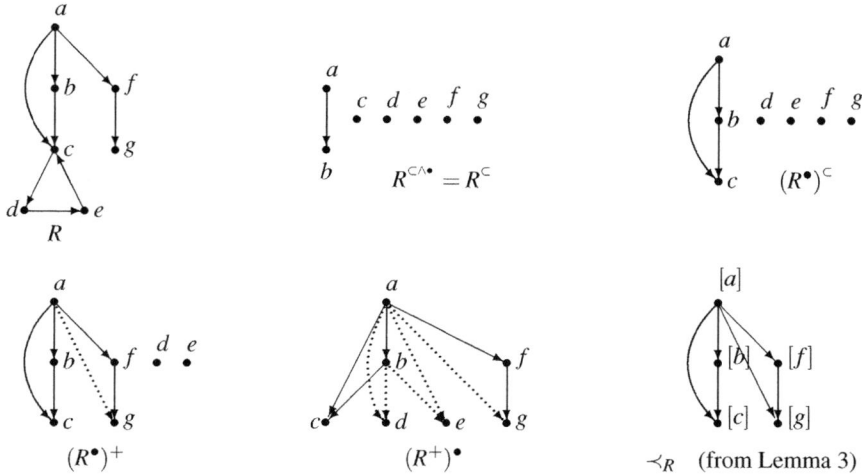

Fig. 2. An example of a relation R, its partial order approximations $R^{c\wedge\bullet} = R^c$, $(R^\bullet)^c$, $(R^\bullet)^+$, $(R^+)^\bullet$, and its relation $\prec_R$ from Lemma 3. Dotted lines in $(R^\bullet)^+$ and $(R^+)^\bullet$ indicate the relationship that is not in R and was added by transitivity operation. For the relation $\prec_R$, $[x]$ denotes $[x]_{R_{id}^{cyc}}$ for $x \in \{a,b,c,d,e,f,g\}$, and $[a] = \{a\}$, $[b] = \{b\}$, $[c] = \{c,d,e\}$, $[f] = \{f\}$, $[g] = \{g\}$.

Assertion (5) follows from (4) and Lemma 2.

If $R = R^\bullet$ then clearly $(R^\bullet)^c = R^c$. We also heve $R^{c\wedge\bullet} = R^c \cap R^\bullet = R^c \cap R = R^c$ as, by Lemma 1, $R^c \subseteq R$. From (3) it follows $(R^\bullet)^+ \subseteq (R^+)^\bullet$. If $R = R^\bullet$, then $(R^+)^\bullet \subseteq R^+ = (R^\bullet)^+$, i.e. $(R^\bullet)^+ = (R^+)^\bullet$, so we have proved (6).

Assertion (7) follows from (1), (3) and Definition 1.

Assertion (8) is a consequence of Lemma 3. Recall that we have

$$a \equiv_{(R^+)^\bullet} b \iff \{x \mid x(R^+)^\bullet a\} = \{x \mid x(R^+)^\bullet b\} \wedge \{x \mid a(R^+)^\bullet x\} = \{x \mid b(R^+)^\bullet x\}.$$

If $aR^{cyc}b$ then $[a]_{R_{id}^{cyc}} = [b]_{R_{id}^{cyc}}$. Hence we have

$$x(R^+)^\bullet a \iff [x]_{R_{id}^{cyc}} \prec_{(R^+)^\bullet} [a]_{R_{id}^{cyc}} \iff [x]_{R_{id}^{cyc}} \prec_{(R^+)^\bullet} [b]_{R_{id}^{cyc}} \iff x(R^+)^\bullet b,$$

which means $\{x \mid x(R^+)^\bullet a\} = \{x \mid x(R^+)^\bullet b\}$. Similarly we can prove $\{x \mid a(R^+)^\bullet x\} = \{x \mid b(R^+)^\bullet x\}$. Thus the assertion (8) does hold as well.

To show (9) first notice that, $(R^c)^\bullet = (R^c)^+ = R^c$ (as R^c is a partial order), $R^+ \cap R^\bullet = (R^+)^\bullet$ (from the definition of acyclic refinement), and $R^+ \cap R^c = R^c$ (since $R^c \subseteq R \subseteq R^+$). Since $R^+ = (R^+)^+$, from (4) we have $(R^+)^\bullet = ((R^+)^+)^\bullet = ((R^+)^\bullet = (R^+)^\bullet$. From (1), (3) and (5) it follows that additional applications of '$\cap$', '^{c}', '$^+$' and '$^\bullet$' do not produce new realtions.

Assertion (10) follows from Definition 2 and Lemma 5(3).

The proof of (11) is virtually the same as the proof of (9). □

Theorem 1 is illustrated by an example in Figure 2. There is no universal inclusion relationship between R^c and $(R^\bullet)^c, (R^\bullet)^+, (R^+)^\bullet$. For the relation R from Figure 1, $R^{c\wedge\bullet} = (R^\bullet)^c = (R^\bullet)^+ = (R^+)^\bullet = \emptyset \subset R^c$, for the relation R from Figure 2, $R^c = R^{c\wedge\bullet} \subset (R^\bullet)^c \subset (R^\bullet)^+ \subset (R^+)^\bullet$.

6 Approximation with Partially Ordered Kernel

Quite often the approximations discussed in the previous section are not sufficient. In some cases we have a partial order $\lhd$ that is a subset of R and we want it also to be a subset of a partial order approximation of R.

Take for example the case of Non-numerical Ranking discussed in Section 2 and a relation R that describe the results of Pairwise Comparisons. Even if R may in general be imprecise, in most cases *some parts* of R describe the precise ranking. For instance if R is the result of expert voting, if all experts agree that aRb, then we may assume that $a <^{rank} b$ (see *Pereto's principle* [6]). In this section we will formally treat such case.

Let R be a relation and let $\lhd$ be a partial order satisfying $\lhd \subseteq R$. We are looking for a partial order approximation of R that includes $\lhd$. The relation $\lhd$ will be called a *partially ordered kernel* of R. In general it may happen that $\lhd$ is not included in any partial order approximation discussed in the previous section (Figure 1 in [11] shows the case of $a \lhd b$ and $\neg a(R^+)^\bullet b$). In general the union of partial orders may not be a partial order at all, however we may use the following lemma.

Lemma 6. *Let R be a relation, $<_1$ and $<_2$ be partial orders satisfying:*

1. $a <_1 b \implies aR^+b$, and
2. $a <_2 b \implies aR^+b \wedge \neg(bR^+a)$.
Then $(<_1 \cup <_2)^+$ is the smallest partial order containing $<_1 \cup <_2$.

Proof. $(<_1 \cup <_2)^+$ is evidently the smallest transitive relation containing $<_1 \cup <_2$. It suffices to show that $(<_1 \cup <_2)^+$ is irreflexive. Suppose it is not irreflexive, i.e. there exoists x_0 such that $x_0(<_1 \cup <_2)^+x_0$. This means $x_0Q_1x_1Q_2x_2...x_{n-1}Q_nx_n$, with $x_n = x_0$, where Q_i is either $<_1$ or $<_2$. Since $<_1$ and $<_2$ are sharp partial orders, then at least one of Q_i's, sat Q_k, must be equal to $<_2$. Since $<_1 \subseteq R^+$ and $<_2 \subseteq R^+$, then for each $i,j \leq n$, we have $x_iR^+x_j \wedge x_jR^+x_i$. In particular $x_kR^+x_{k-1}$, a contradiction as $x_{k-1} <_2 x_k \implies \neg x_{k-1}R^+x_k$. Hence $(<_1 \cup <_2)^+$ is irreflexive. $\qquad\square$

The above result can easily be applied to all partially ordered approximations from the previous section.

Corollary 3. *For each $< \in \{R^c, R^{c \wedge \bullet}, (R^\bullet)^c, (R^\bullet)^+, (R^+)^\bullet\}$, and each partial order $\lhd \subseteq R$, $(\lhd \cup <)^+$ is the smallest partial order containing $\lhd \cup <$.* $\qquad\square$

A special case of Corollary 3 was used in the ranking algorithms proposed in [9,11].

7 Classical Rough Relations

How can the problem of approximating an arbitrary relation by a partial order be solved with Rough Sets? This topic will be discussed in this and the following sections.

In the spirit of Rough Sets [16,17], the relations $(R^\bullet)^+$ and $(R^+)^\bullet$ can be seen as upper partial order approximations of R, while $R^{c \wedge \bullet}$ or $(R^\bullet)^c$ can be seen as lower partial order approximations of R. But this is true in spirit only, as in general case R

may be included in neither $(R^\bullet)^+$ nor $(R^+)^\bullet$, and neither $R^{\subset\wedge\bullet}$ nor $(R^\bullet)^\subset$ satisfy all of the properties required from lower (Rough Sets) approximations.

The principles of Rough Rets [16,17] can be formulated as follows. Let U be a finite and nonempty universe of elements, and let $E \subseteq U \times U$ be an *equivalence relation*. The elements of U/E are called elementary sets and they are interpreted as basic observable, measurable, or definable sets. The pair (U,E) is referred to as a Pawlak approximation space. A set $X \subseteq U$ is approximated by two subsets of U, $\underline{A}(X)$ - called the lower approximation of X, and $\overline{A}(X)$ - called the upper approximation of X, where:

$$\underline{A}(X) = \bigcup\{[x]_E \mid x \in U \wedge [x]_E \subseteq X\}, \qquad \overline{A}(X) = \bigcup\{[x]_E \mid x \in U \wedge [x]_E \cap X \neq \emptyset\}.$$

Rough set approximations satisfy the following properties:

Proposition 1 (Pawlak [17])

1. $X \subseteq Y \implies \underline{A}(X) \subseteq \underline{A}(Y)$,   *6. $X \subseteq Y \implies \overline{A}(X) \subseteq \overline{A}(Y)$,*
2. $\underline{A}(X \cap Y) = \underline{A}(X) \cap \underline{A}(Y)$,   *7. $\overline{A}(X \cup Y) = \overline{A}(X) \cup \overline{A}(Y)$,*
3. $\underline{A}(X) \subseteq X$,   *8. $X \subseteq \overline{A}(X)$,*
4. $\underline{A}(X) = \underline{A}(\underline{A}(X))$,   *9. $\overline{A}(X) = \overline{A}(\overline{A}(X))$,*
5. $\overline{A}(X) = \underline{A}(\overline{A}(X))$,   *10. $\underline{A}(X) = \overline{A}(\underline{A}(X))$.* $\square$

Since every relation is a set of pairs, this approach can be used for relations as well [21]. Unfortunately, in such cases as ours we want approximations to have some specific properties like irreflexivity, transitivity etc., and most of those properties are not closed under the set union operator. As was pointed out in [25], in general one cannot expect approximations to have the desired properties (see [25] for details). It is also unclear how to define the relation E for cases such as ours.

However the Rough Sets can also be defined in an orthogonal (sometimes called 'topological') manner [17,22,24]. For a given (U,E) we may define $\mathcal{D}(U)$ as the smallest set containing $\emptyset$, all of the elements of U/E and that is closed under set union. Clearly U/E is the set of all components generated by $\mathcal{D}(U)$ [15]. We may start with defining a space as $(U,\mathcal{D})$ where $\mathcal{D}$ is a family of sets that contains $\emptyset$ and for each $x \in U$ there is $X \in \mathcal{D}$ such that $x \in X$ (i.e. $\mathcal{D}$ is a cover of U [18]). We may now define $E_\mathcal{D}$ as the equivalence relation generated by the set of all components defined by $\mathcal{D}$ (see for example [15]). Hence both approaches are equivalent [17,22,25], however now for each $X \subseteq U$ we have:

$$\underline{A}(X) = \bigcup\{Y \mid Y \subseteq X \wedge Y \in \mathcal{D}\}, \qquad \overline{A}(X) = \bigcap\{Y \mid X \subseteq Y \wedge Y \in \mathcal{D}\}.$$

We can now define $\mathcal{D}$ as a set of relations having the desired properties and then calculate $\underline{A}(R)$ and/or $\overline{A}(R)$ with respect to a given $\mathcal{D}$. Such an approach was proposed and analysed in [25], however it seems to have only limited applications. It assumes that the set $\mathcal{D}$ is closed under both union and intersection, and few properties of relations do this. For instance, transitivity is not closed under union and having a cycle is not closed under intersection. Some properties, like "having exactly one cycle" are preserved by neither union nor intersection. This problem was discussed in [25] and they proposed that perhaps a different $\mathcal{D}$ could be used for the lower and upper approximations. But this solution again seems to have rather limited applications. The approach of [25] assumes additionally that, for the upper approximation there is at least one element of $\mathcal{D}$ that contains R, and, for the lower approximation there exists at least one element of $\mathcal{D}$ that is included in R. These are assumptions that are too strong for cases such as ours. If R contains a cycle, then there is no partial order that contains R! Very often $R \setminus (R^+)^\bullet \neq \emptyset$ and $R \setminus (R^\bullet)^+ \neq \emptyset$. A possible solution to this problem involves complicated generalisations of the concepts of both lower and upper approximations and the use of mixed approximations. For example, $R^\bullet$ can be interpreted as a lower acyclic approximation of R, and then $(R^\bullet)^+$ can be interpreted as an upper transitive approximation of $R^\bullet$. Similarly, R^+ can be interpreted an upper transitive approximation of R, and $(R^+)^\bullet$ as a lower acyclic approximation of R^+. Note that $(R^\bullet)^+ \subseteq (R^+)^\bullet$, as expected from the point of view of the Rough Sets paradigm; however, here we are mixing different approximations.

8   Property-Driven Rough Approximations of Binary Relations

Let X be a set, $R \subseteq X \times X$ be a binary relation and let α be any first-order predicate (c.f. [7]) with one atomic formula R and all variables over the set X. Any predicate α of this kind will be called *a property over R*. The predicate α is called a *property of R* if (X, R) is a model of α, i.e. α holds for any assignment (c.f. [7]). Obvious examples of properties are transitivity, reflexivity, etc. We would like to point out the difference between *a property over R*, i.e. just a statement that may or may not be true, and a *property of R*, a statement that is true for all assignments.

Let $R \subseteq X \times X$ be a non-empty relation and let $\mathcal{P}$ be a set of properties *over R*. Any element $\alpha \in \mathcal{P}$ is called an *elementary property*. We assume that for each $\alpha \in \mathcal{P}$ there is a non-empty family of relations $P_\alpha \subseteq 2^{X \times X}$ such that $P_\alpha \neq \{\emptyset\}$ and for every $Q \subseteq X \times X$, if α is a property of Q then $Q \in P_\alpha$. In other words P_α is the set of relations over X that satisfy the property α.

Let $\mathcal{P}^\cap$ be a subset of $\mathcal{P}$ such that $\alpha \in \mathcal{P}^\cap$ iff P_α is closed under intersection, and $\mathcal{P}^\cup$ be a subset of $\mathcal{P}$ such that $\alpha \in \mathcal{P}^\cup$ iff P_α is closed under union.

We assume that $\mathcal{P} = \mathcal{P}^\cap \cup \mathcal{P}^\cup$ and the pair $(X, \mathcal{P})$ (or $(X, \{P_\alpha \mid \alpha \in \mathcal{P}\})$) will be called an *propert-driven approximation space for the binary relations over X.*

Let $R \subseteq X \times X$ be a non-empty relation and $\alpha \in \mathcal{P}$. We say that:

- R has α-lower bound $\iff \exists Q \in P_\alpha.\, Q \subseteq R$,
- R has α-upper bound $\iff \exists Q \in P_\alpha.\, R \subseteq Q$.

Some examples:

- $\alpha = transitivity$, R - any relation, both α-lower bound and α-upper bound do exist,
- $\alpha = partial\ order$, R has a cycle, α-lower bound exists but α-upper bound does not exist,
- $\alpha = having\ a\ cycle$, R is a partial order, α-lower bound does not exist, α-upper bound exists,
- $\alpha = (a,b) \in R \wedge (c,d) \notin R$, R any relation such that $(a,b) \notin R$ and $(c,d) \in R$, neither α-lower bound nor α-upper bound exist.

We also define

- $lb_\alpha(R) = \{Q \mid Q \in P_\alpha \wedge Q \subseteq R\}$, the set of all α-lower bounds of R, and
- $ub_\alpha(R) = \{Q \mid Q \in P_\alpha \wedge R \subseteq Q\}$, the set of all α-upper bounds of R.

For every family of relations $\mathcal{F} \subseteq 2^{X \times X}$, we define

- $min(\mathcal{F}) = \{R \mid \forall Q \in \mathcal{F}. Q \subseteq R \Rightarrow R = Q\}$, the set of all minimal elements of $\mathcal{F}$, and
- $max(\mathcal{F}) = \{R \mid \forall Q \in \mathcal{F}. R \subseteq Q \Rightarrow R = Q\}$, the set of all maximal elements of $\mathcal{F}$.

We are now able to provide the two main definitions of this section:

- If R has α-lower bound then we define its α-*lower approximation* as:

$$\underline{\mathbf{A}}_\alpha(R) = \bigcap \{Q \mid Q \in max(lb_\alpha(R))\}.$$

- If R has α-upper bound then we define its α-*upper approximation* as:

$$\overline{\mathbf{A}}_\alpha(R) = \bigcup \{Q \mid Q \in min(ub_\alpha(R))\}.$$

If R *does not have* α-lower bound (α-upper bound) then its α-lower approximation (α-upper approximation) *does not exist*. The result below shows that the above two definitions are sound when $R \in P_\alpha$.

Proposition 2. *If $R \in P_\alpha$ then $\underline{\mathbf{A}}_\alpha(R) = \overline{\mathbf{A}}_\alpha(R) = R$.*

Proof. If $R \in P_\alpha$ then $lb_\alpha(R) = ub_\alpha(R) = \{R\}$.   □

Directly from the definitions it follows that $\underline{\mathbf{A}}_\alpha(R)$ is well defined if $\alpha \in \mathcal{P}^\cap$ and $\overline{\mathbf{A}}_\alpha(R)$ is well defined if $\alpha \in \mathcal{P}^\cup$. The result below shows that both concepts are well defined for all $\alpha \in \mathcal{P} = \mathcal{P}^\cup \cup \mathcal{P}^\cap$.

Proposition 3

1. *If $\alpha \in \mathcal{P}^\cup$ and R has α-lower bound, then*

$$\underline{\mathbf{A}}_\alpha(R) = \bigcup \{Q \mid Q \in lb_\alpha(R)\} = \bigcup \{Q \mid Q \subseteq R \wedge Q \in P_\alpha\}.$$

2. *If $\alpha \in \mathcal{P}^\cap$ and R has α-upper bound, then*

$$\overline{\mathbf{A}}_\alpha(R) = \bigcap \{Q \mid Q \in ub_\alpha(R)\} = \bigcap \{Q \mid R \subseteq Q \wedge Q \in P_\alpha\}.$$

Proof
(1) If $\alpha \in \mathcal{P}^\cup$ and R has α-lower bound, then $max(lb_\alpha(R)) = \{\bigcup \{Q \mid Q \in lb_\alpha(R)\}\}$.
(2) If $\alpha \in \mathcal{P}^\cap$ and R has α-upper bound, then $min(ub_\alpha(R)) = \{\bigcap \{Q \mid Q \in ub_\alpha(R)\}\}$.□

The next result shows *when this model is exactly the same as the classical Rough Sets approach to relations* (the version from [24,25]).

Corollary 4

1. If $\alpha \in \mathcal{P}^{\cup} \cap \mathcal{P}^{\cap}$ and R has α-lower bound, then $\underline{\mathbf{A}}_\alpha(R) = \underline{\mathbf{A}}(R)$, and,
2. if $\alpha \in \mathcal{P}^{\cup} \cap \mathcal{P}^{\cap}$ and R has α-upper bound, then $\overline{\mathbf{A}}_\alpha(R) = \overline{\mathbf{A}}(R)$,

where $\underline{\mathbf{A}}(R)$ and $\overline{\mathbf{A}}(R)$ are classical upper and lower rough approximations over the space $(X \times X, \{P_\alpha \mid \alpha \in \mathcal{P}\})$.

Proof. (1) From the second equality in Proposition 3(1).
(2) From the second equality in Proposition 3(2). □

The next two results will show that our definitions of α-lower approximation and α-upper upproximation are sound, and their properties pretty close (but not identical) to those of standard rough set approximations as presented in Proposition 1. We start with the properties of α-lower approximation (compare with Proposition 1(1–4) for standard rough set lower approximation).

Proposition 4. *If $R, Q \subseteq X \times X$ have α-lower bound then:*

1. $R \subseteq Q \implies \underline{\mathbf{A}}_\alpha(R) \subseteq \underline{\mathbf{A}}_\alpha(Q)$,
2. $\underline{\mathbf{A}}_\alpha(R) \subseteq R$,
3. $\underline{\mathbf{A}}_\alpha(R) = \underline{\mathbf{A}}_\alpha(\underline{\mathbf{A}}_\alpha(R))$,
4. $\underline{\mathbf{A}}_\alpha(R \cap Q) = \underline{\mathbf{A}}_\alpha(\underline{\mathbf{A}}_\alpha(R) \cap \underline{\mathbf{A}}_\alpha(Q))$,
5. if $\alpha \in \mathcal{P}^{\cap}$ then $\underline{\mathbf{A}}_\alpha(R \cap Q) = \underline{\mathbf{A}}_\alpha(R) \cap \underline{\mathbf{A}}_\alpha(Q)$,
6. if R has α-upper bound then $\overline{\mathbf{A}}_\alpha(R) = \underline{\mathbf{A}}_\alpha(\overline{\mathbf{A}}_\alpha(R))$.

Proof. (1) Since $R \subseteq Q \implies lb_\alpha(R) \subseteq lb_\alpha(Q) \implies max(lb_\alpha(R)) \subseteq lb_\alpha(Q)$, then for each $S \in max(lb_\alpha(R))$ there is $S' \in max(lb_\alpha(Q))$ such that $S \subseteq S'$; and intersection preserves inclusion.
(2) Since $S \in lb_\alpha(R) \implies S \subseteq R$, and intersection preserves inclusion.
(3) From Proposition 2 because $\underline{\mathbf{A}}_\alpha(R) \in P_\alpha$.
(4) By (1) we have $\underline{\mathbf{A}}_\alpha(R \cap Q) \subseteq \underline{\mathbf{A}}_\alpha(R)$ and $\underline{\mathbf{A}}_\alpha(R \cap Q) \subseteq \underline{\mathbf{A}}_\alpha(Q)$, so $\underline{\mathbf{A}}_\alpha(R \cap Q) \subseteq \underline{\mathbf{A}}_\alpha(R) \cap \underline{\mathbf{A}}_\alpha(Q)$. Hence by (2) and (3) $\underline{\mathbf{A}}_\alpha(R \cap Q) \subseteq \underline{\mathbf{A}}_\alpha(\underline{\mathbf{A}}_\alpha(R) \cap \underline{\mathbf{A}}_\alpha(Q))$.
By the definition we have $\underline{\mathbf{A}}_\alpha(\underline{\mathbf{A}}_\alpha(R) \cap \underline{\mathbf{A}}_\alpha(Q)) = \bigcap\{S \mid S \in max(lb_\alpha(\underline{\mathbf{A}}_\alpha(Q) \cap \underline{\mathbf{A}}_\alpha(Q)))\}$.
Let $T \in lb_\alpha(\underline{\mathbf{A}}_\alpha(Q) \cap \underline{\mathbf{A}}_\alpha(Q)))$. This means $T \in P_\alpha \wedge T \subseteq \underline{\mathbf{A}}_\alpha(R) \cap \underline{\mathbf{A}}_\alpha(Q)$, hence $T \in P_\alpha \wedge T \subseteq R \wedge T \subseteq Q$, i.e. $T \in P_\alpha \wedge T \subseteq R \cap Q$. Therefore $T \in lb_\alpha(R \cap Q)$.
In this way we proved that $lb_\alpha(\underline{\mathbf{A}}_\alpha(Q) \cap \underline{\mathbf{A}}_\alpha(Q)) \subseteq lb_\alpha(R \cap Q)$. Hence $max(lb_\alpha(\underline{\mathbf{A}}_\alpha(Q) \cap \underline{\mathbf{A}}_\alpha(Q))) \subseteq lb_\alpha(X \cap Q)$, i.e. for each $S \in max(lb_\alpha(\underline{\mathbf{A}}_\alpha(Q) \cap \underline{\mathbf{A}}_\alpha(Y)))$ there exists $S' \in max(lb_\alpha(R \cap Q))$, such that $S \subseteq S'$. Since intersection preserves inclusion this means that $\underline{\mathbf{A}}_\alpha(\underline{\mathbf{A}}_\alpha(R) \cap \underline{\mathbf{A}}_\alpha(Q)) \subseteq \underline{\mathbf{A}}_\alpha(R \cap Q)$.
(5) If $\alpha \in \mathcal{P}^{\cap}$ then $\underline{\mathbf{A}}_\alpha(R) \cap \underline{\mathbf{A}}_\alpha(Q) \in P_\alpha$ so by Proposition 2 we have $\underline{\mathbf{A}}_\alpha(R) \cap \underline{\mathbf{A}}_\alpha(Q) = \underline{\mathbf{A}}_\alpha(\underline{\mathbf{A}}_\alpha(R) \cap \underline{\mathbf{A}}_\alpha(Q))$.
(6) If R has α-upper bound then $\overline{\mathbf{A}}_\alpha(R) \in P_\alpha$ so from Proposition 2, it follows, $\overline{\mathbf{A}}_\alpha(X) = \underline{\mathbf{A}}_\alpha(\overline{\mathbf{A}}_\alpha(X))$. □

The difference from the classical case is that intersection splits into two cases and mixing lower with upper α-approximation is conditional.

We will now present the properties of α-upper approximation (compare with Proposition 1(5–10) for standard rough set upper approximation).

Proposition 5. *If $R, Q \subseteq U$ have α-upper bound then:*

1. $R \subseteq Q \implies \overline{\mathbf{A}}_\alpha(R) \subseteq \overline{\mathbf{A}}_\alpha(Q)$,
2. $R \subseteq \overline{\mathbf{A}}_\alpha(R)$,
3. $\overline{\mathbf{A}}_\alpha(R) = \overline{\mathbf{A}}_\alpha(\overline{\mathbf{A}}_\alpha(R))$,
4. $\overline{\mathbf{A}}_\alpha(R \cup Q) = \overline{\mathbf{A}}_\alpha(\overline{\mathbf{A}}_\alpha(R) \cup \overline{\mathbf{A}}_\alpha(Q))$,
5. *if* $\alpha \in \mathcal{P}^\cup$ *then* $\overline{\mathbf{A}}_\alpha(R \cup Q) = \overline{\mathbf{A}}_\alpha(R) \cup \overline{\mathbf{A}}_\alpha(Q)$,
6. *if R has α-lower bound then* $\underline{\mathbf{A}}_\alpha(R) = \overline{\mathbf{A}}_\alpha(\underline{\mathbf{A}}_\alpha(R))$.

Proof. (1)   Since $R \subseteq Q \implies ub_\alpha(Q) \subseteq ub_\alpha(R) \implies min(ub_\alpha(Q)) \subseteq ub_\alpha(R)$, then for each $S' \in min(ub_\alpha(Q))$ there is $S \in min(ub_\alpha(R))$ such that $S \subseteq S'$; and union preserves inclusion.
(2)   Since $S \in ub_\alpha(R) \implies R \subseteq S$, and and union preserves inclusion.
(3)   From Proposition 2 because $\overline{\mathbf{A}}_\alpha(R) \in P_\alpha$.
(4)   By (1) we have $\overline{\mathbf{A}}_\alpha(R) \subseteq \underline{\mathbf{A}}_\alpha(R \cup Q)$ and $\overline{\mathbf{A}}_\alpha(Q) \subseteq \underline{\mathbf{A}}_\alpha(R \cup Q)$, so $\overline{\mathbf{A}}_\alpha(R) \cup \underline{\mathbf{A}}_\alpha(Q) \subseteq \underline{\mathbf{A}}_\alpha(R \cup Q)$. Hence by (2) and (3) $\underline{\mathbf{A}}_\alpha(\underline{\mathbf{A}}_\alpha(R) \cup \underline{\mathbf{A}}_\alpha(Q)) \subseteq \overline{\mathbf{A}}_\alpha(R \cup Q)$.
Since $R \subseteq \overline{\mathbf{A}}_\alpha(R)$ and $Q \subseteq \overline{\mathbf{A}}_\alpha(Q)$ then $R \cup Q \subseteq \overline{\mathbf{A}}_\alpha(R) \cup \overline{\mathbf{A}}_\alpha(Q)$, i.e.
$up_\alpha(\overline{\mathbf{A}}_\alpha(R) \cup \overline{\mathbf{A}}_\alpha(Q)) \subseteq up_\alpha(R \cup Q)$, and consequently $min(up_\alpha(\overline{\mathbf{A}}_\alpha(R) \cup \overline{\mathbf{A}}_\alpha(Q))) \subseteq up_\alpha(R \cup Q)$. Hence for each $S' \in min(up_\alpha(\overline{\mathbf{A}}_\alpha(R) \cup \overline{\mathbf{A}}_\alpha(Q)))$, there exists $S \in min(up_\alpha(R \cup Q))$ such that $S \subseteq S'$. Since union preserves inclusion, we obtained $\overline{\mathbf{A}}_\alpha(R \cup Q) \subseteq \overline{\mathbf{A}}_\alpha(\overline{\mathbf{A}}_\alpha(R) \cup \overline{\mathbf{A}}_\alpha(Q))$.
(5)   If $\alpha \in \mathcal{P}^\cup$ then $\overline{\mathbf{A}}_\alpha(R) \cup \overline{\mathbf{A}}_\alpha(Q) \in P_\alpha$ so by Proposition 2 we have $\overline{\mathbf{A}}_\alpha(R) \cup \overline{\mathbf{A}}_\alpha(Q) = \overline{\mathbf{A}}_\alpha(\overline{\mathbf{A}}_\alpha(R) \cup \overline{\mathbf{A}}_\alpha(Q))$.
(6)   If R has α-lower bound then $\underline{\mathbf{A}}_\alpha(R) \in P_\alpha$ so by Proposition 2 we have, $\underline{\mathbf{A}}_\alpha(X) = \overline{\mathbf{A}}_\alpha(\underline{\mathbf{A}}_\alpha(X))$.   □

Here the difference from the classical case is that union splits into two cases and mixing upper with lower α-approximation is conditional.

The theory proposed above can be applied to any kind of binary relations. We will now apply it to the approximation by partial orders.

9   Property-Driven Rough Partial Order Approximations of Arbitrary Relations

Let X be a set and $R \subseteq X \times X$ be any relation, and let $\lhd \subseteq X \times X$ be a partial order. Let $\mathcal{P}$ be the following set of properties *over the relation R* (i.e. $\lhd$ *is treated as a constant*), $\mathcal{P} = \{\alpha_1, \alpha_2, \alpha_3, \alpha_4, \alpha_5, \alpha_6\}$, where:

- $\alpha_1 \stackrel{df}{=} \forall a, b \in X.\ bR^\circ \subset aR^\circ \wedge R^\circ a \subset R^\circ b$, i.e. $\alpha_1 = $ *inclusion property*,
- $\alpha_2 \stackrel{df}{=} \forall a, b, c \in X.\ aRb \wedge bRa \implies aRc$, i.e. $\alpha_2 = $ *transitivity*,
- $\alpha_3 \stackrel{df}{=} \forall a, b \in X.\ \neg(aR^{cyc}b)$, i.e. $\alpha_3 = $ *acyclicity*,
- $\alpha_4 \stackrel{df}{=} \alpha_1 \wedge \alpha_3$,
- $\alpha_5 \stackrel{df}{=} (\forall a \in X.\ \neg(aRa)) \wedge \alpha_2$, i.e. $\alpha_5 = $ *partial ordering*,
- $\alpha_6 \stackrel{df}{=} \forall a, b \in X.\ a \lhd b \implies aRb$, i.e. $\alpha_6 = \lhd$ *is included in R.*

Consider the property-driven rough set approximation space $(X, \mathcal{P})$.

Directly from the definitions we may conclude that an arbitrary relation R has:

- α_1-lower bound, but not α_1-upper bound,
- α_2-lower bound and α_2-upper bound,
- α_3-lower bound, but not α_3-upper bound,
- α_4-lower bound, but not α_4-upper bound,
- α_5-lower bound, but not α_5-upper bound,
- α_6-upper bound, but not α_6-lower bound,

We have here $\mathcal{P}^\cap = \{\alpha_1, \alpha_2, \alpha_3, \alpha_4, \alpha_5\}$ and $\mathcal{P}^\cup = \{\alpha_6\}$.

From Theorem 1 if also follows that if a relation $S \in P_{\alpha_1}$ or $S \in P_{\alpha_4}$ then S is a partial order and $S \subseteq R$. If $\lhd \subseteq R$, then from Proposition 2 we immediately have $\underline{\mathbf{A}}_{\alpha_6}(R) = \overline{\mathbf{A}}_{\alpha_6}(R) = R$.

We will now show how the partially ordered approximations R^C, $R^{\mathsf{C}\wedge\bullet}$, $(R^\bullet)^\mathsf{C}$, $(R^\bullet)^+$ and $(R^+)^\bullet$ defined in Section 5 can naturally be obtain using the Rough Set approach proposed in the previous section.

Proposition 6

1. $R^\mathsf{C} = \underline{\mathbf{A}}_{\alpha_1}(R)$,
2. $R^+ = \overline{\mathbf{A}}_{\alpha_2}(R)$,
3. $R^\bullet = \underline{\mathbf{A}}_{\alpha_3}(R)$,
4. $R^{\mathsf{C}\wedge\bullet} = \underline{\mathbf{A}}_{\alpha_4}(R) = \underline{\mathbf{A}}_{\alpha_1 \wedge \alpha_3}(R)$,
5. $(R^\bullet)^\mathsf{C} = \underline{\mathbf{A}}_{\alpha_1}(\underline{\mathbf{A}}_{\alpha_3}(R))$,
6. $(R^\bullet)^+ = \overline{\mathbf{A}}_{\alpha_2}(\underline{\mathbf{A}}_{\alpha_3}(R))$,
7. $(R^+)^\bullet = \underline{\mathbf{A}}_{\alpha_3}(\overline{\mathbf{A}}_{\alpha_2}(R))$,
8. $R \cup \lhd = \overline{\mathbf{A}}_{\alpha_6}(R)$,
9. For any $Q \in \{R^\mathsf{C}, R^{\mathsf{C}\wedge\bullet}, (R^\bullet)^\mathsf{C}, (R^\bullet)^+, (R^+)^\bullet\}$, $(Q \cup \lhd)^+ = \overline{\mathbf{A}}_{\alpha_2}(\overline{\mathbf{A}}_{\alpha_6}(Q))$.

Proof. (1) Since $lb_{\alpha_1}(R) = \{R^\mathsf{C}\}$.

(2) Since R has α_2-upper bound and $\alpha_2 \in \mathcal{P}^\cap$, by Corollary 4(2), $\overline{\mathbf{A}}_{\alpha_2}(R)$ is just the smallest transitive relation containing R, i.e. R^+ (see for instance [18]).

(3) By the definition we have $\underline{\mathbf{A}}_{\alpha_3}(R) = \bigcap\{Q \mid Q \in max(lb_{\alpha_3}(R))\}$. Let $aR^{cyc}b$. This means $a = a_1 R a_2 R ... R a_{k-1} R a_k = b$, where $i \neq j \Rightarrow a_i \neq a_j$. Let $Q \in lb_{\alpha_1}(R)$. Note that $Q \in max(lb_{\alpha_1}(R))$ if and only if there is a_r such that $(a_{r-1}, a_r) \notin Q$ but $a = a_0 Q a_1 ... Q a_{r-1}$ and $a_r Q ... Q a_k$. Hence $R^\bullet = \bigcap\{Q \mid Q \in max(lb_{\alpha_3}(R))\} = \underline{\mathbf{A}}_{\alpha_3}(R)$.

(4) We have $\underline{\mathbf{A}}_{\alpha_4}(R) = \bigcap\{Q \mid Q \in max(lb_{\alpha_1 \wedge \alpha_3}(R))\} = \bigcap\{Q \mid Q \in max(\{R^\mathsf{C} \cap S \mid S \in lb_{\alpha_3}(R)\})\} = R^\mathsf{C} \cap \bigcap\{Q \mid Q \in max(lb_{\alpha_3}(R))\}$. From (3) it follows $\bigcap\{Q \mid Q \in max(lb_{\alpha_3}(R))\} = R^\bullet$, so $R^{\mathsf{C}\wedge\bullet} = R^\mathsf{C} \cap R^\bullet = \underline{\mathbf{A}}_{\alpha_1 \wedge \alpha_3}(R) = \underline{\mathbf{A}}_{\alpha_4}(R)$.

(5) From (1), (3) and the fact that $\underline{\mathbf{A}}_{\alpha_3}(R)$ has α_1-lower bound.

(6) From (2), (3) and the fact that $\underline{\mathbf{A}}_{\alpha_3}(R)$ has α_2-upper bound.

(7) From (2), (3) and the fact that $\overline{\mathbf{A}}_{\alpha_2}(R)$ has α_3-lower bound.

(8) Since R has α_6-upper bound and $\alpha_2 \in \mathcal{P}^\cup$, by Corollary 4(2), $\overline{\mathbf{A}}_{\alpha_6}(R)$ is just the smallest set containing both R and $\lhd$, i.e. $R \cup \lhd$.

(9) Note that each Q has α_6-upper bound and $\overline{\mathbf{A}}_{\alpha_6}(Q)$ has α_2-upper bound. The rest follows from (2) and (8). □

Proposition 6 not only provides a link between classical realtional algebra approach and rough set approach, but also well illustrate basic properties of property-driven rough set approximations of binary relations. Below we provide some important observations and the questions they raise:

- Property α_5 does not appear in Proposition 6 at all. It is actually a rather useless property. No upper bound exist in a general case and $\underline{\mathbf{A}}_{\alpha_5}(R)$ are not very interesting. For the examples in Figures 1 and 2, $\underline{\mathbf{A}}_{\alpha_5}(R) = \emptyset$! Property α_5 (being a partial order) is just too strong to be efficiently handled as a whole. We can get much better results when we treat the components of α_5, for instance acyclicity and transitivity, separately and then compose the results obtained.
- We have $\alpha_4 = \alpha_1 \wedge \alpha_3$, but $R^{\subset \wedge \bullet} = \underline{\mathbf{A}}_{\alpha_4}(R) = \underline{\mathbf{A}}_{\alpha_1 \wedge \alpha_3}(R) \neq (R^\bullet)^\subset = \underline{\mathbf{A}}_{\alpha_1}(\underline{\mathbf{A}}_{\alpha_3}(R))$, and $R^{\subset \wedge \bullet} = \underline{\mathbf{A}}_{\alpha_1 \wedge \alpha_3}(R) \neq \underline{\mathbf{A}}_{\alpha_3 \wedge \alpha_1}(R) = (R^\subset)^\bullet = R^\subset$. Moreover $\underline{\mathbf{A}}_{\alpha_1 \wedge \alpha_3}(R) \subseteq \underline{\mathbf{A}}_{\alpha_1}(\underline{\mathbf{A}}_{\alpha_3}(R))$, $\underline{\mathbf{A}}_{\alpha_1 \wedge \alpha_3}(R) \subseteq \underline{\mathbf{A}}_{\alpha_3}(\underline{\mathbf{A}}_{\alpha_1}(R))$ and $\underline{\mathbf{A}}_{\alpha_1 \wedge \alpha_3}(R) \neq \underline{\mathbf{A}}_{\alpha_3 \wedge \alpha_1}(R)$.
  This means that starting with compound properties like $\alpha_1 \wedge \alpha_3$ may lead to different approximations that approximating each property separately and that the order of composition does matter. Are there some general rules regulating this process?
- We also have $(R^\bullet)^+ = \overline{\mathbf{A}}_{\alpha_2}(\underline{\mathbf{A}}_{\alpha_3}(R)) \subseteq \underline{\mathbf{A}}_{\alpha_3}(\overline{\mathbf{A}}_{\alpha_2}(R)) = (R^+)^\bullet$. Is this a coincidence or a general pattern?

We will try to answer some of the above questions in the next section.

10   Compound Properties and Mixed Approximations

Most of the interesting properties are compound properties, like transitivity *and* reflexivity, and they can be imposed in various orders, for instance as $(R^\bullet)^+$ or $(R^+)^\bullet$. In this section we will propose a framework for doing this in a systematic way.

Let $\alpha, \beta \in \mathcal{P}$. We say that β *is consistent with* α iff for every $R \in P_\alpha$:

- if R has β-lower bound then $\underline{\mathbf{A}}_\beta(R) \in P_\alpha$,
- if R has β-upper bound then $\overline{\mathbf{A}}_\beta(R) \in P_\alpha$.

We will say that α and β are *consistent* iff β *is consistent with* α and α *is consistent with* β.

We will also *assume* that for all $\alpha, \beta \in \mathcal{P}$, α and β are *consistent*.

From now on when writing a formula like $\underline{\mathbf{A}}_\beta(\overline{\mathbf{A}}_\alpha(R))$ we will assume that all necessary conditions are satisfied, i.e. in this case, R has α-upper bound and $\overline{\mathbf{A}}_\alpha(R)$ has β-lower bound.

Proposition 7

1. $\mathbf{A}_\alpha(\mathbf{A}_\beta(R)) \in P_\alpha \cap P_\beta$ for $\mathbf{A}_\alpha \in \{\underline{\mathbf{A}}_\alpha, \overline{\mathbf{A}}_\alpha\}, \mathbf{A}_\beta \in \{\underline{\mathbf{A}}_\beta, \overline{\mathbf{A}}_\beta\}$,
2. $\overline{\mathbf{A}}_\alpha(\underline{\mathbf{A}}_\beta(R)) \subseteq \underline{\mathbf{A}}_\beta(\overline{\mathbf{A}}_\alpha(R))$.

Proof. (1)   Because all α and β from $\mathcal{P}$ are consistent.

(2)   By Proposition 5(2), $R \subseteq \overline{\mathbf{A}}_\alpha(R)$, so $\underline{\mathbf{A}}_\beta(R) \subseteq \underline{\mathbf{A}}_\beta(\overline{\mathbf{A}}_\alpha(R))$, and $\overline{\mathbf{A}}_\alpha(\underline{\mathbf{A}}_\beta(R)) \subseteq \overline{\mathbf{A}}_\alpha(\underline{\mathbf{A}}_\beta(\overline{\mathbf{A}}_\alpha(R)))$. By (1) of this proposition, $\underline{\mathbf{A}}_\beta(\overline{\mathbf{A}}_\alpha(R)) \in P_\alpha$, so by Proposition 2, $\overline{\mathbf{A}}_\alpha(\underline{\mathbf{A}}_\beta(\overline{\mathbf{A}}_\alpha(R))) = \underline{\mathbf{A}}_\beta(\overline{\mathbf{A}}_\alpha(R))$. Therefore $\overline{\mathbf{A}}_\alpha(\underline{\mathbf{A}}_\beta(R)) \subseteq \underline{\mathbf{A}}_\beta(\overline{\mathbf{A}}_\alpha(R))$. □

The case $(R^\bullet)^+ = \overline{\mathbf{A}}_{\alpha_2}(\underline{\mathbf{A}}_{\alpha_3}(R)) \subseteq \underline{\mathbf{A}}_{\alpha_3}(\overline{\mathbf{A}}_{\alpha_2}(R)) = (R^+)^\bullet$ illustrates the above proposition. Note that the above result is consistent with Propositions 4(6) and 5(6).

Proposition 8. *Assume that α, β and $\alpha \wedge \beta$ belong to $\mathcal{P}$.*

1. $\underline{\mathbf{A}}_{(\alpha \wedge \beta)}(R) \subseteq \underline{\mathbf{A}}_\alpha(\underline{\mathbf{A}}_\beta(R))$,

2. $\overline{\mathbf{A}}_\alpha(\overline{\mathbf{A}}_\beta(R)) \subseteq \overline{\mathbf{A}}_{(\alpha \wedge \beta)}(R)$.

Proof. (1)   Since obviously $lb_{(\alpha \wedge \beta)}(R) \subseteq lb_\beta(R)$ then $\underline{\mathbf{A}}_{(\alpha \wedge \beta)}(R) \subseteq \underline{\mathbf{A}}_\beta(R)$. Hence $\underline{\mathbf{A}}_\alpha(\underline{\mathbf{A}}_{(\alpha \wedge \beta)}(R)) \subseteq \underline{\mathbf{A}}_\alpha(\underline{\mathbf{A}}_\beta(R))$. Since $\underline{\mathbf{A}}_{(\alpha \wedge \beta)}(R) \in P_\alpha$, then due to Proposition 2, $\underline{\mathbf{A}}_\alpha(\underline{\mathbf{A}}_{(\alpha \wedge \beta)}(R)) = \underline{\mathbf{A}}_{(\alpha \wedge \beta)}(R)$, which ends the proof of (1).

(2)   Since obviously $ub_{(\alpha \wedge \beta)}(R) \subseteq ub_\beta(R)$ then $min(ub_{(\alpha \wedge \beta)}(R)) \subseteq ub_\beta(R)$. This means $\overline{\mathbf{A}}_\beta(R) \subseteq \overline{\mathbf{A}}_{(\alpha \wedge \beta)}(R)$. Hence $\overline{\mathbf{A}}_\alpha(\overline{\mathbf{A}}_\beta(R)) \subseteq \overline{\mathbf{A}}_\alpha(\overline{\mathbf{A}}_{(\alpha \wedge \beta)}(R))$. Since $\overline{\mathbf{A}}_{(\alpha \wedge \beta)}(R) \in P_\alpha$, then due to Proposition 2, $\overline{\mathbf{A}}_\alpha(\overline{\mathbf{A}}_{(\alpha \wedge \beta)}(R)) = \overline{\mathbf{A}}_{(\alpha \wedge \beta)}(R)$, which ends the proof of (2).   □

The case $\underline{\mathbf{A}}_{\alpha_1 \wedge \alpha_3}(R) \subseteq \underline{\mathbf{A}}_{\alpha_1}(\underline{\mathbf{A}}_{\alpha_3}(R))$ illustrates Proposition 8(1).

Proposition 8 suggest an important technique for the design of approximation schema. It in principle says that using a complex predicate as a property result is a *worse* approximation than when the property is decomposed into simpler ones, and then we approximate all simpler properties. This means before starting an approximation process we should think carefully how the given property could be decomposed into the simpler ones.

Define $\widehat{\mathcal{P}} = \mathcal{P} \times \{0, 1\}$. The elements of $\widehat{\mathcal{P}}$ will be called *labelled elementary properties*. We will also write $\alpha^{(0)}$ or $\underline{\alpha}$ instead of $(\alpha, 0)$ and $\alpha^{(1)}$ or $\overline{\alpha}$ instead of $(\alpha, 1)$.

A sequence $s = \alpha_1^{(i_1)} \alpha_2^{(i_2)} .... \alpha_k^{(i_k)}$ of elements of $\widehat{\mathcal{P}}$ such that $\alpha_i \neq \alpha_{i+1}$, for $i = 1, ..., k - 1$, is called a *schedule*.

For will also use $\mathbf{A}^{(0)}$ instead of $\underline{\mathbf{A}}$ and $\mathbf{A}^{(1)}$ instead of $\overline{\mathbf{A}}$.

A schedule $s = \alpha_1^{(i_1)} \alpha_2^{(i_2)} .... \alpha_k^{(i_k)}$ is *proper* if for each $R \subseteq X \times X$ the following *mixed approximation*

$$\mathbf{A}^s(R) = \mathbf{A}_{\alpha_1}^{(i_1)}(\mathbf{A}_{\alpha_2}^{(i_2)}(...(\mathbf{A}_{\alpha_k}^{(i_k)}(R))...))$$

is well defined. For example for $\mathcal{P} = \{\alpha_1, \alpha_2, \alpha_3, \alpha_4, \alpha_5, \alpha_6\}$ from the previous section, the sequences $\alpha_2^{(1)} \alpha_3^{(0)}$, $\alpha_3^{(0)} \alpha_2^{(1)}$, $\alpha_3^{(0)} \alpha_1^{(0)}$, $\alpha_2^{(1)} \alpha_6^{(1)} \alpha_2^{(1)} \alpha_3^{(0)}$, $\alpha_2^{(1)} \alpha_6^{(1)} \alpha_1^{(0)}$, etc., are proper schedules.

Let $\mathbf{PS}_\mathcal{P}$ denote the set of all proper schedules.

Each schedule $s = \alpha_1^{(i_1)} \alpha_2^{(i_2)} .... \alpha_k^{(i_k)}$ defines a *composite property*

$$\pi(s) = \alpha_1 \wedge \alpha_2 \wedge .... \wedge \alpha_k.$$

For the case from the previous section, $\pi(\alpha_2^{(1)} \alpha_3^{(0)}) = \pi(\alpha_3^{(0)} \alpha_2^{(1)}) = \alpha_2 \wedge \alpha_3$, $\pi(\alpha_3^{(0)} \alpha_1^{(0)}) = \alpha_1 \wedge \alpha_3$, $\pi(\alpha_2^{(1)} \alpha_6^{(1)} \alpha_2^{(1)} \alpha_3^{(0)}) = \alpha_2 \wedge \alpha_3 \wedge \alpha_6$, $\pi(\alpha_2^{(1)} \alpha_6^{(1)} \alpha_1^{(0)}) = \alpha_1 \wedge \alpha_2 \wedge \alpha_6$, etc. Obviously $\pi(\alpha_1^{(0)}) = \alpha_1$ and $\pi(\alpha_3^{(0)}) = \alpha_3$.

A composite property α is *approximable* if there exists a proper sequence $s \in \mathbf{PS}_\mathcal{P}$ such that $\alpha = \pi(s)$.

The property of being a partial order may be defined in various ways, for example by composite properties, $\alpha_2 \wedge \alpha_3$, $\alpha_1 \wedge \alpha_3$, or by elemetary properties α_1, α_4 and α_5. The composite properties $\alpha_2 \wedge \alpha_3$, $\alpha_1 \wedge \alpha_3$ are approximable, as $\pi(\alpha_2^{(1)} \alpha_3^{(0)}) = \pi(\alpha_3^{(0)} \alpha_2^{(1)}) = \alpha_2 \wedge \alpha_3$ and $\pi(\alpha_3^{(0)} \alpha_1^{(0)}) = \alpha_1 \wedge \alpha_3$ belong to $\mathbf{PS}_\mathcal{P}$ where $\mathcal{P} = \{\alpha_1, \alpha_2, \alpha_3, \alpha_4, \alpha_5, \alpha_6\}$. The 'natural' partial order property α_5 is useless for approximation reasons, the properties α_1 and α_4 give some approximations that might be reasonable in some circumstances.

The proper schedules could be interpreted as different "metrics" used for approximation purposes.

11 Final Comment

Two systematic approaches to finding partial order approximations of arbitrary relations have been proposed. The first one uses the standard theory of relations, the second one is rooted in the Rough Sets paradigm. While in both cases we obtained the same set of "natural" approximations, namely the partial orders R^C, $R^{C \wedge \bullet}$, R^C, $(R^\bullet)^C$, $(R^\bullet)^+$ and $(R^+)^\bullet$, the *Rough Sets approach is much more general*, it could be used for approximation by any kind of relations; and also gives much better intuitive justification of the results obtained.

It is usually assumed that ranking is a *weak order* [6], and none of the relations R^C, $R^{C \wedge \bullet}$, R^C, $(R^\bullet)^C$, $(R^\bullet)^+$ and $(R^+)^\bullet$ guarantee this, so they must eventually be extended to appropriate weak orders using one of the methods proposed in [5]. This process is not discussed in this paper, an interested reader is referred to [9,11]. By modifying an example from the Introduction one may show that each of the five partial order approximations of R is better than the others in given circumstances, however some experiments made to justify some claims of [8] indicate that often $(R^+)^\bullet$ could be interpreted as the "best" partial order approximation. This appears to be especially true when cycles of R are naturally interpreted as indifference (see Theorem 1(8)).

References

1. Arrow, K.J.: Social Choice and Individual Values. J. Wiley, New York (1951)
2. Burris, S., Sankappanavar, H.P.: A Course in Universal Algebra. Springer, New York (1981)
3. Cowan, N.: The magical number 4 in short-term memory. A reconsideration of mental storage capacity. Behavioural and Brain Sciences 24, 87–185 (2001)
4. Dyer, J.S.: Remarks on the Analytic Hierarchy Process. Management Sci. 36, 244–258 (1990)
5. Fishburn, P.C.: Interval Orders and Interval Graphs. J. Wiley, New York (1985)
6. French, S.: Decision Theory. Ellis Horwood, New York (1986)
7. Huth, M.R.A., Ryan, M.D.: Logic in Computer Science. Cambridge University Press, Cambridge (2000)
8. Janicki, R.: Pairwise Comparisons, Incomparability and Partial Orders. In: Proc. of Int. Conf. on Enterprise Information Systems, ICEIS 2007, Funchal, Portugal, vol. 2, pp. 297–302 (2007)

9. Janicki, R.: Ranking with Partial Orders and Pairwise Comparisons. In: Wang, G., Li, T., Grzymala-Busse, J.W., Miao, D., Skowron, A., Yao, Y. (eds.) RSKT 2008. LNCS (LNAI), vol. 5009, pp. 442–451. Springer, Heidelberg (2008)

10. Janicki, R.: Some Remarks on Approximations of Arbitrary Binary Relations by Partial Orders. In: Chan, C.-C., Grzymala-Busse, J.W., Ziarko, W.P. (eds.) RSCTC 2008. LNCS (LNAI), vol. 5306, pp. 81–91. Springer, Heidelberg (2008)

11. Janicki, R., Koczkodaj, W.W.: Weak Order Approach to Group Ranking. Computers Math. Applic. 32(2), 51–59 (1996)

12. Janicki, R., Koutny, M.: Structure of Concurrency. Theoretical Computer Science 112, 5–52 (1993)

13. de Condorcet, M.: Essai sur l'application de l'analyse a la probabilite des decisions rendue a la pluralite des voix, Paris (1785) (see [1])

14. Kleinberg, J., Tardos, E.: Algorithm Design. Addison-Wesley, Reading (2006)

15. Kuratowski, K., Mostowski, A.: Set Theory, 2nd edn. North-Holland, Amsterdam (1976)

16. Pawlak, Z.: Rought Sets. International Journal of Computer and Information Sciences 34, 557–590 (1982)

17. Pawlak, Z.: Rough Sets. Kluwer, Dordrecht (1991)

18. Rosen, K.H.: Discrete Mathematics and Its Applications. McGraw-Hill, New York (1999)

19. Saaty, T.L.: A Scaling Methods for Priorities in Hierarchical Structure. Journal of Mathematical Psychology 15, 234–281 (1977)

20. Schröder, E.: Algebra der Logik. Teuber, Leipzig (1895); 2nd edition published by Chelsea (1966)

21. Skowron, A., Stepaniuk, J.: Approximation of relations. In: Ziarko, W. (ed.) Rough Sets, Fuzzy Sets and Knowledge Discovery, pp. 161–166. Springer, London (1994)

22. Skowron, A., Stepaniuk, J.: Tolerence approximation spaces. Fundamenta Informaticae 27, 245–253 (1996)

23. Stewart, J.: Calculus. Concepts and Contexts. Brooks/Cole, Pacific Grove (1997)

24. Yao, Y.Y.: Two views of the theory of rough sets in finite universes. International Journal of Approximate Reasoning 15, 291–317 (1996)

25. Yao, Y.Y., Wang, T.: On Rough Relations: An Alternative Formulations. In: Zhong, N., Skowron, A., Ohsuga, S. (eds.) RSFDGrC 1999. LNCS (LNAI), vol. 1711, pp. 82–91. Springer, Heidelberg (1999)

Hybridization of Rough Sets and Statistical Learning Theory

Wojciech Jaworski

Faculty of Mathematics, Computer Science and Mechanics
Warsaw University, Banacha 2, 02-097 Warsaw, Poland
`wjaworski@mimuw.edu.pl`

Abstract. In this paper we propose the hybridization of the rough set concepts and statistical learning theory. We introduce new estimators for rule accuracy and coverage, which base on the assumptions of the statistical learning theory. These estimators allow us to select rules describing statistically significant dependencies in data. Then we construct classifier which uses these estimators for rule induction. In order to make our solution applicable for information systems with missing values and multiple valued attributes, we propose axiomatic representation of information systems and we redefine the indiscernibility relation as a relation on objects characterized by axioms. Finally, we test our classifier on benchmark datasets.

Keywords: Rough sets, quality measures, accuracy, coverage, significance, rule induction, rule selection, missing values, multiple valued attributes.

1 Introduction

Rough set theory [1,2] and statistical learning theory [3] provide two different methodologies for reasoning from data.

The rough set concept theory is a theoretical framework for describing and inferring knowledge. Examined knowledge is imperfect. It is imprecise due to vague concepts involved in knowledge representation and it is based on incomplete data. The central point of the theory is the idea of concept approximation by the set of objects that certainly belong to the concept and the set of those which may belong to the concept on the basis of possessed data. Then these two sets are described in terms of available attributes.

The main goal of statistical learning theory is to provide a framework for studying the problem of inference. For this purpose, there are introduced statistical assumptions about the way the data is generated. A probabilistic model of data generation process, which is the core of the theory, establishes the formalisation of relationships between past and future observations.

While rough set theory provides an intuitive description of relationships in data and approximations for dependencies that cannot be defined in an exact

J.F. Peters et al. (Eds.): Transactions on Rough Sets XIII, LNCS 6499, pp. 39–55, 2011.

way, statistical learning theory measures the significance and correctness of discovered dependencies.

The combination of both approaches provides us tools for building simple, human understandable classifiers, whose quality will be guaranteed by the statistical assumptions.

In this paper we propose the hybridization of the rough set approach and statistical learning theory. We define the probabilistic model of data generation process, which allow us to explain the process of data acquisition and to infer knowledge that would be applied for all existing objects, not only for the ones that are mentioned in data.

We recall rough set concepts in this new setting. However, we introduce axiomatic representation of information systems, which we developed in [4]. We define rough set concepts such as indiscernibility, definability and set approximations in this setting. In the case of complete information systems, the proposed approach is equivalent to the approach used so far in rough set theory [1,2]. Yet, it allows us to incorporate information systems with missing values and multiple valued attributes into the theory 'seamlessly' — without the need of any modification of the rough set concepts.

Then we show how to extend set approximations from a sample to the set of all objects. Our attitude is similar to the idea of inductive extensions of approximation spaces presented, for example, in [5,6].

We introduce measures of approximation quality: accuracy and coverage. Taking advantage of the underlying probabilistic model we estimate values of the above indices on the set of all objects using a sample. We propose two estimators: one based on Hoeffding inequality [7], and second based on the optimal probability bound presented in [8,9].

The statistical nature of estimators leads us to the index, the measure called significance. Significance measures how often sample-based accuracy and coverage estimations are correct. The trade-off relation between these three measures allow us to balance the approximation between fitting to the sample and generalisation.

The properties of accuracy and coverage were thoroughly studied in [10]. The author proposed a probabilistic definition of the indices, yet he neither defined any underlying probability model nor showed the trade-off between accuracy or coverage and significance. Quality measures were also examined from the statistical point of view in [11], but without placing them in the rough set context.

[12] propose an application of statistical techniques in rough set data analysis, yet they did not incorporate the assumptions on the data generating process required by these techniques into the presented model.

In order to show how the estimators behave in practice we developed a simple rule-based classifier. Estimated indices guarantee the quality of each rule, determine the required accuracy level for rule to be accepted and decide how many objects have to match the rule in order to make it significant. We test the classifier on benchmark datasets obtained from [13] and we apply it in analysis of Neo-Sumerian economic documents (for details see [14]).

Test results reveal that the obtained classifier generates highly relevant rules. Each rule is assigned with its accuracy and coverage estimations. Rules cover only that part of universe for which it is possible to predict decision with high accuracy. As a consequence the classifier is able to judge whether it has enough knowledge to classify a certain object.

2   Probabilistic Model

We propose the following definition of the problem of induction. Let $\mathbb{U}$ be a finite set of objects for a given domain. We denote $\mathbb{U}$ as *universe*. We introduce a probability measure $P_\#$ on $2^{\mathbb{U}}$ according to the following formula:

$$P_\#(X) := \frac{|X|}{|\mathbb{U}|},$$

where $|\cdot|$ denotes the number of elements in a set.

Statistical learning theory [3] assumes that the phenomena underlying generated data have statistical nature and the observed objects are independent, identically distributed random variables.

Formally we introduce a probability space $(\Omega, 2^{\Omega}, P)$. Observed objects $u_1, u_2, \ldots, u_i, \ldots$ are values of independent random variables $U_1, U_2, \ldots, U_i, \ldots$. Each U_i is a function $U_i : \Omega \to \mathbb{U}$. The distribution of U_i is identical to $P_\#$, i.e.:

$$\forall_i \forall_{X \subseteq \mathbb{U}} P_\#(X) = P(\{\omega \in \Omega \mid U_i(\omega) \in X\}) = P(U_i^{-1}(X))$$

We do not know $\mathbb{U}$ and $P_\#$.

Let $U \subseteq \mathbb{U}$ be a non-empty, finite set of observed objects called a *sample*. U is the only part of the domain $\mathbb{U}$ which is known to us. We denote elements of U by $u_1, \ldots, u_n$, where u_i is a realisation (or value) of the random variable U_i. The information which we posses about the domain is usually represented in terms of information system.

3   Complete Information Systems

In this section, we define information systems [15] and we recall their axiomatic representation, which we introduced in [4].

Information systems are based on the assumption that examined domain is organised in terms of *objects* possessing *attributes*. Depending on the nature of domain, objects are interpreted as, e.g. cases, states, processes, patients, observations. Attributes are interpreted as features, variables, characteristics, conditions, etc.

Let $U \subset \mathbb{U}$ be a sample — a non-empty, finite set of known objects. Let A be a non-empty finite set of known attributes. Each attribute $a \in A$ has its domain V_a. An information system defines attribute values for given objects. Let

$$m(u, a)$$

denote the set of values of the attribute a for the object u in the information system.

Usually information systems are presented in a form of tables whose rows represent objects and columns are labelled by attributes.

In case when each attribute has exactly one and known value for each object, i.e $m(u, a)$ contains one element for every u and a, information system is called *complete*.

We consider an information system as a set of axioms. The information system provides the structural information about the domains of the attributes. We represent this information by means of axioms that set constraints on a set of possible worlds. For each attribute a we state

$$\forall_{x,y} \, a(x, y) \Longrightarrow x \in \mathbb{U} \wedge y \in V_a.$$

The complete information system also states that every attribute has exactly one value for each object: for each attribute a we write the following axiom

$$\forall_{x \in \mathbb{U}} \, \exists!_y \, a(x, y).$$

We encode the contents of the information system as a set formulae in the following way: For each $u \in U$, for each $a \in A$ such that $v \in m(u, a)$ in the information system we add the following axiom:

$$a(u, v).$$

The above transformation treats both an object etiquette and an attribute value as constants. The attributes are considered as binary relations.

Real-life data is frequently incomplete , i.e. values for some attributes are missing (see e.g. [16,17,18,19]). We will assume three different interpretations of missing values:

- missing attribute values that are *lost*, i.e they are specified, yet their values are unknown
- attributes *not applicable* in a certain case, e.g. the date of death of a person who is still alive.
- *do not care values*: the attribute may have any value from its domain.

We will extend the definition of $m(u, a)$. $m(u, a) = ?$ will mean that the value of attribute a for object u is lost, $m(u, a) = \star$ that it is 'do not care' and $m(u, a) = -$ that it is not applicable.

We express the various types of missing value semantics using axioms:

- for each $u \in U$, for each $a \in A$ we state

$$a(u, v),$$

where $v \in m(u, a)$ in the information system.

- 'lost' values are defined as follows: for each $u \in U$, for each $a \in A$ we state

$$a(u, v_1) \vee \cdots \vee a(u, v_n),$$

where $v_1, \ldots, v_n$ are all possible values of attribute a.
- for each $u \in U$, for each $a \in A$ whose value is not applicable we state

$$\forall_x \neg a(u, x),$$

- for each $u \in U$, for each $a \in A$, for each v from the domain of a we state

$$a(u, v),$$

when the value of a is 'do not care' for object u.

Multiple valued attributes (introduced in [15] and studied in [20]) may reflect our incomplete knowledge about their values, what makes them similar to 'lost' missing values. The may also represent attributes that have a few values simultaneously, in which case the are like 'do not care' missing values.

- 'lost' multiple values we define as follows: for each $u \in U$, for each $a \in A$ we state

$$a(u, v_1) \vee \cdots \vee a(u, v_n),$$

where $v_1, \ldots, v_n$ are all possible values of attribute a for object u mentioned in the information system.
- for each $u \in U$, for each $a \in A$, for each value v of attribute a for object u in information system

$$a(u, v),$$

when the value of a is 'do not care' multiple value for object u.

4   Rough Set Theory

The rough set theory [1,2] is based on the idea of an indiscernibility relation. In this section, we define indiscernibility and set approximations.

In this and the following sections we assume that we are given an information system $\mathcal{A}$. U denote the finite set of objects described in $\mathcal{A}$, A is the finite set of attributes in $\mathcal{A}$ and $\mathbb{A}$ is a set of axioms derived from $\mathcal{A}$.

Let B be a nonempty subset of A. The indiscernibility relation $IND(B)$ is a relation on objects in a complete information system defined for $x, y \in U$ as follows

$$(x, y) \in IND(B) \text{ iff } \forall a \in B\big(m(x, a) = m(x, a)\big).$$

IND is an equivalence relation. We will denote its equivalence class generated by object u as

$$[u]_{IND(B)}.$$

The notion of indiscernibility is used to define set approximations. A given set $X \subseteq U$ may be approximated using only the information contained in $B \subset A$ by constructing the *B-lower* and *B-upper approximations of* X, denoted $\underline{B}X$ and $\overline{B}X$ respectively, where

$$\underline{B}X = \bigcup\{[u]_B \mid [u]_B \subseteq X\}$$

and

$$\overline{B}X = \bigcup\{[u]_B \mid [u]_B \cap X \neq \emptyset\}.$$

The above theory was designed or complete information systems. However, in [4], we proved that the concepts of indiscernibility and set approximations in a way that they could cover also information systems with missing values and multivariate attributes.

Now we recall this definition. First we introduce auxiliary concepts of descriptor, query and conditional formula.

Definition 1. *For a given set of attributes* $B \subseteq A$, *formulae of the form*

$$a(x, v),$$

where $a \in A$, $v \in V_a$ *and* x *is a free variable, are called* descriptors *over* B.

Definition 2. *By a* query *over the set of attributes* B *we denote any formula*

$$\bigwedge_{i=1}^{n} \varphi_i(x),$$

where each φ_i *is a descriptor over* B *and* $n \leq |B|$. x *is a free variable ranging over objects.*

Definition 3. *The set of* conditional formulae *over* B *is defined as the least set containing all descriptors over* B *and closed with respect to the propositional connectives* $\wedge$ *(conjunction),* $\vee$ *(disjunction) and* $\neg$ *(negation).*

Note that every query is a conditional formula.

Definition 4. *Let* $\varphi(x)$ *be a conditional formula. By* $||\varphi(x)||_{U,\mathbb{A}}$ *we will denote the set of all elements from* U *for which* φ *is a semantic consequence of* $\mathbb{A}$, *i.e.:*

$$||\varphi(x)||_{U,\mathbb{A}} = \{x \in U \mid \mathbb{A} \models \varphi(x)\}.$$

We postulate the following definition of indiscernibility:

Definition 5. *Let* $\varphi(x)$ *be a query with free variable* x. *Let* u_1 *and* u_2 *be constants. We say that* u_1 *and* u_2 *are indiscernible by the query* $\varphi(x)$ *if*

$$\left(\mathbb{A} \models \varphi(u_1)\right) \Longleftrightarrow \left(\mathbb{A} \models \varphi(u_2)\right).$$

Theorem 1. *Let* $\mathcal{A}$ *be a complete information system. Let* B *be a subset of* A. *Objects* $u_1 \in U$ *and* $u_2 \in U$ *are indiscernible with respect to attribute set* B *iff they are indiscernible with respect to every query over the set of attributes* B.

Proof. See [4].

A given set $X \subset U$ is either *definable* or *indefinable* by attributes in the information system depending on the existence of conditional formula that recognizes its elements:

Definition 6. *Let X be a subset of U. We say that X is* definable *by $\mathbb{A}$ iff there exist queries $\varphi_1(x), \ldots, \varphi_n(x)$ such that*

$$X = ||\varphi_1(x) \vee \cdots \vee \varphi_n(x)||_{U,\mathbb{A}}$$

Each definable set is a sum of objects that satisfy at least one of a given queries.

Any set $X \subset U$ may be approximated by two definable sets. The first one is called the *lower approximation* of X, denoted by $\underline{\mathbb{A}}X$, and is defined by

$$\bigcup \{Y \mid Y \subset X \wedge Y \text{ is definable by } \mathbb{A}\}.$$

The second set is called the *upper approximation* of X, denoted by $\overline{\mathbb{A}}X$, and is defined by

$$\bigcap \{Y \mid X \subset Y \wedge Y \text{ is definable by } \mathbb{A}\}.$$

$\overline{\mathbb{A}}X \subset U$ because every definable set is a subset of U.

Theorem 2

$$\underline{A}X = \underline{\mathbb{A}}X \text{ and } \overline{A}X = \overline{\mathbb{A}}X.$$

For the proof of the above theorem and further results concerning comparison of our concept of set approximations with the one proposed by other authors see [4].

Classification is a process of finding dependencies between values of attributes. Let $\mathbb{A}$ be a given set of axioms which define attributes A for objects from the set U. We select one of attributes from A which we denote as d — decision attribute. Let $B = A \setminus \{d\}$. Our goal is to estimate the value of attribute d on the basis of other attribute values for a given object. For each value v of the decision attribute, there exist conditional formulae over B that define the lower and upper approximation of $||d(x, v)||_{U,\mathbb{A}}$. We denote them $\underline{\varphi}_v(x)$ and $\overline{\varphi}_v(x)$ respectively.

$$||\underline{\varphi}_v(x)||_{U,\mathbb{A}} \subseteq ||d(x, v)||_{U,\mathbb{A}} \subseteq ||\overline{\varphi}_v(x)||_{U,\mathbb{A}}$$

Set approximations for all decision values compose a classifier.

5   Extended Approximations

In the above section we considered set approximations that described the dependence between the attribute values and the value of decision for objects in U. Now, we extend set approximations on the whole universe $\mathbb{U}$.

The assumption that past and future observations are both sampled independently from the same distribution provides us with tools for extending the approximations. However, the extension will be correct only with some probability.

Inductive reasoning is based on the assumption that the definition generated for the sample data is still valid in the general case. For a given set of attributes B, *extended approximations* are represented by means of conditional formulae over B interpreted in the universe $\mathbb{U}$. Let φ be a conditional formula over B and let $||\varphi||_{\mathbb{U},\mathbb{A}}$ denote the subset of elements of the universe $\mathbb{U}$ that satisfy the formula.

For every U_i we obtain from its definition[1]

$$P_{\#}(||a(x,v)||_{\mathbb{U},\mathbb{A}}) = P_{\#}(\{x \in \mathbb{U} \mid \mathbb{A} \models a(x,v)\}) =$$

$$= P(\{\omega \in \Omega | a(U_i(\omega), v)\}) = P(a(U_i, v)).$$

This correspondence may be easily extended on all conditional formulae.

Now, we define extended approximations using conditional formulae interpreted in the universe $\mathbb{U}$:

Definition 7. *Let $X \subseteq \mathbb{U}$ and B be a set of attributes and let $Y \subseteq \mathbb{U}$ be such that*

$$Y = ||\varphi||_{\mathbb{U},\mathbb{A}},$$

where φ is a conditional formula over B. Let $\alpha, \kappa \in [0,1]$. The set $Y \subseteq \mathbb{U}$ is called B-α-κ-approximation of X when

$$P_{\#}(X \mid Y) \geq \alpha \text{ and } P_{\#}(Y \mid X) \geq \kappa.$$

We call α as the approximation accuracy *and we denote κ as the approximation* coverage.

As opposed to the standard approximations defined in a decision system, this definition does not construct a set Y, it only states whether a given set possesses a property of being an α-κ-approximation.

Accuracy and coverage are indices of the approximation quality. Accuracy measures the probability that an object belonging to the approximation belongs also to the approximated set. Coverage measures the fraction of objects in a set that are included in its approximation. When the approximation accuracy is equal to 1 and the coverage is maximised the approximation may be considered as *lower* one and when the approximation coverage is equal to 1 and the accuracy is maximised the approximation may be considered as *upper* one.

Accuracy and coverage are defined by means of the underlying probability distribution, according to which the sample is drawn. Since we are given only a sample and we do not know the probability distribution, we must estimate values of the indices using the sample and probabilistic inequalities of the form

$$P\big(P_{\#}(X \mid Y) \geq f_n(U_1, \ldots, U_n)\big) \geq \gamma_n.$$

---

[1] The latter equality introduces a standard probabilistic notation in which 'ω', '{' and '}' are omitted in expressions with random variables.

The above inequality may be interpreted in the following way: if we draw $\{(u_1^i, u_2^i, \ldots, u_n^i)\}_{i=1}^{\infty}$, an infinite sequence of n-element samples, where u_j^i is a realisation of U_j^i, then according to the law of large numbers

$$P\big(P_\#(X \mid Y) \geq f_n(U_1, \ldots, U_n)\big) = P\big(P_\#(X \mid Y) \geq f_n(U_1^i, \ldots, U_n^i)\big) =$$

$$= \lim_{k \to \infty} \frac{1}{k} \cdot |\{i \leq k \mid P_\#(X \mid Y) \geq f_n(u_1^i, \ldots, u_n^i)\}|.$$

Hence γ_n describes how frequent it is true that $P_\#(X \mid Y) \geq f_n(u_1^i, \ldots, u_n^i)$ or, in other words how likely $P_\#(X \mid Y) \geq f_n(u_1^i, \ldots, u_n^i)$ is to happen in one occurrence. γ_n is a measure called *significance*.

We propose two methods of deriving estimators of the accuracy and the coverage on the basis of sample. The first bases on the Hoeffding inequality [7]:

Theorem 3. *Let $Z_1, \ldots, Z_n$ be identically distributed independent random variables. Assume that each $Z_i : \Omega \to [0,1]$. Then, for every $\varepsilon > 0$, the following inequality holds:*

$$P(EZ_1 \leq \frac{1}{n} \sum_{i=1}^{n} Z_i + \varepsilon) \geq 1 - e^{-2n\varepsilon^2}. \tag{1}$$

$\square$

We derive estimator from this theorem as follows: assume that Y is an α-κ-approximation for a set X. Let U be a sample and let $\{U_1, \ldots, U_n\} = U \cap Y$. For the purpose of accuracy estimation we declare that

$$Z_i = \begin{cases} 0, & \text{when } U_i \in X \\ 1, & \text{when } U_i \notin X \end{cases}.$$

Since

$$EZ_1 = P(Z_1 = 1) = P(U_1 \notin X \mid U_1 \in Y) = 1 - P_\#(X \mid Y),$$

we obtain the following inequality

$$P((1 - P_\#(X \mid Y)) \leq \frac{1}{n} \sum_{i=1}^{n} Z_i + \varepsilon) \geq 1 - e^{-2n\varepsilon^2}$$

Now, we take the advantage of the law of large numbers and the fact that we know the realisation of the sample U. We calculate a realisation for each Z_i in the following way

$$z_i = \begin{cases} 0, & \text{when } u_i \in X \\ 1, & \text{when } u_i \notin X \end{cases},$$

where u_i is i-th u_k such that $u_k \in Y$. The statement

$$(1 - P_\#(X \mid Y)) - \frac{1}{n} \sum_{i=1}^{n} z_i \leq \varepsilon$$

is likely to happen with significance $1 - e^{-2n\varepsilon^2}$.

n denotes the number of variables Z_i. It is equal, by definition, to the number of elements in the sample that belong to Y. On the other hand $Z_i = 1$ if and only if the corresponding U_i does not belong to X. Since U_i have to belong to U and Y we obtain

$$n = |U \cap Y| \quad \text{and} \quad \frac{1}{n} \sum_{i=1}^{n} z_i = \frac{|(U \cap Y) \setminus X|}{|U \cap Y|} = 1 - \frac{|U \cap Y \cap X|}{|U \cap Y|}.$$

If we assume that significance is equal to γ we obtain

$$\varepsilon = \sqrt{\frac{\ln(1 - \gamma)}{-2|U \cap Y|}}$$

and the approximation accuracy is estimated from (1) with the significance γ according to the formula

$$P_\#(X \mid Y) \geq \frac{|U \cap Y \cap X|}{|U \cap Y|} - \sqrt{\frac{\ln(1 - \gamma)}{-2|U \cap Y|}}.$$

The coverage estimator is developed in the analogous way from (1), and the following estimator is obtained

$$P_\#(Y \mid X) \geq \frac{|U \cap Y \cap X|}{|U \cap X|} - \sqrt{\frac{\ln(1 - \gamma)}{-2|U \cap X|}}.$$

Table 1. Exemplary decision system

	a	d
u_0	1	1
u_1	0	0
u_2	0	0
$\vdots$	$\vdots$	$\vdots$
u_{100}	0	0

We illustrate the trade-off between these three numerical factors using the following example. Consider decision system presented in Table 1. We obtain the following lower approximation for the objects in the system:

$$\underline{\{a\}}||d(x,0)||_{U,\mathbb{A}} = ||a(x,0)||_{U,\mathbb{A}}, \quad \underline{\{a\}}||d(x,1)||_{U,\mathbb{A}} = ||a(x,1)||_{U,\mathbb{A}}.$$

Yet we cannot state that $||a(x,0)||_{U,\mathbb{A}}$ is an approximation of $||d(x,0)||_{U,\mathbb{A}}$ with a 100% accuracy, since there may exist an object u_{101} in $\mathbb{U} \setminus U$ such that $a(u_{101}) = 0$ and $d(u_{101}) = 1$. The given decision system suggests that such an event is unlikely, yet still it is possible.

We estimate the approximation accuracy with significance 95%:

$$P_\#(||d(x,0)||_{U,A} \mid ||a(x,0)||_{U,A}) \geq \frac{|\;||d(x,0) \wedge a(x,0)||_{U,A}|}{|\;||a(x,0)||_{U,A}|} - \sqrt{\frac{\ln(1-0.95)}{-2|\;||a(x,0)||_{U,A}|}} =$$

$$= \frac{100}{100} - \sqrt{\frac{\ln(0.05)}{-200}} = 0.88.$$

Hence, the accuracy of the approximation of the set $||d=0||_{U,A}$ by means of $||a=0||_{U,A}$ is greater than 88% with significance 95%. On the other hand, for the approximation $\{a\}||d(x,1)||_{U,A} = ||a(x,1)||_{U,A}$, we do not obtain any significant accuracy estimation.

Hoeffding inequality provides us with a simple analytic formula for the approximation accuracy, yet the obtained estimator is not optimal. That is why we propose the second estimator based on the bound proposed in [8]. It results in an optimal estimator.

Theorem 4. *Let $Z_1, \ldots, Z_n$ be identically distributed independent random variables such that $Z_i : \Omega \to \{0,1\}$, $i = 1, \ldots, n$. Then, the following inequality holds:*

$$P\left(EZ_1 > g_{n,\gamma}(\frac{1}{n}\sum_{i=1}^{n} Z_i)\right) < \gamma,$$

where, for a given $k < n$, $g_{n,\gamma}$ satisfies the equation and $g_{n,\gamma}(1) = 1$. $g_{n,\gamma}$ provides the optimal bound of EZ_1.

The second estimator does not provide any analytic formula for the estimator value, yet $g_{n,\gamma}(\frac{k}{m})$ may be calculated using the algorithm proposed in [8].

According to the second estimator the accuracy of the approximation of the set $||d=0||_{U,A}$ by means of $||a=0||_{U,A}$ is greater than 97% with significance 95%.

6   Rule Induction Algorithm

Extended approximations of all decision classes compose a classifier. Unfortunately an extended approximation for a given set is not uniquely defined. Many algorithms for calculating approximations have been developed. Often approximations are represented by means of decision rules.

A *decision rule* for a given decision system is any expression of the form $\varphi(x) \to d(x,v)$, where φ is a conditional formula, d is a decision attribute, $v \in V_d$ and $||\varphi(x)||_{U,A} \neq \emptyset$. A decision rule $\varphi(x) \to d(x,v)$ is *true* in the decision system if, and only if, $||\varphi||_{U,A} \subseteq ||d=v||_{U,A}$. A decision rule describes the dependence between a decision class and its approximation.

In order to illustrate the link of theory with practical results we propose a simple algorithm for rule induction. The algorithm generates a classifier calculating extended approximations for all decision classes. Each approximation is represented as a set of decision rules whose predecessors are conjunctions of descriptors. For each rule, the accuracy, the coverage and the significance are calculated. The algorithm is parametrised by minimal levels of significance and accuracy and it induces all the rules that satisfy these minimal levels of indices. As a consequence induced rules do not cover all objects, and the classifier has not enough knowledge to recognise some objects. On the other hand all the classified objects are certified to be classified correctly with a very high probability.

The algorithm works as follows:

- In the 0th step it checks using the estimator whether there is a decision value v such that the rule with empty predecessor and decision value v would have the desired accuracy and significance. If the answer is positive, then the rule is generated and the rule induction process ends. Otherwise the algorithm moves to the 1st step.
- In the 1st step the set P_1 of all the possible rule predecessors with one descriptor are generated. Each element of P_1 is checked using the estimator. If the answer is positive, then the rule is generated. Then we remove from P_1 all elements used to generate rules and we denote the remaining set as P_1'.
- In the k-th step, $k > 1$, the generates the set P_k on the basis of P_{k-1}' in the following way: each element $\varphi(x)$ of P_{k-1}' and for each descriptor $a(x, v)$ such that a does not appear in $\varphi(x)$ we add add $\varphi(x) \wedge a(x, v)$ to P_k. Each element of P_k is checked using the estimator. If the answer is positive, then the rule is generated. Then we remove from P_k all elements used to generate rules and we denote the remaining set as P_k'.
- The algorithm uses two heuristics that speed it up: it does not try to generate a rule that is more specific than any existing rule and it checks whether there sufficiently many objects matching the rule predecessor to make it significant.
- The algorithm ends when no more rules may be created.

The algorithm generates short and relevant rules that cover only a part of universe.

In the case when during classification several rules may be applied to a given object, we choose the rule with the greatest accuracy.

Many more effective algorithms for rule generation that the one described above were developed (for example, in RSES [21] system). However, our objective was to illustrate the theory with a practical application and to show the link between set approximations and induced rules only.

7   Tests

To evaluate the performance of the algorithm, 3 benchmark data sets were selected: *chess, nursery, census94*. The data sets are obtained from the repository of University of California at Irvine [13].

Each data set is split into a training and a test set. For *census94* data sets the original partition into 32561 training samples and 16281 test samples available in the repository was used in the experiments. The remaining data sets (*chess*

Table 2. Test results obtained using the estimator based on Thm. 1

dataset	min accuracy	number of rules	classifier accuracy	classifier coverage
nursery	0.900000	42	0.985617	0.778395
chess	0.900000	80	0.952963	0.954944
census94	0.950000	32	0.951100	0.502610
census94	0.900000	83	0.899346	0.758307
census94	0.800000	107	0.812987	0.998894

Table 3. Test results obtained using the estimator based on Thm. 2

dataset	min accuracy	number of rules	classifier accuracy	classifier coverage
nursery	0.900000	112	0.989269	0.884722
chess	0.900000	310	0.957419	0.968085
census94	0.950000	92	0.951274	0.590873

Table 4. Part of 53 rules induced from *census94* dataset with significance 0.95 and minimal accuracy 0.85

Accuracy	Coverage	Rule
0.874541	0.388500	sex=Female → class=<=50K
0.938863	0.417531	marital-status=Never-married → class=<=50K
0.883111	0.310253	relationship=Not-in-family → class=<=50K
0.958077	0.198552	relationship=Own-child → class=<=50K
0.967552	0.195818	age=17-23 → class=<=50K
0.893863	0.143788	age=24-28 → class=<=50K
0.899943	0.064978	hours-per-week=18-24 → class=<=50K
0.940021	0.168487	capital-gain=7000-99999 → class=>50K
0.843932	0.050181	occupation=Machine-op-inspct, hours-per-week=40 → class=<=50K
0.879734	0.052272	occupation=Handlers-cleaners → class=<=50K
0.827732	0.040772	occupation=Adm-clerical, education=Some-college → class=<=50K
0.901273	0.048653	education=11th → class=<=50K

and *nursery*) consisted of 3196 and 12960 samples and were randomly split into a training and a test part with the split ratio 2 to 1.

All the selected sets are the data sets from UCI repository that have data objects represented as vectors of attributes values and have the size between a few thousand and several tens thousand of objects. To make the results from different experiments comparable the random partition was done once for each data set and the same partition was used in all the performed experiments.

Chess has 36 nominal attributes and 2 decision classes. One attribute has 3 values while the other have 2 values. *nursery* has 8 attributes and 5 decision classes. All attributes are nominal and have from 2 till 5 values. *Census94* possess 10 nominal, 6 numeric attributes and 2 decision classes. The numeric attributes were discretised. Attribute domains range from 2 till 30 values. There are also missing values present in *Census94* data set.

Table 2 presents test results obtained using the estimator based on Thm. 1. Table 3 presents test results obtained using the estimator based on Thm. 2. In both cases rules were induced with significance 95%.

The tests results show that the algorithm generates a small number of highly relevant rules which makes it useful for knowledge discovery. The fact that it estimates accuracy and coverage for each rule provide us with an insight into the internal structure of data. Table 4 illustrates the above statements presenting a part of rules induced from *census94* dataset.

8   Decision Rules Extraction

We used our methodology in practical task of decision rules extraction from Neo-Sumerian economic documents [22,23,24,25].

Sumerians lived from prehistoric times until late 3rd millennium BC in lower Mesopotamia (modern Iraq). Sumer was the first highly developed urban civilisation, which used cuneiform script. During the reign of the 3rd dynasty of Ur (2100 BC-2000 BC), whose power extended as far as present Iraq and western Iran, the state introduced a centrally planned economy with an extensive bureaucratic apparatus.

Civil servants reported on clay tablets agriculture and factory production, lists of worker salaries, summaries of executed work, distribution of commodities, goods, animals etc., lists of sacrificed animals, travel diets and other economical information.

Economic documents are an essential source of information about ancient Sumer. The corpus contains crucial information about economic, social and political history of the state, as well as its political system and administration structure. The sources of this type provide the most complete information about the daily life of those days.

For our studies, we have selected a subcorpus of 11891 documents concerning distribution of domestic animals. The distribution was organised in the form of transactions. During each **Transaction** one person, called **Supplier**, transfers a **Number** of **Animals** to another Person, called **Receiver**. Animal description consists of information like: species, age, gender, fodder, etc. Person is described by means of his/her name, filiation, job and/or nationality. Apart from the Supplier and Receiver, other persons might have assisted in the transaction:

Giri. Middleman between Supplier and and Receiver;
MuDu. Person on whose account the transaction took place, Receiver was probably Mu Du's representative;
MuSze. Person in whose name the Receiver or Supplier acted;
Kiszib. Person who sealed the document
Maszkim. Overseer of the transaction;
Bala. Person who provided goods as royal tax.

The roles are named after the Sumerian phrases used to introduce them, their meaning is still studied by sumerologists. The date of each Transaction, composed out of **Year, Month** and **Day**, is also provided.

We in [14] we described, how to process Neo-Sumerian economic documents into an information system. Transactions are objects in this system while concepts written in bold in the above paragraph are their attributes.

We extracted decision rules from the information system. These rules bring to light dependencies between Sumerian officials, animal types and transaction dates. The following table presents a few rules (generated with significance at least 95%):

Accuracy	Coverage	Rule
0.762606	0.895833	Receiver=lu2-{d}gesz-bar-e3 → Kiszib=a-kal-la sipa
0.866110	1.000000	MuDu=lum-ma → Kiszib=ensi2 u3-da
0.888281	0.873563	MuSze=ur-ra, Giri={d}en-lil2-la2 → Kiszib={d}szul-gi-a-a-mu
0.863621	0.739130	Year=SS09, Supplier=du-du → Kiszib=u4-de3-nig2-sag10
0.799641	0.226601	Year=SZ41, Animal=udu → Maszkim=en-{d}nansze-ki-ag2
0.775528	0.928571	Year=AS07, Kiszib=ab-ba-sa6-ga → Maszkim=du-du
0.770440	0.857143	MuSze=ensi2 zimbir{ki} → Maszkim=ur-{d}gubalag nar ta2-hi-isz-a-tal
0.827830	0.273504	Receiver=lu2-mah → MuDu={d}szara2

The first rule states that if `lu2-{d}gesz-bar-e3` is a receiver of goods in transaction then with a probability 76% `a-kal-la sipa` seals the document. And this rule covers 89% of cases when `a-kal-la sipa` seals any document. The remaining rules are interpreted analogically.

We extracted 12841 rules. These rules help to direct Sumerological research pointing out interesting dependencies. Sumerologists may, for example, try to deduct from documents the reasons for a given dependence. Also, analysis of the whole set of rules is interesting, because it provide broad picture of Sumerian economy.

9   Conclusions

The hybridization of roughs sets and statistical learning theory resulted in the concept of extended approximation and statistical estimators for rule accuracy and coverage.

These estimators may be used with any rule induction algorithm. They guarantee the relevance of induced rules.

Extended approximations create a theoretical background for the classification. They indicate the connection between lower and upper approximations and rules induced from sample.

The theory and algorithms may be further developed to make them suitable for handling numerical attributes and other types of data.

Acknowledgment. The research has been partially supported by grants N N516 368334 and N N206 400234 from Ministry of Science and Higher Education of the Republic of Poland.

References

1. Pawlak, Z.: Rough sets. International Journal of Computer and Information Sciences 11(5), 341–356 (1982)
2. Pawlak, Z.: Rough Sets: Theoretical Aspects of Reasoning about Data. Kluwer Academic Publishers, Dordrecht (1991)
3. Vapnik, V.N.: Statistical Learning Theory. John_Wiley, New York (1998)
4. Jaworski, W.: Generalized indiscernibility relations: Applications for missing values and analysis of structural objects. Transactions of Rough Sets 8, 116–145 (2008)
5. Pawlak, Z., Skowron, A.: Rough sets: Some extensions. Inf. Sci. 177(1), 28–40 (2007)
6. Skowron, A., Świniarski, R.W., Synak, P.: Approximation spaces and information granulation. In: Peters, J.F., Skowron, A. (eds.) Transactions on Rough Sets III. LNCS, vol. 3400, pp. 175–189. Springer, Heidelberg (2005)
7. Hoeffding, W.: Probability inequalities for sums of bounded random variables. Journal of the American Statistical Association 58, 13–30 (1963)
8. Jaworski, W.: Model selection and assessment for classification using validation. In: Ślęzak, D., Wang, G., Szczuka, M.S., Düntsch, I., Yao, Y. (eds.) RSFDGrC 2005. LNCS (LNAI), vol. 3641, pp. 481–490. Springer, Heidelberg (2005)
9. Jaworski, W.: Bounds for validation. Fundam. Inform. 70(3), 261–275 (2006)
10. Tsumoto, S.: Accuracy and coverage in rough set rule induction. In: Alpigini, J.J., Peters, J.F., Skowron, A., Zhong, N. (eds.) RSCTC 2002. LNCS (LNAI), vol. 2475, pp. 373–380. Springer, Heidelberg (2002)
11. Guillet, F., Hamilton, H.J. (eds.): Quality Measures in Data Mining. SCI, vol. 43. Springer, Heidelberg (2007)
12. Gediga, G., Düntsch, I.: Rough Set Data Analysis — A Road to Non-Invasive Knowledge Discovery. Methodos Publishers, UK (2000)
13. Asuncion, A., Newman, D.J.: UCI machine learning repository. Technical report, University of California, Irvine, School of Information and Computer Sciences (2007)
14. Jaworski, W.: Contents modelling of Neo-Sumerian Ur III economic text corpus. In: Proc. of the 22nd International Conference on Computational Linguistics (COLING 2008), Manchester, UK, pp. 369–376. Coling 2008 Organizing Committee (2008)
15. Pawlak, Z.: Information systems — theoretical foundations. Information Systems 6(3), 205–218 (1981)
16. Grzymała-Busse, J.W., Grzymala-Busse, W.J.: An experimental comparison of three rough set approaches to missing attribute values. In: Peters, J.F., Skowron, A., Düntsch, I., Grzymała-Busse, J.W., Orłowska, E., Polkowski, L. (eds.) Transactions on Rough Sets VI. LNCS, vol. 4374, pp. 31–50. Springer, Heidelberg (2007)
17. Grzymała-Busse, J.W.: A rough set approach to data with missing attribute values. In: Wang, G.-Y., Peters, J.F., Skowron, A., Yao, Y. (eds.) RSKT 2006. LNCS (LNAI), vol. 4062, pp. 58–67. Springer, Heidelberg (2006)
18. Kryszkiewicz, M.: Rough set approach to incomplete information systems. Inf. Sci. 112(1-4), 39–49 (1998)
19. Kryszkiewicz, M.: Properties of incomplete information systems in the framework of rough sets. In: Polkowski, L., Skowron, A. (eds.) Rough Sets in Knowledge Discovery 1. Methodology and Applications. Studies in Fuzziness and Soft Computing, pp. 422–450. Physica-Verlag, Heidelberg (1998)

20. Lipski, W.J.: On Databases with Incomplete Information. Journal of the Association of Computing Machinery 28(1), 41–70 (1981)
21. Bazan, J., Szczuka, M.: RSES and RSESlib - a collection of tools for rough set computations. In: Ziarko, W.P., Yao, Y. (eds.) RSCTC 2000. LNCS (LNAI), vol. 2005, pp. 106–113. Springer, Heidelberg (2001)
22. Stępień, M.: Animal Husbandry in the Ancient Near East: A Prosopographic Study of Third-Millennium Umma. CDL Press, Bethesda (1996)
23. Stępień, M.: Ensi w czasach III dynastii z Ur: aspekty ekonomiczne i administracyjne pozycji namiestnika prowincji w świetle archiwum z Ummy [Ensi in the Third Dynasty of Ur: Economic and Administrative Aspects of the Province Governor Position on the Basis of Umma Archive]. Wydawnictwa Uniwersytetu Warszawskiego (2006)
24. Sharlach, T.: Provincial Taxation and the Ur III State, Leiden-Boston (2004)
25. Steinkeller, P.: The administrative and economic organization of the ur iii state: The core and the periphery. In: Biggs, R.D., Gibson, M.G. (eds.) The Organization of Power: Aspect of Bureaucracy in the Ancient Near East, Chicago. SAOC, vol. 46, pp. 19–41 (1987)

Fuzzy-Rough Nearest Neighbour Classification

Richard Jensen[1] and Chris Cornelis[2]

[1] Dept. of Comp. Sci., Aberystwyth University, Ceredigion, SY23 3DB, Wales, UK
rkj@aber.ac.uk
[2] Dept. of Appl. Math. and Comp. Sci., Ghent University, Gent, Belgium
Chris.Cornelis@UGent.be

Abstract. A new fuzzy-rough nearest neighbour (FRNN) classification algorithm is presented in this paper, as an alternative to Sarkar's fuzzy-rough ownership function (FRNN-O) approach. By contrast to the latter, our method uses the nearest neighbours to construct lower and upper approximations of decision classes, and classifies test instances based on their membership to these approximations. In the experimental analysis, we evaluate our approach with both classical fuzzy-rough approximations (based on an implicator and a t-norm), as well as with the recently introduced vaguely quantified rough sets. Preliminary results are very good, and in general FRNN outperforms FRNN-O, as well as the traditional fuzzy nearest neighbour (FNN) algorithm.

Keywords: Fuzzy-rough sets, nearest neighbour algorithms, classification.

1 Introduction

Lately there has been great interest in developing methodologies which are capable of dealing with imprecision and uncertainty, and the resounding amount of research currently being done in the areas related to fuzzy [30] and rough sets [18] is representative of this. The success of rough set theory is due in part to three aspects of the theory. Firstly, only the facts hidden in data are analysed. Secondly, no additional information about the data is required for data analysis such as thresholds or expert knowledge on a particular domain. Thirdly, it finds a minimal knowledge representation for data. As rough set theory handles only one type of imperfection found in data, it is complementary to other concepts for the purpose, such as fuzzy set theory. The two fields may be considered analogous in the sense that both can tolerate inconsistency and uncertainty - the difference being the type of uncertainty and their approach to it; fuzzy sets are concerned with vagueness, rough sets are concerned with indiscernibility.

Many relationships have been established and more so, most of the recent studies have concluded at this complementary nature of the two methodologies, especially in the context of granular computing. Therefore, it is desirable to extend and hybridize the underlying concepts to deal with additional aspects of data imperfection. Such developments offer a high degree of flexibility and provide robust solutions and advanced tools for data analysis.

J.F. Peters et al. (Eds.): Transactions on Rough Sets XIII, LNCS 6499, pp. 56–72, 2011.
© Springer-Verlag Berlin Heidelberg 2011

The K-nearest neighbour (KNN) algorithm [9] is a well-known classification technique that assigns a test object to the decision class most common among its K nearest neighbours, i.e., the K training objects that are closest to the test object. An extension of the KNN algorithm to fuzzy set theory (FNN) was introduced in [17]. It allows partial membership of an object to different classes, and also takes into account the relative importance (closeness) of each neighbour w.r.t. the test instance. However, as Sarkar correctly argued in [22], the FNN algorithm has problems dealing adequately with insufficient knowledge. In particular, when every training pattern is far removed from the test object, and hence there are no suitable neighbours, the algorithm is still forced to make clear-cut predictions. This is because the predicted membership degrees to the various decision classes always need to sum up to 1.

To address this problem, Sarkar [22] introduced a so-called fuzzy-rough ownership function that, when plugged into the conventional FNN algorithm, produces class confidence values that do not necessarily sum up to 1. However, this method (called FRNN-O throughout this paper) does not refer to the main ingredients of rough set theory, i.e., lower and upper approximation. In this paper, therefore, we present an alternative approach, which uses a test object's nearest neighbours to construct the lower and upper approximation of each decision class, and then computes the membership of the test object to these approximations. The method is very flexible, as there are many options to define the fuzzy-rough approximations, including the traditional implicator/t-norm based model [21], as well as the vaguely quantified rough set (VQRS) model [6], which is more robust in the presence of noisy data.

This paper is structured as follows. Section 2 provides necessary details for fuzzy rough set theory, while Section 3 is concerned with the existing fuzzy (-rough) NN approaches. Section 4 outlines our algorithm, while comparative experimentation on a series of classification and prediction problems is provided in Section 5. The paper is concluded in section 6.

2   Hybridization of Rough Sets and Fuzzy Sets

2.1   Rough Set Theory

Rough set theory (RST) [18] provides a tool by which knowledge may be extracted from a domain in a concise way; it is able to retain the information content whilst reducing the amount of knowledge involved. Central to RST is the concept of indiscernibility. Let $(\mathbb{U}, \mathbb{A})$ be an information system, where $\mathbb{U}$ is a non-empty set of finite objects (the universe of discourse) and $\mathbb{A}$ is a non-empty finite set of attributes such that $a : \mathbb{U} \to V_a$ for every $a \in \mathbb{A}$. V_a is the set of values that attribute a may take. With any $B \subseteq \mathbb{A}$ there is an associated equivalence relation R_B:

$$R_B = \{(x, y) \in \mathbb{U}^2 | \forall a \in B, \ a(x) = a(y)\} \tag{1}$$

If $(x, y) \in R_B$, then x and y are indiscernible by attributes from B. The equivalence classes of the B-indiscernibility relation are denoted $[x]_B$. Let $A \subseteq \mathbb{U}$. A can be approximated using the information contained within B by constructing the B-*lower* and B-*upper* approximations of A:

$$R_B \downarrow A = \{x \in \mathbb{U} \mid [x]_B \subseteq A\} \tag{2}$$

$$R_B \uparrow A = \{x \in \mathbb{U} \mid [x]_B \cap A \neq \emptyset\} \tag{3}$$

The tuple $\langle R_B \downarrow A, R_B \uparrow A \rangle$ is called a rough set.

A *decision system* $(X, \mathcal{A} \cup \{d\})$ is a special kind of information system, used in the context of classification, in which d ($d \notin \mathcal{A}$) is a designated attribute called the decision attribute. Its equivalence classes $[x]_{R_d}$ are called decision classes. The set of decision classes is denoted $\mathcal{C}$ in this paper. Given $B \subseteq \mathcal{A}$, the B-positive region POS_B contains those objects from X for which the values of B allow to predict the decision class unequivocally:

$$POS_B = \bigcup_{x \in X} R_B \downarrow [x]_{R_d} \tag{4}$$

Indeed, if $x \in POS_B$, it means that whenever an object has the same values as x for the attributes in B, it will also belong to the same decision class as x. The predictive ability w.r.t. d of the attributes in B is then measured by the following value (degree of dependency of d on B):

$$\gamma_B = \frac{|POS_B|}{|X|} \tag{5}$$

$(X, \mathcal{A} \cup \{d\})$ is called *consistent* if $\gamma_{\mathcal{A}} = 1$. A subset B of $\mathcal{A}$ is called a *decision reduct* if it satisfies $POS_B = POS_{\mathcal{A}}$, i.e., B preserves the decision making power of $\mathcal{A}$, and moreover it cannot be further reduced, i.e., there exists no proper subset B' of B such that $POS_{B'} = POS_{\mathcal{A}}$. If the latter constraint is lifted, i.e., B is not necessarily minimal, we call B a decision superreduct.

2.2   Fuzzy Set Theory

Fuzzy set theory [30] allows that objects belong to a set, or couples of objects belong to a relation, to a given degree. Recall that a fuzzy set in X is an $X \to [0, 1]$ mapping, while a fuzzy relation in X is a fuzzy set in $X \times X$. For all y in X, the R-foreset of y is the fuzzy set Ry defined by

$$Ry(x) = R(x, y) \tag{6}$$

for all x in X. If R is a reflexive and symmetric fuzzy relation, that is,

$$R(x, x) = 1 \tag{7}$$

$$R(x, y) = R(y, x) \tag{8}$$

hold for all x and y in X, then R is called a fuzzy tolerance relation. For a fuzzy tolerance relation R, we call Ry the fuzzy tolerance class of y.

For fuzzy sets A and B in X, $A \subseteq B \iff (\forall x \in X)(A(x) \leq B(x))$. If X is finite, the cardinality of A is calculated by

$$|A| = \sum_{x \in X} A(x). \tag{9}$$

Fuzzy logic connectives play an important role in the development of fuzzy rough set theory. We therefore recall some important definitions. A triangular norm (t-norm for short) $\mathcal{T}$ is any increasing, commutative and associative $[0,1]^2 \to [0,1]$ mapping satisfying $\mathcal{T}(1,x) = x$, for all x in $[0,1]$. In this paper, we use $\mathcal{T}_M$ and $\mathcal{T}_L$ defined by $\mathcal{T}_M(x,y) = \min(x,y)$ and $\mathcal{T}_L(x,y) = \max(0, x+y-1)$ (Łukasiewicz t-norm), for x,y in $[0,1]$. On the other hand, an implicator is any $[0,1]^2 \to [0,1]$-mapping $\mathcal{I}$ satisfying $\mathcal{I}(0,0) = 1, \mathcal{I}(1,x) = x$, for all x in $[0,1]$. Moreover we require $\mathcal{I}$ to be decreasing in its first, and increasing in its second component. The implicators used in this paper are $\mathcal{I}_M$ and $\mathcal{I}_L$ defined by $\mathcal{I}_M(x,y) = \max(1-x,y)$ (Kleene-Dienes implicator) and $\mathcal{I}_L(x,y) = \min(1, 1-x+y)$ (Łukasiewicz implicator) for x,y in $[0,1]$.

2.3 Fuzzy-Rough Set Theory

The process described above can only operate effectively with datasets containing discrete values. As most datasets contain real-valued attributes, it is necessary to perform a discretization step beforehand. A more intuitive and flexible approach, however, is to model the approximate equality between objects with continuous attribute values by means of a fuzzy relation R in $\mathbb{U}$, i.e., a $\mathbb{U} \to [0,1]$ mapping that assigns to each couple of objects their degree of similarity. In general, it is assumed that R is at least a fuzzy tolerance relation, that is, $R(x,x) = 1$ and $R(x,y) = R(y,x)$ for x and y in $\mathbb{U}$. Given y in $\mathbb{U}$, its foreset Ry is defined by $Ry(x) = R(x,y)$ for every x in $\mathbb{U}$.

Given a fuzzy tolerance relation R and a fuzzy set A in $\mathbb{U}$, the lower and upper approximation of A by R can be constructed in several ways. A general definition [7,21] is the following:

$$(R{\downarrow}A)(x) = \inf_{y \in \mathbb{U}} \mathcal{I}(R(x,y), A(y)) \tag{10}$$

$$(R{\uparrow}A)(x) = \sup_{y \in \mathbb{U}} \mathcal{T}(R(x,y), A(y)) \tag{11}$$

Here, I is an implicator and T a t-norm. When A is a crisp (classical) set and R is an equivalence relation in $\mathbb{U}$, the traditional lower and upper approximation are recovered.

Just like their crisp counterparts, formulas (10) and (11) (henceforth called the FRS approximations) are quite sensitive to noisy values. That is, a change in a single object can result in drastic changes to the approximations (due to the use of sup and inf, which generalize the existential and universal quantifier, respectively). In the context of classification tasks, this behaviour may affect accuracy adversely. Therefore, in [6], the concept of vaguely quantified rough sets (VQRS) was introduced. It uses the linguistic quantifiers "most" and "some", as opposed to the traditionally used crisp quantifiers "all" and "at least one", to decide to what extent an object belongs to the lower and upper approximation. Given a couple (Q_u, Q_l) of fuzzy quantifiers[1] that model "most" and "some", the lower and upper approximation of A by R are defined by

$$(R\!\downarrow^{Q_u}\!A)(y) = Q_u\left(\frac{|Ry \cap A|}{|Ry|}\right) = Q_u\left(\frac{\sum\limits_{x \in X} \min(R(x,y), A(x))}{\sum\limits_{x \in X} R(x,y)}\right) \qquad (12)$$

$$(R\!\uparrow^{Q_l}\!A)(y) = Q_l\left(\frac{|Ry \cap A|}{|Ry|}\right) = Q_l\left(\frac{\sum\limits_{x \in X} \min(R(x,y), A(x))}{\sum\limits_{x \in X} R(x,y)}\right) \qquad (13)$$

where the fuzzy set intersection is defined by the min t-norm and the fuzzy set cardinality by the sigma-count operation. As an important difference to (10) and (11), the VQRS approximations do not extend the classical rough set approximations, in a sense that when A and R are crisp, the lower and upper approximations may still be fuzzy.

2.4   Fuzzy-Rough Classification

Due to its recency, there have been very few attempts at developing fuzzy-rough set theory for the purpose of classification. Previous work has focused on using crisp rough set theory to generate fuzzy rulesets [14,23] but mainly ignores the direct use of fuzzy-rough concepts.

The induction of gradual decision rules, based on fuzzy-rough hybridization, is given in [12]. For this approach, new definitions of fuzzy lower and upper approximations are constructed that avoid the use of fuzzy logical connectives altogether. Decision rules are induced from lower and upper approximations defined for positive and negative relationships between credibility of premises and conclusions. Only the ordinal properties of fuzzy membership degrees are used. More recently, a fuzzy-rough approach to fuzzy rule induction was presented in [27], where fuzzy reducts are employed to generate rules from data. This method also employs a fuzzy-rough feature selection preprocessing step.

Also of interest is the use of fuzzy-rough concepts in building fuzzy decision trees. Initial research is presented in [2] where a method for fuzzy decision tree construction is given that employs the fuzzy-rough ownership function. This is used to define both an index of fuzzy-roughness and a measure of fuzzy-rough entropy as a node splitting criterion. Traditionally, fuzzy entropy (or its extension) has been used for this purpose. In [16], a fuzzy decision tree algorithm is proposed, based on fuzzy ID3, that incorporates the fuzzy-rough dependency function as a splitting criterion. A fuzzy-rough rule induction method is proposed in [13] for generating certain and possible rulesets from hierarchical data.

---

[1] By a fuzzy quantifier, we mean an increasing $[0,1] \rightarrow [0,1]$ mapping such that $Q(0) = 0$ and $Q(1) = 1$.

3   Fuzzy Nearest Neighbour Classification

The fuzzy K-nearest neighbour (FNN) algorithm [17] was introduced to classify test objects based on their similarity to a given number K of neighbours (among the training objects), and these neighbours' membership degrees to (crisp or fuzzy) class labels. For the purposes of FNN, the extent $C'(y)$ to which an unclassified object y belongs to a class C is computed as:

$$C'(y) = \sum_{x \in N} R(x,y)C(x) \tag{14}$$

where N is the set of object y's K nearest neighbours, obtained by calculating the fuzzy similarity between y and all training objects, and choosing the K objects that have highest similarity degree. $R(x,y)$ is the [0,1]-valued similarity of x and y. In the traditional approach, this is defined in the following way:

$$R(x,y) = \frac{||y-x||^{-2/(m-1)}}{\sum\limits_{j \in N} ||y-j||^{-2/(m-1)}} \tag{15}$$

where $|| \cdot ||$ denotes the Euclidean norm, and m is a parameter that controls the overall weighting of the similarity. Assuming crisp classes, Fig. 1 shows an application of the FNN algorithm that classifies a test object y to the class with the highest resulting membership. The idea behind this algorithm is that the degree of closeness of neighbours should influence the impact that their class membership has on deriving the class membership for the test object. The complexity of this algorithm for the classification of one test pattern is $O(|\mathbb{U}| + K \cdot |C|)$.

FNN($\mathbb{U}, \mathcal{C}, y, K$).
$\mathbb{U}$, the training data; $\mathcal{C}$, the set of decision classes;
y, the object to be classified; K, the number of nearest neighbours.

(1) $N \leftarrow$ getNearestNeighbours(y, K);
(2) $\forall C \in \mathcal{C}$
(3) $C'(y) = \sum_{x \in N} R(x,y)C(x)$
(4)   **output** $\arg\max\limits_{C \in \mathcal{C}} (C'(y))$

Fig. 1. The fuzzy KNN algorithm

Initial attempts to combine the FNN algorithm with concepts from fuzzy rough set theory were presented in [22,26]. In these papers, a fuzzy-rough ownership function is constructed that attempts to handle both "fuzzy uncertainty" (caused by overlapping classes) and "rough uncertainty" (caused by insufficient knowledge, i.e., attributes, about the objects). The fuzzy-rough ownership function τ_C of class C was defined as, for an object y,

$$\tau_C(y) = \frac{\sum\limits_{x \in \mathbb{U}} R(x,y)C(x)}{|\mathbb{U}|} \tag{16}$$

In this, the fuzzy relation R is determined by:

$$R(x,y) = \exp\left(-\sum_{a \in \mathbb{C}} \kappa_a(a(y) - a(x))^{2/(m-1)}\right) \qquad (17)$$

where m controls the weighting of the similarity (as in FNN) and κ_a is a parameter that decides the bandwidth of the membership, defined as

$$\kappa_a = \frac{|\mathbb{U}|}{2\sum_{x \in \mathbb{U}} ||a(y) - a(x)||^{2/(m-1)}} \qquad (18)$$

$\tau_C(y)$ is interpreted as the confidence with which y can be classified to class C. The corresponding crisp classification algorithm, called FRNN-O in this paper, can be seen in Fig. 2. Initially, the parameter κ_a is calculated for each attribute and all memberships of decision classes for test object y are set to 0. Next, the weighted distance of y from all objects in the universe is computed and used to update the class memberships of y via equation (16). Finally, when all training objects have been considered, the algorithm outputs the class with highest membership. The algorithm's complexity is $O(|\mathbb{C}||\mathbb{U}| + |\mathbb{U}| \cdot (|\mathbb{C}| + |\mathcal{C}|))$.

By contrast to the FNN algorithm, the fuzzy-rough ownership function considers all training objects rather than a limited set of neighbours, and hence no decision is required as to the number of neighbours to consider. The reasoning behind this is that very distant training objects will not influence the outcome (as opposed to the case of FNN). For comparison purposes, the K-nearest neighbours version of this algorithm is obtained by replacing line (3) with $N \leftarrow$ getNearestNeighbours(y,K).

It should be noted that the algorithm does not use fuzzy lower or upper approximations to determine class membership. A very preliminary attempt to do so was described in [3]. However, the authors did not state how to use the upper and lower approximations to derive classifications.

FRNN-O$(\mathbb{U},\mathbb{C},\mathcal{C},y)$.
$\mathbb{U}$, the training data; $\mathbb{C}$, the set of conditional features;
$\mathcal{C}$, the set of decision classes; y, the object to be classified.

(1) $\forall a \in \mathbb{C}$
(2) $\kappa_a = |\mathbb{U}|/2\sum_{x \in \mathbb{U}} ||a(y) - a(x)||^{2/(m-1)}$
(3) $N \leftarrow |\mathbb{U}|$
(4) $\forall C \in \mathcal{C}, \tau_C(y) = 0$
(5) $\forall x \in N$
(6) $d = \sum_{a \in \mathbb{C}} \kappa_a(a(y) - a(x))^2$
(7) $\forall C \in \mathcal{C}$
(8) $\tau_C(y) += \frac{C(x) \cdot \exp(-d^{1/(m-1)})}{|N|}$
(9)   **output** $\arg\max_{C \in \mathcal{C}} \tau_C(y)$

Fig. 2. The fuzzy-rough ownership nearest neighbour algorithm

4   Fuzzy-Rough Nearest Neighbour (FRNN) Algorithm

Figure 3 outlines our proposed algorithm, combining fuzzy-rough approximations with the ideas of the classical FNN approach. In what follows, FRNN-FRS and FRNN-VQRS denote instances of the algorithm where traditional, and VQRS, approximations are used, respectively. The rationale behind the algorithm is that the lower and the upper approximation of a decision class, calculated by means of the nearest neighbours of a test object y, provide good clues to predict the membership of the test object to that class.

In particular, if $(R{\downarrow}C)(y)$ is high, it reflects that all (most) of y's neighbours belong to C, while a high value of $(R{\uparrow}C)(y)$ means that at least one (some) neighbour(s) belong(s) to that class, depending on whether the FRS or VQRS approximations are used. A classification will always be determined for y due to the initialisation of τ to zero in line (2). To perform crisp classification, the algorithm outputs the decision class with the resulting best combined fuzzy lower and upper approximation memberships, seen in line (4) of the algorithm. This is only one way of utilising the information in the fuzzy lower and upper approximations to determine class membership, other ways are possible but are not investigated in this paper. The complexity of the algorithm is $O(|\mathcal{C}| \cdot (2|\mathbb{U}|))$.

FRNN($\mathbb{U}$,$\mathcal{C}$,y).
$\mathbb{U}$, the training data; $\mathcal{C}$, the set of decision classes;
y, the object to be classified.

$$
\begin{array}{ll}
(1) & N \leftarrow \text{getNearestNeighbors}(y,K) \\
(2) & \tau \leftarrow 0,\ Class \leftarrow \emptyset \\
(3) & \forall C \in \mathcal{C} \\
(4) & \quad \textbf{if } (((R{\downarrow}C)(y) + (R{\uparrow}C)(y))/2 \geq \tau) \\
(5) & \quad\quad Class \leftarrow C \\
(6) & \quad\quad \tau \leftarrow ((R{\downarrow}C)(y) + (R{\uparrow}C)(y))/2 \\
(7) & \textbf{output } Class
\end{array}
$$

Fig. 3. The fuzzy-rough nearest neighbour algorithm - classification

Furthermore, the algorithm is dependent on the choice of the fuzzy tolerance relation R A general way of constructing R is as follows: given the set of conditional attributes $\mathbb{C}$, R is defined by

$$
R(x,y) = \min_{a \in \mathbb{C}} R_a(x,y) \tag{19}
$$

in which $R_a(x,y)$ is the degree to which objects x and y are similar for attribute a. Possible options include

$$
R_a^1(x,y) = \exp\left(-\frac{(a(x) - a(y))^2}{2\sigma_a{}^2}\right) \tag{20}
$$

$$
R_a^2(x,y) = 1 - \frac{|a(x) - a(y)|}{|a_{\max} - a_{\min}|} \tag{21}
$$

FRNN2($\mathbb{U}$,d,y).
$\mathbb{U}$, the training data; d, the decision feature;
y, the object to be classified.

$\quad$ (1) $\quad$ $N \leftarrow$ getNearestNeighbors(y,K)
$\quad$ (2) $\quad$ $\tau_1 \leftarrow 0$, $\tau_2 \leftarrow 0$
$\quad$ (3) $\quad$ $\forall z \in N$
$\quad$ (4) $\quad\quad$ $M \leftarrow ((R{\downarrow}R_d z)(y) + (R{\uparrow}R_d z)(y))/2$
$\quad$ (5) $\quad\quad$ $\tau_1 \leftarrow \tau_1 + M * d(y)$
$\quad$ (6) $\quad\quad$ $\tau_2 \leftarrow \tau_2 + M$
$\quad$ (7) $\quad$ **output** τ_1/τ_2

Fig. 4. The fuzzy-rough nearest neighbour algorithm - prediction

where $\sigma_a{}^2$ is the variance of attribute a, and $a_{\max}$ and $a_{\min}$ are the maximal and minimal occurring value of that attribute.

When using FRNN-FRS, the use of K is not required in principle: as $R(x,y)$ gets smaller, x tends to have only have a minor influence on $(R{\downarrow}C)(y)$ and $(R{\uparrow}C)(y)$. For FRNN-VQRS, this may generally not be true, because $R(x,y)$ appears in the numerator as well as the denominator of (12) and (13).

When dealing with real-valued decision features, the above algorithm can be modified to that found in Fig. 4. This is a zero order Takagi-Sugeno controller, with each neighbour acting as a rule. Here, the lower and upper approximations are defined as:

$$(R{\downarrow}R_d z)(x) = \inf_{y \in N} \mathcal{I}(R(x,y), R_d(y,z)) \tag{22}$$

$$(R{\uparrow}R_d z)(x) = \sup_{y \in N} \mathcal{T}(R(x,y), R_d(y,z)) \tag{23}$$

where R_d is the fuzzy tolerance relation for the decision feature d. In this paper, we use the same relation as that used for the conditional features. This need not be the case in general; indeed, it is conceivable that there may be situations where the use of a different similarity relation is sensible for the decision feature.

5 Experimentation

This section details the experimentation performed for the evaluation of the proposed algorithms for both classification and prediction tasks.

5.1 Classification

To demonstrate the power of the proposed fuzzy-rough NN approach, two sets of classification experiments were conducted. In the first set, the performance of the fuzzy and fuzzy-rough NN approaches were compared. The second set of experiments compared the proposed NN approaches (FRNN-FRS and FRNN-VQRS) with a variety of leading classification algorithms. Both sets of experiments were conducted over eight benchmark datasets from [4] and [22]. The details of the datasets used can be found in table 1. All of them have a crisp decision attribute.

Table 1. Dataset details

Dataset	Objects	Attributes
Cleveland	297	14
Glass	214	10
Heart	270	14
Letter	3114	17
Olitos	120	26
Water 2	390	39
Water 3	390	39
Wine	178	14

Fuzzy NN approaches. This section presents the initial experimental evaluation of the classification methods FNN, FRNN-O, FRNN-FRS and FRNN-VQRS for the task of pattern classification[2].

For FNN and FRNN-O, m is set to 2. For the new approaches, the fuzzy relation given in equation (21) was chosen. In the FRNN-FRS approach, we used the min t-norm and the Kleene-Dienes implicator I defined by $I(x,y) = \max(1-x, y)$. The FRNN-VQRS approach was implemented using $Q_l = Q_{(0.1,0.6)}$ and $Q_u = Q_{(0.2,1.0)}$, according to the general formula

$$Q_{(\alpha,\beta)}(x) = \begin{cases} 0, & x \leq \alpha \\ \frac{2(x-\alpha)^2}{(\beta-\alpha)^2}, & \alpha \leq x \leq \frac{\alpha+\beta}{2} \\ 1 - \frac{2(x-\beta)^2}{(\beta-\alpha)^2}, & \frac{\alpha+\beta}{2} \leq x \leq \beta \\ 1, & \beta \leq x \end{cases}$$

Initially, the impact of the number of neighbours K was investigated for the nearest neighbour approaches. K was initialized to $|\mathbb{U}|$, the number of objects in the training dataset, and then decremented by $1/30^{th}$ of $|\mathbb{U}|$ each time, resulting in 30 experiments for each dataset. For each choice of parameter K, $2\times$ 10-fold cross-validation was performed. The results of this for two datasets can be seen in Fig. 5 and Fig. 6.

It can be seen that FRNN-FRS is indeed unaffected by the choice of K for nominal-valued decision features. FRNN-O also appears to be relatively unaffected by K. For the Letter dataset, FRNN-VQRS and FNN exhibit degradation in classification performance as the number of nearest neighbours increases beyond 10. Therefore, for these methods the choice of K is an important consideration, with a value of around 10 neighbours being a sensible choice.

Based on this, further experimentation was conducted on a range of datasets. For this experimentation, each NN approach is run twice, the first time setting $K = 10$ and the second time with K set to the full set of training objects. Again, this is evaluated via $2\times$10-fold cross-validation.

The results of the experiments are shown in Table 2, where the average classification accuracy for the methods is recorded. For clarity, the method names have

---

[2] These methods and many more have been integrated into the WEKA package [29] and can be downloaded from: http://users.aber.ac.uk/rkj/book/programs.php

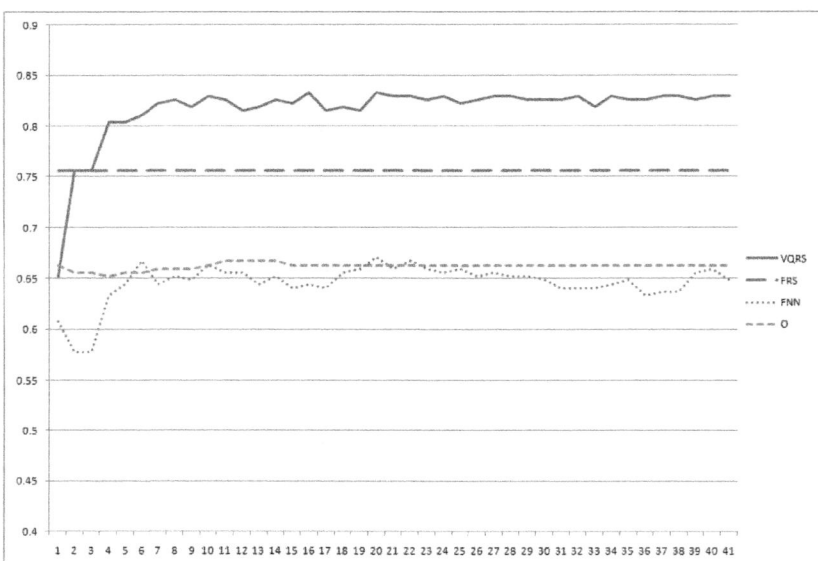

Fig. 5. Classification accuracy for the four methods and different values of K for the Heart dataset

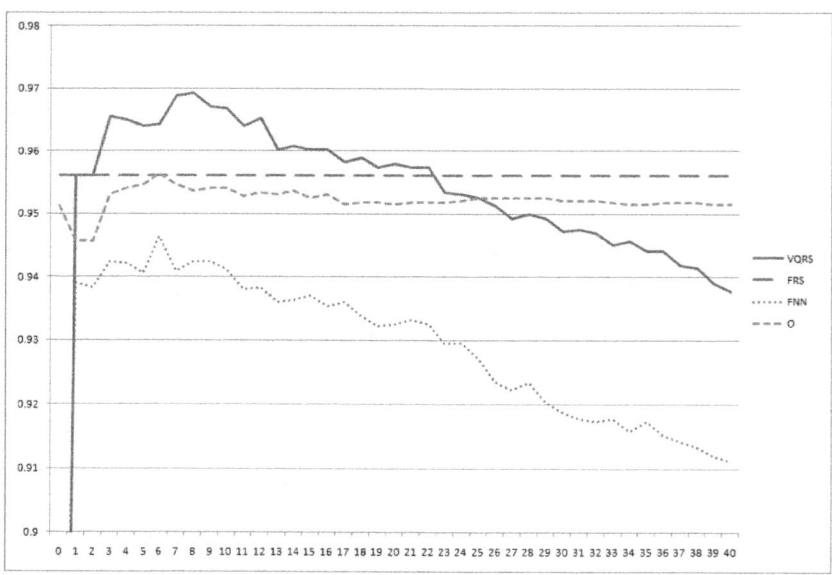

Fig. 6. Classification accuracy for the four methods and different values of K for the Letter dataset

Table 2. Nearest neighbour results

Dataset	FRS(10)	FRS	VQRS(10)	VQRS	FNN(10)	FNN	O(10)	O
Cleveland	53.21	53.21	59.41	53.89	50.19	53.89	47.50	47.50
Glass	73.13	73.13	69.36	38.06*	69.15	62.85*	71.22	71.22
Heart	76.30	76.30	82.04v	65.19*	66.11*	61.48*	66.48	66.30
Letter	95.76	95.76	96.69v	71.25*	94.25*	80.21*	95.45	95.26
Olitos	78.33	78.33	78.75	41.67*	63.75*	43.33*	65.83*	65.83*
Water 2	83.72	83.72	85.26	80.00	77.18*	80.00	79.62	79.62
Water 3	80.26	80.26	81.41	73.59*	74.49*	73.59*	73.08*	73.08*
Wine	98.02	98.02	97.75	63.79*	96.05	93.25*	95.78	95.78
Summary	(v/ /*)	(0/8/0)	(2/6/0)	(0/2/6)	(0/3/5)	(0/2/6)	(0/6/2)	(0/6/2)

been condensed in the table to: FRS (denoting FRNN-FRS), VQRS (denoting FRNN-VQRS), FNN (the standard fuzzy nearest neighbours algorithm), and O (denoting FRNN-O). A paired t-test was used to determine the statistical significance of the results at the 0.05 level when compared to FRNN-FRS. A 'v' next to a value indicates that the performance was statistically better than FRNN-FRS, and a '*' indicates that the performance was worse statistically. This is summarised by the final line in the table which shows the count of the number of statistically better, equivalent and worse results for each method in comparison to FRNN-FRS. For example (0/2/6) in the FNN column indicates that this method performed better than FRNN-FRS in zero datasets, equivalently to FRNN-FRS in two datasets, and worse than FRNN-FRS in six datasets.

For all datasets, either FRNN-FRS or FRNN-VQRS(10) yields the best results. Overall, FRNN-FRS produces the most consistent results. This is particularly remarkable considering the inherent simplicity of the method. FRNN-VQRS is best for **heart** and **letter**, which might be attributed to the comparative presence of noise in those datasets.

It is also interesting to consider the influence of the number of nearest neighbours. Both FRNN-FRS and FRNN-O remain relatively unaffected by changes in K. This could be explained in that, for FRNN-FRS, an infimum and supremum are used which can be thought of as a worst case and best case respectively. When more neighbours are considered, $R(x, y)$ values decrease as these neighbours are less similar, hence $I(R(x, y), C(x))$ increases, and $T(R(x, y), C(x))$ decreases. In other words, the more distant a neighbour is, the more unlikely it is to change the infimum and supremum value. For FRNN-O, again $R(x, y)$ decreases when more neighbours are added, and hence the value $R(x, y)C(x)$ that is added to the numerator is also small. Since each neighbour has the same weight in the denominator, the ratios stay approximately the same when adding new neighbours.

For FNN and FRNN-VQRS, increasing K can have a significant effect on classification accuracy. This is most clearly observed in the results for the **olitos** data, where there is a clear downward trend. For FRNN-VQRS, the ratio $|Ry \cap C|/|Ry|$ has to be calculated. Each neighbour has a different weight in the denominator, so the ratios can fluctuate considerably even when adding distant neighbours.

Table 3. Comparison results

Dataset	FRS	VQRS	IBk	JRip	PART	J48	SMO	NB
Cleveland	53.21	59.41	51.53	54.22	50.34	52.89	57.77	56.78
Glass	73.13	69.36	69.83	68.63	67.25	67.49	57.24*	49.99*
Heart	76.30	82.04v	76.11	80.93	74.26	78.52	84.07v	83.7v
Letter	95.76	96.69v	94.94	92.88*	93.82*	92.84*	89.05*	78.57*
Olitos	78.33	78.75	75.00	67.92*	63.33*	66.67*	87.5	76.67
Water 2	83.72	85.26	84.74	81.79	83.72	82.44	82.95	70.77*
Water 3	80.26	81.41	81.15	82.31	84.10	83.08	87.05v	85.51v
Wine	98.02	97.75	94.93	94.05	93.27	94.12	98.61	97.19
Summary (v/ /*)		(2/6/0)	(0/8/0)	(0/6/2)	(0/6/2)	(0/6/2)	(2/4/2)	(2/3/3)

Comparison with leading approaches. In order to demonstrate the efficacy of the proposed methods, further experimentation was conducted involving several leading classifiers: IBk, JRip, PART, J48, SMO (a support vector-based method) and NB (naive bayes). The same datasets as above were used and 2×10-fold cross validation was performed. For FRNN-FRS and FRNN-VQRS, K was set to 10. The results can be seen in Table 3, with statistical comparisons again between each method and FRNN-FRS.

IBk [1] is a simple (non-fuzzy) K-nearest neighbour classifier that uses Euclidean distance to compute the closest neighbour (or neighbours if more than one object has the closest distance) in the training data, and outputs this object's decision as its prediction. JRip [5] learns propositional rules by repeatedly growing rules and pruning them. During the growth phase, features are added greedily until a termination condition is satisfied. Features are then pruned in the next phase subject to a pruning metric. Once the ruleset is generated, a further optimization is performed where classification rules are evaluated and deleted based on their performance on randomized data. PART [28,29] generates rules by means of repeatedly creating partial decision trees from data. The algorithm adopts a divide-and-conquer strategy such that it removes instances covered by the current ruleset during processing. Essentially, a classification rule is created by building a pruned tree for the current set of instances; the leaf with the highest coverage is promoted to a rule. J48 [20] creates decision trees by choosing the most informative features and recursively partitioning the data into subtables based on their values. Each node in the tree represents a feature with branches from a node representing the alternative values this feature can take according to the current subtable. Partitioning stops when all data items in the subtable have the same classification. A leaf node is then created, and this classification assigned. SMO [24] implements a sequential minimal optimization algorithm for training a support vector classifier. Pairwise classification is used to solve multi-class problems.

Both FRNN-FRS and FRNN-VQRS perform well. There are two datasets (Water 3 and Heart) for which the methods are bettered by SMO and NB, but for the remainder their performance is equivalent to or better than all classifiers.

This is interesting, given the comparative algorithmic simplicity of FRNN-FRS and FRNN-VQRS.

5.2   Prediction

For the task of prediction, eight datasets were chosen that possess real-valued decision features (Table 4). The algae data sets[3] are provided by ERUDIT [11] and describe measurements of river samples for each of seven different species of alga, including river size, flow rate and chemical concentrations. The decision feature is the corresponding concentration of the particular alga. The housing dataset is taken from the Machine Learning Repository.

Seven methods were compared, namely the four nearest neighbour methods, IBk, SMOreg (support vector-based regression), LR (linear regression) and Pace. For the nearest neighbour methods, K was set to 10. Again, 2×10-fold cross validation was performed and the average root mean squared error (RMSE) was recorded.

Table 4. Dataset details

Dataset	Objects	Attributes
Algae A→G	187	11
Housing	506	13

The linear regression model [10] is applicable for numeric classification and prediction provided that the relationship between the input attributes and the output attribute is almost linear. The relation is then assumed to be a linear function of some parameters - the task being to estimate these parameters given training data. This is often accomplished by the method of least squares, which consists of finding the values that minimize the sum of squares of the residuals. Once the parameters are established, the function can be used to estimate the output values for unseen data. Projection adjustment by contribution estimation (Pace) regression [25] is a recent approach to fitting linear models, based on considering competing models. Pace regression improves on classical ordinary least squares regression by evaluating the effect of each variable and using a clustering analysis to improve the statistical basis for estimating their contribution to the overall regression. SMOreg is a sequential minimal optimization algorithm for training a support vector regression using polynomial or Radial Basis Function kernels [19,24]. It reduces support vector machine training down to a series of smaller quadratic programming subproblems that have an analytical solution. This has been shown to be very efficient for prediction problems using linear support vector machines and/or sparse data sets.

The results for the prediction experimentation can be seen in Table 5. It can be seen that FRNN-O and IBk perform poorly, and the other methods perform similarly to FRNN-FRS. The average RMSEs for FRNN-FRS and FRNN-VQRS are generally lower than those obtained for the other algorithms.

---

[3] See http://archive.ics.uci.edu/ml/datasets/Coil+1999+Competition+Data

Table 5. Prediction results (RMSE)

Dataset	FRS	VQRS	FNN	O	IBk	SMOreg	LR	Pace
Algae A	17.15	16.81	15.79	24.55*	24.28*	17.97	18.00	18.18
Algae B	10.77	10.57	10.68	13.04*	17.18*	10.08	10.30	10.06
Algae C	6.81	6.68	6.99	8.16*	9.07*	7.12	7.11	7.26
Algae D	2.91	2.88	3.04	3.47*	4.62*	2.99	3.86	3.95
Algae E	6.88	6.85	7.38	9.10*	9.02*	7.18	7.61	7.59
Algae F	10.40	10.33	11.24	12.60*	13.51*	10.09	10.33	9.65
Algae G	4.97	4.84	5.23	5.38	6.48	4.96	5.21	4.96
Housing	4.72	4.85	6.62*	24.27*	4.59	4.95	4.80	4.79
Summary	(v/ /*)	(0/8/0)	(0/7/1)	(0/1/7)	(0/2/6)	(0/8/0)	(0/8/0)	(0/8/0)

6   Conclusion and Future Work

This paper has presented two new techniques for fuzzy-rough classification based on the use of lower and upper approximations w.r.t. fuzzy tolerance relations. The difference between them is in the definition of the approximations: while FRNN-FRS uses "traditional" operations based on a t-norm and an implicator, FRNN-VQRS uses a fuzzy quantifier-based approach. The results show that these methods are effective, and that they are very competitive with existing methods for both classification and prediction. Further investigation is still needed to adequately explain the impact of the choice of fuzzy relations, connectives and quantifiers. Of particular importance is the choice of relation composition operator as this determines the overall similarity of objects based on the full set of data features. The use of a t-norm for this operation is sensible from a theoretical viewpoint, but may introduce problems from a practical perspective as the overall similarity of a pair of objects will be zero if these objects have zero similarity for just one of their features. Therefore, an alternative method of combining relations is desirable.

Also, the impact of a feature selection preprocessing step upon classification accuracy needs to be investigated. It is expected that feature selectors that incorporate fuzzy relations expressing closeness of objects (see e.g. [8,15]) should be able to further improve the effectiveness of the classification methods presented here.

Acknowledgment.   Chris Cornelis would like to thank the Research Foundation—Flanders for funding his research.

References

1. Aha, D.: Instance-based learning algorithm. Machine Learning 6, 37–66 (1991)
2. Bhatt, R.B., Gopal, M.: FRID: Fuzzy-Rough Interactive Dichotomizers. In: IEEE International Conference on Fuzzy Systems (FUZZ-IEEE 2004), pp. 1337–1342 (2004)

3. Bian, H., Mazlack, L.: Fuzzy-Rough Nearest-Neighbor Classification Approach. In: Proceeding of the 22nd International Conference of the North American Fuzzy Information Processing Society (NAFIPS), pp. 500–505 (2003)
4. Blake, C.L., Merz, C.J.: UCI Repository of machine learning databases. University of California, Irvine (1998), http://archive.ics.uci.edu/ml/
5. Cohen, W.W.: Fast effective rule induction. In: Machine Learning: Proceedings of the 12th International Conference, pp. 115–123 (1995)
6. Cornelis, C., De Cock, M., Radzikowska, A.M.: Vaguely Quantified Rough Sets. In: An, A., Stefanowski, J., Ramanna, S., Butz, C.J., Pedrycz, W., Wang, G. (eds.) RSFDGrC 2007. LNCS (LNAI), vol. 4482, pp. 87–94. Springer, Heidelberg (2007)
7. Cornelis, C., De Cock, M., Radzikowska, A.M.: Fuzzy Rough Sets: from Theory into Practice. In: Pedrycz, W., Skowron, A., Kreinovich, V. (eds.) Handbook of Granular Computing. Wiley, Chichester (2008)
8. Cornelis, C., Hurtado Martín, G., Jensen, R., Slezak, D.: Feature Selection with fuzzy decision reducts. In: Wang, G., Li, T., Grzymala-Busse, J.W., Miao, D., Skowron, A., Yao, Y. (eds.) RSKT 2008. LNCS (LNAI), vol. 5009, pp. 284–291. Springer, Heidelberg (2008)
9. Duda, R., Hart, P.: Pattern Classification and Scene Analysis. Wiley, New York (1973)
10. Edwards, A.L.: An Introduction to Linear Regression and Correlation. W.H. Freeman, San Francisco (1976)
11. European Network for Fuzzy Logic and Uncertainty Modelling in Information Technology (ERUDIT), Protecting rivers and streams by monitoring chemical concentrations and algae communities, Computational Intelligence and Learning (CoIL) Competition (1999)
12. Greco, S., Inuiguchi, M., Slowinski, R.: Fuzzy rough sets and multiple-premise gradual decision rules. International Journal of Approximate Reasoning 41, 179–211 (2005)
13. Hong, T.P., Liou, Y.L., Wang, S.L.: Learning with Hierarchical Quantitative Attributes by Fuzzy Rough Sets. In: Proceedings of the Joint Conference on Information Sciences, Advances in Intelligent Systems Research(2006)
14. Hsieh, N.-C.: Rule Extraction with Rough-Fuzzy Hybridization Method. In: Washio, T., Suzuki, E., Ting, K.M., Inokuchi, A. (eds.) PAKDD 2008. LNCS (LNAI), vol. 5012, pp. 890–895. Springer, Heidelberg (2008)
15. Jensen, R., Shen, Q.: Fuzzy-Rough Sets Assisted Attribute Selection. IEEE Transactions on Fuzzy Systems 15(1), 73–89 (2007)
16. Jensen, R., Shen, Q.: Computational Intelligence and Feature Selection: Rough and Fuzzy Approaches. Wiley-IEEE Press (2008)
17. Keller, J.M., Gray, M.R., Givens, J.A.: A fuzzy K-nearest neighbor algorithm. IEEE Trans. Systems Man Cybernet. 15(4), 580–585 (1985)
18. Pawlak, Z.: Rough Sets: Theoretical Aspects of Reasoning About Data. Kluwer Academic Publishing, Dordrecht (1991)
19. Platt, J.: Fast Training of Support Vector Machines using Sequential Minimal Optimization. In: Schölkopf, B., Burges, C., Smola, A. (eds.) Advances in Kernel Methods - Support Vector Learning. MIT Press, Cambridge (1998)
20. Quinlan, J.R.: C4.5: Programs for Machine Learning. The Morgan Kaufmann Series in Machine Learning. Morgan Kaufmann Publishers, San Mateo (1993)
21. Radzikowska, A.M., Kerre, E.E.: A comparative study of fuzzy rough sets. Fuzzy Sets and Systems 126(2), 137–155 (2002)
22. Sarkar, M.: Fuzzy-Rough nearest neighbors algorithm. Fuzzy Sets and Systems 158, 2123–2152 (2007)

23. Shen, Q., Chouchoulas, A.: A rough-fuzzy approach for generating classification rules. Pattern Recognition 35(11), 2425–2438 (2002)
24. Smola, A.J., Schölkopf, B.: A Tutorial on Support Vector Regression, NeuroCOLT2 Technical Report Series - NC2-TR-1998-030 (1998)
25. Wang, Y.: A new approach to fitting linear models in high dimensional spaces, PhD Thesis, Department of Computer Science, University of Waikato (2000)
26. Wang, X., Yang, J., Teng, X., Peng, N.: Fuzzy-Rough Set Based Nearest Neighbor Clustering Classification Algorithm. In: Wang, L., Jin, Y. (eds.) FSKD 2005. LNCS (LNAI), vol. 3613, pp. 370–373. Springer, Heidelberg (2005)
27. Wang, X., Tsang, E.C.C., Zhao, S., Chen, D., Yeung, D.S.: Learning fuzzy rules from fuzzy samples based on rough set technique. Information Sciences 177(20), 4493–4514 (2007)
28. Witten, I.H., Frank, E.: Generating Accurate Rule Sets Without Global Optimization. In: Proceedings of the 15th International Conference on Machine Learning. Morgan Kaufmann Publishers, San Francisco (1998)
29. Witten, I.H., Frank, E.: Data Mining: Practical machine learning tools with Java implementations. Morgan Kaufmann, San Francisco (2000)
30. Zadeh, L.A.: Fuzzy sets. Information and Control 8, 338–353 (1965)

Efficient Mining of Jumping Emerging Patterns with Occurrence Counts for Classification

Łukasz Kobyliński and Krzysztof Walczak

Institute of Computer Science, Warsaw University of Technology
ul. Nowowiejska 15/19, 00-665 Warszawa, Poland
{L.Kobylinski,K.Walczak}@ii.pw.edu.pl

Abstract. In this paper we propose an efficient method of discovering Jumping Emerging Patterns with Occurrence Counts for the use in classification of data with numeric or nominal attributes. This new extension of Jumping Emerging Patterns proved to perform well when classifying image data and here we experimentally compare it to other methods, by using generalized border-based pattern mining algorithm to build the classifier.

Keywords: data mining, emerging patterns, image representation, classification.

1 Introduction

Recently there has been a strong progress in the area of rule- and pattern-based classification algorithms, following the very fruitful research in the area of association rules and emerging patterns. One of the most recent and promising methods is classification using jumping emerging patterns (JEPs). It is based on the idea that JEPs, as their support changes sharply from one dataset to another, carry highly discriminative information that allows creating classifiers, which associate previously unseen records of data to one of these datasets. As JEPs have been originally conceived for transaction databases, where each data record is a set of items, a JEP-based classifier is not usually directly applicable to relational databases, i.e. containing numeric or nominal attributes. In such case an additional discretization step is required to transform the available data to transactional form.

In this article we address the problem of efficiently discovering JEPs and using them directly for supervised learning in databases, where the data can be described as multi-sets of features. This is an enhancement of the transactional database representation, where instead of a binary relation between items and database records, an occurrence count is associated with every item in a set. Example real-world problems that could be approached in this way include market-basket analysis (quantities of bought products), as well as text and multimedia data mining (numbers of occurrences of particular features). We use a new type of JEPs to accomplish this task – the jumping emerging patterns with occurrence counts (occJEPs) – show both the original semi-naïve algorithm and

J.F. Peters et al. (Eds.): Transactions on Rough Sets XIII, LNCS 6499, pp. 73–88, 2011.

a new border-based algorithm for finding occJEPs and compare their discriminative value with other recent classification methods.

The rest of the paper is organized as follows. Section 2 outlines previous work done in the field, while Section 3 gives an overview of the concept of emerging patterns in transaction databases. In Sections 4–7 we introduce jumping emerging patterns with occurrence counts (occJEPs), present their discovery algorithms and describe the chosen method of performing classification with a set of found occJEPs. Section 8 presents experimental results of classification and a comparison of some of the most current classifiers. Section 9 closes with a conclusion and discussion on possible future work.

2   Previous Work

The number of papers concerning emerging patterns and the more general concept known as contrast patterns is growing rapidly since the original introduction of EPs by Dong and Li in [1]. Emerging patterns have been defined there as itemsets, for which supports increase significantly from one dataset to another. This idea has become the subject of interest of many researchers, as the patterns proved to be a very accurate alternative to previously proposed rule- and tree-based classifiers. The first classification algorithm based on EP mining – Classification based on Aggregating Emerging Patterns (CAEP) – has been proposed in [2].

Efficient mining of emerging patterns has been studied in [3], while more accurate classification algorithm has been proposed in [4]. In [5] DeEPS, a new, lazy learning scheme for EP-based classification has been presented. A particularly good performance of classification has been achieved using only a subset of EPs, the jumping emerging patterns [6]. Mining efficiency of such patterns has been further improved in [7,8,9].

More recently, a rough set theory approach to pattern mining has been presented in [10] and a method based on the concept of equivalence classes in [11].

Many applications of EPs have been proposed to date, with a particularly fruitful research in the area of bioinformatics, specifically classification and finding relationships in gene data. The first EP-based algorithms concerning analysis of such data have been proposed in [12,13].

The concept of recurrent items in transactional systems has been presented in the area of multimedia data analysis in [14] in the context of association rules, while general and efficient algorithms for discovering rules with recurrent items have been studied in [15] and [16]. The extension of the definition of jumping emerging patterns to include recurrent items and using them for building classifiers has been proposed in [17].

3   Emerging Patterns

Emerging patterns may be briefly described as patterns, which occur frequently in one set of data and seldom in another. We now give a formal definition of emerging patterns in transaction systems.

Let a transaction system be a pair $(\mathcal{D}, \mathcal{I})$, where $\mathcal{D}$ is a finite sequence of transactions $(T_1, \ldots, T_n)$ (database), such that $T_i \subseteq \mathcal{I}$ for $i = 1, \ldots, n$ and $\mathcal{I}$ is a non-empty set of items (itemspace). A support of an itemset $X \subset \mathcal{I}$ in a sequence $D = (T_i)_{i \in K \subseteq \{1, \ldots, n\}} \subseteq \mathcal{D}$ is defined as:

$$\text{supp}_D(X) = \frac{|\{i \in K : X \subseteq T_i\}|}{|K|}. \tag{1}$$

Given two databases $D_1, D_2 \subseteq \mathcal{D}$ the growth rate of an itemset $X \subset \mathcal{I}$ from D_1 to D_2 is defined as:

$$GR_{D_1 \to D_2}(X) = \begin{cases} 0 & \text{if } \text{supp}_{D_1}(X) = 0 \text{ and } \text{supp}_{D_2}(X) = 0, \\ \infty & \text{if } \text{supp}_{D_1}(X) = 0 \text{ and } \text{supp}_{D_2}(X) \neq 0, \\ \frac{\text{supp}_{D_2}(X)}{\text{supp}_{D_1}(X)} & \text{otherwise.} \end{cases} \tag{2}$$

Given a minimum growth rate ρ, we define an itemset $X \subset \mathcal{I}$ to be a ρ-emerging pattern (ρ-EP) from D_1 to D_2 if $GR_{D_1 \to D_2}(X) > \rho$. Furthermore, we say that an itemset X is a jumping emerging pattern (JEP), when its growth rate is infinite, that is $GR_{D_1 \to D_2}(X) = \infty$. Having a minimum support threshold ξ, we define a strong ξ-jumping emerging pattern to be a JEP from D_1 to D_2 for which $\text{supp}_{D_1}(X) = 0$ and $\text{supp}_{D_2}(X) > \xi$. A set of all JEPs from D_1 to D_2 is called a JEP space and denoted by $JEP(D_1, D_2)$.

Example 1. For the example database given by Table 1, having $D_1 = (T_{1,2,3})$, $D_2 = (T_{4,5})$, the set of minimal JEPs from D_1 to D_2 is equal to: $\{\{black\}, \{brown\}, \{red, white\}\}$. A minimal JEP is a jumping emerging pattern X, such that no proper subset of X is a JEP, e.g. in the case of $\{red, white\}$ neither $\{red\}$ nor $\{white\}$ are JEPs.

Table 1. Transaction database example. T_i – transactions with binary items, T_i^r – transactions with recurrent items.

	i	T_i	T_i^r
D_1	1	blue, green, white, yellow	$8 \cdot blue$, $4 \cdot green$, $3 \cdot white$, $1 \cdot yellow$
	2	beige, red, yellow	$10 \cdot beige$, $3 \cdot red$, $3 \cdot yellow$
	3	white, magenta	$12 \cdot white$, $4 \cdot magenta$
D_2	4	blue, brown, white	$6 \cdot blue$, $2 \cdot brown$, $8 \cdot white$
	5	black, white, red, yellow	$9 \cdot black$, $2 \cdot white$, $3 \cdot red$, $2 \cdot yellow$

4 Jumping Emerging Patterns with Occurrence Counts

Let a transaction system with recurrent items be a pair $(\mathcal{D}^r, \mathcal{I})$, where $\mathcal{D}^r$ is a database and $\mathcal{I}$ is an itemspace (the definition of itemspace remains unchanged). We define database $\mathcal{D}^r$ as a finite sequence of transactions $(T_1^r, \ldots, T_n^r)$ for $i = 1, \ldots, n$. Each transaction is a set of pairs $T_i^r = \{(t_i, q_i); t_i \in \mathcal{I}\}$, where $q_i : \mathcal{I} \to \mathbb{N}$ is a function, which assigns the number of occurrences to each item of the transaction. Similarly, a multiset of items X^r is defined as a set of pairs $\{(x,\ p);\ x \in \mathcal{I}\}$, where $p : \mathcal{I} \to \mathbb{N}$. We say that $x \in X^r \iff p(x) \geq 1$ and define $X = \{x : x \in X^r\}$. We will write $X^r = (X,\ P)$ to distinguish X as the set of items contained in a multiset X^r and P as the set of functions, which assign occurrence counts to particular items.

The support of a multiset of items X^r in a sequence $D^r = (T_i^r)_{i \in K \subseteq \{1,\dots,n\}} \subseteq \mathcal{D}^r$ is defined as:

$$\mathrm{supp}_D(X^r, \theta) = \frac{|\{i \in K :\ X^r \overset{\theta}{\subseteq} T_i^r\}|}{|K|}, \tag{3}$$

where $\overset{\theta}{\subseteq}$ is an inclusion relation between a multiset $X^r = (X,\ P)$ and a transaction $T^r = (T,\ Q)$ with an occurrence threshold $\theta \geq 1$:

$$X^r \overset{\theta}{\subseteq} T^r \iff \forall_{x \in \mathcal{I}}\ q(x) \geq \theta \cdot p(x). \tag{4}$$

We will assume that the relation $\subseteq$ is equivalent to $\overset{1}{\subseteq}$ in the context of two multisets.

Example 2. The support of a multiset of items $X^r = \{1 \cdot white,\ 2 \cdot yellow\}$ for threshold $\theta = 1$ in transaction sequence $D_1 = (T_{1,2,3}^r)$ of database given by Table 1 is equal to: $\mathrm{supp}_{D_1}(X^r, 1) = 0$. Similarly, for $D_2 = (T_{4,5}^r)$, $\mathrm{supp}_{D_2}(X^r, 1) = 1$. Having the threshold $\theta = 2$, $\mathrm{supp}_{D_1}(X^r, 2) = 0$ and $\mathrm{supp}_{D_2}(X^r, 2) = 0$.

Let a decision transaction system be a tuple $(\mathcal{D}^r, \mathcal{I}, \mathcal{I}_d)$, where $(\mathcal{D}^r, \mathcal{I} \cup \mathcal{I}_d)$ is a transaction system with recurrent items and $\forall_{T^r \in \mathcal{D}^r} |T \cap \mathcal{I}_d| = 1$. Elements of $\mathcal{I}$ and $\mathcal{I}_d$ are called condition and decision items, respectively. A support for a decision transaction system $(\mathcal{D}^r, \mathcal{I}, \mathcal{I}_d)$ is understood as a support in the transaction system $(\mathcal{D}^r, \mathcal{I} \cup \mathcal{I}_d)$.

For each decision item $c \in \mathcal{I}_d$ we define a decision class sequence $C_c = (T_i^r)_{i \in K}$, where $K = \{k \in \{1,\dots,n\} : c \in T_k\}$. Notice that each of the transactions from $\mathcal{D}^r$ belongs to exactly one class sequence. In addition, for a database $D = (T_i^r)_{i \in K \subseteq \{1,\dots,n\}} \subseteq \mathcal{D}^r$, we define a complement database $D' = (T_i^r)_{i \in \{1,\dots,n\}-K}$.

Given two databases $D_1, D_2 \subseteq \mathcal{D}^r$ we call a multiset of items X^r a jumping emerging pattern with occurrence counts (occJEP) from D_1 to D_2, if $\mathrm{supp}_{D_1}(X^r, 1) = 0 \wedge \mathrm{supp}_{D_2}(X^r, \theta) > 0$, where θ is the occurrence threshold. A set of all occJEPs with a threshold θ from D_1 to D_2 is called an occJEP space and denoted by $occJEP(D_1, D_2, \theta)$. We distinguish the set of all minimal occJEPs as $occJEP_m$, $occJEP_m(D_1, D_2, \theta) \subseteq occJEP(D_1, D_2, \theta)$. Notice also that $occJEP(D_1, D_2, \theta) \subseteq occJEP(D_1, D_2, \theta - 1)$ for $\theta \geq 2$. In the rest of the paper we will refer to multisets of items as itemsets and use the symbol X^r to avoid confusion.

Example 3. Taking into consideration $D_1 = (T_{1,2,3}^r)$ and $D_2 = (T_{4,5}^r)$ from Table 1, the set of minimal occJEPs from D_1 to D_2 with threshold $\theta = 1$ is equal to: $\{\{1 \cdot black\}, \{1 \cdot brown\}, \{1 \cdot blue, 4 \cdot white\}, \{1 \cdot red, 1 \cdot white\}, \{1 \cdot white, 2 \cdot yellow\}\}$.

Changing the threshold to $\theta = 2$ results in reducing the set of patterns to: $\{\{1 \cdot black\}, \{1 \cdot brown\}, \{1 \cdot blue, 4 \cdot white\}, \{1 \cdot red, 1 \cdot white\}\}$. This is because $\text{supp}_{D_1}(\{1 \cdot white, 2 \cdot yellow\}, 1) = 0$ and $\text{supp}_{D_2}(\{1 \cdot white, 2 \cdot yellow\}, 1) > 1$, but $\text{supp}_{D_2}(\{1 \cdot white, 2 \cdot yellow\}, 2) = 0$.

Occurrence Threshold. The introduction of an occurrence threshold θ allows for differentiating transactions containing the same sets of items with a specified tolerance margin of occurrence counts. It is thus possible to define a difference in the number of occurrences, which is necessary to consider such a pair of transactions as distinct sets of items.

For the example image database given by Table 1 we can see that the differences between counts of such items as *white* and *yellow* may be too small to assume they represent a general pattern present in the database that would allow building a classifier. Setting the threshold to a higher value results in a smaller number of patterns, but the discovered ones have a greater confidence.

5   A Semi-Naïve Mining Algorithm

Our previous method of discovering occJEPs, introduced in [17], is based on the observation that only minimal patterns need to be found to perform classification. Furthermore, it is usually not necessary to mine patterns longer than a few items, as their support is very low and thus their impact on classification accuracy is negligible. This way we can reduce the problem to: (a) finding only such occJEPs, for which no patterns with a lesser number of items and the same or lower number of item occurrences exist; (b) discovering patterns of less than δ items.

Let C_c be a decision class sequence of a database $\mathcal{D}^{r}$ for a given decision item c and C'_c a complement sequence to C_c. We define $D_1 = C'_c$, $D_2 = C_c$ and the aim of the algorithm to discover $occJEP_m(D_1, D_2, \theta)$. We begin by finding the patterns, which are not supported in D_1, as possible candidates for occJEPs. In case of multi-item patterns at least one of the item counts of the candidate pattern has to be larger than the corresponding item count in the database. We can write this as:

$$X^{r} = (X, P) \text{ is an occJEP candidate} \iff \forall_{T^{r}=(T,Q) \in D_1} \exists_{x \in X} \; p(x) > q(x).$$

Table 2 shows an example set of conditions for single and multi-item occJEP candidates.

The first step of the algorithm is then to create a set of conditions in the form of $[p(i_j) > q_1(i_j) \lor \ldots \lor p(i_k) > q_1(i_k)] \land \ldots \land [p(i_j) > q_n(i_j) \lor \ldots \lor p(i_k) > q_n(i_k)]$ for each of the candidate itemsets $X^{r} = (X, P)$, $X \subseteq 2^{\mathcal{I}}$, where j and k are subscripts of items appearing in a particular X^{r} and n is the number of transactions in D_1. Solving this set of inequalities results in its transformation to the form of $[p(i_j) > r_j \land \ldots \land p(i_k) > r_k] \lor \ldots \lor [p(i_j) > s_j \land \ldots \land p(i_k) > s_k]$, where r and s are the occurrence counts of respective items. The counts have to be incremented by 1, to fulfill the condition of $\text{supp}_{D_1}(X^{r}, \theta) = 0$.

Table 2. Finding occJEPs in a transaction database with recurrent items. Example conditions for single-item patterns and patterns consisting of two items.

D_1	$q(i_1)$	$q(i_2)$	$q(i_3)$	$cond(p(i_1))$	$cond(p(i_2), p(i_3))$
T_1^{r}	3	12	3	$p(i_1) > 3$	$p(i_2) > 12 \vee p(i_3) > 3$
T_2^{r}	0	16	11	$p(i_1) > 0$	$p(i_2) > 16 \vee p(i_3) > 11$
T_3^{r}	5	19	4	$p(i_1) > 5$	$p(i_2) > 19 \vee p(i_3) > 4$
T_4^{r}	2	14	13	$p(i_1) > 2$	$p(i_2) > 14 \vee p(i_3) > 13$

Example 4. For the previously introduced example of D_1 from Table 2, we can see that $cond(p(i_1))$ resolves to $p(i_1) > 5$ and $cond(p(i_1), p(i_2))$ to $p(i_2) > 19 \vee p(i_3) > 13 \vee (p(i_2) > 14 \wedge p(i_3) > 11) \vee (p(i_2) > 16 \wedge p(i_3) > 4)$. Notice that resolved conditions for pattern length l also contain all conditions for $l - 1$. The resulting candidates for minimal patterns, after incrementing the occurrence counts, are thus the following: $X_1^{\mathrm{r}} = \{6 \cdot i_1\}$, $X_2^{\mathrm{r}} = \{20 \cdot i_2\}$, $X_3^{\mathrm{r}} = \{14 \cdot i_3\}$, $X_4^{\mathrm{r}} = \{15 \cdot i_2, 12 \cdot i_3\}$, $X_5^{\mathrm{r}} = \{17 \cdot i_2, 5 \cdot i_3\}$.

Having found the minimum occurrence counts of items in the candidate itemsets, we then calculate the support of each of the itemsets in D_2 with a threshold θ. The candidates, for which $\mathrm{supp}_{D_2}(X, \theta) > 0$ are the minimal $occJEPs(D_1, D_2, \theta)$.

Example 5. Continuing the above example, we see from Table 3 that the support of candidate patterns $\mathrm{supp}_{D_2}(X_1^{\mathrm{r}}, 3) > 0$, $\mathrm{supp}_{D_2}(X_3^{\mathrm{r}}, 2) = 0$ and $\mathrm{supp}_{D_2}(X_5^{\mathrm{r}}, 1) > 0$. X_1^{r} and X_5^{r} are thus minimal occJEPs with threshold values $\theta = 3$ and $\theta = 1$ respectively. By the definition of occJEPs, X_1^{r} is also an occJEP for $\theta \in [1, 3]$. Other minimal occJEPs are X_3^{r} and X_4^{r} for $\theta = 1$, as their respective supports in D_2 are equal to $1/4$.

Table 3. Finding occJEPs in a transaction database with recurrent items. Calculating the support count of candidate itemsets in complementary database.

D_2	$p(i_1)$	$p(i_2)$	$p(i_3)$	$\phi(X_1^{\mathrm{r}}, T^{\mathrm{r}}, 3)$	$\phi(X_3^{\mathrm{r}}, T^{\mathrm{r}}, 2)$	$\phi(X_5^{\mathrm{r}}, T^{\mathrm{r}}, 1)$
T_1^{r}	11	15	4	0	0	0
T_2^{r}	18	16	12	1	0	0
T_3^{r}	12	17	5	0	0	1
T_4^{r}	23	14	14	1	0	0
				$\mathrm{supp}_{D_2} = 1/2$	$\mathrm{supp}_{D_2} = 0$	$\mathrm{supp}_{D_2} = 1/4$

6   Border-Based Mining Algorithm

The border-based occJEP discovery algorithm is an extension of the EP-mining method described in [3]. Similarly, as proved in [7] for regular emerging patterns, we can use the concept of borders to represent a collection of occJEPs. This is because the occJEP space S is convex, that is it follows: $\forall X^{\mathrm{r}}, Z^{\mathrm{r}} \in S^{\mathrm{r}} \ \forall Y^{\mathrm{r}} \in$

2^{S^r} $X^\mathrm{r} \subseteq Y^\mathrm{r} \subseteq Z^\mathrm{r} \Rightarrow Y^\mathrm{r} \in S^\mathrm{r}$. For the sake of readability we will now onward denote particular items with consecutive alphabet letters, with an index indicating the occurrence count, and skip individual brackets, e.g. $\{a_1 b_2, c_3\}$ instead of $\{\{1 \cdot i_1, 2 \cdot i_2\}, \{3 \cdot i_3\}\}$.

Example 6. $\mathcal{S} = \{a_1, a_1 b_1, a_1 b_2, a_1 c_1, a_1 b_1 c_1, a_1 b_2 c_1\}$ is a convex collection of sets, but $\mathcal{S}' = \{a_1, a_1 b_1, a_1 c_1, a_1 b_1 c_1, a_1 b_2 c_1\}$ is not convex. We can partition it into two convex collections $\mathcal{S}'_1 = \{a_1, a_1 b_1\}$ and $\mathcal{S}'_2 = \{a_1 c_1, a_1 b_1 c_1, a_1 b_2 c_1\}$.

A border is an ordered pair $< \mathcal{L}, \mathcal{R} >$ such that $\mathcal{L}$ and $\mathcal{R}$ are antichains, $\forall X^\mathrm{r} \in \mathcal{L} \; \exists Y^\mathrm{r} \in \mathcal{R} \; X^\mathrm{r} \subseteq Y^\mathrm{r}$ and $\forall X^\mathrm{r} \in \mathcal{R} \; \exists Y^\mathrm{r} \in \mathcal{L} \; Y^\mathrm{r} \subseteq X^\mathrm{r}$. The collection of sets represented by a border $< \mathcal{L}, \mathcal{R} >$ is equal to:

$$[\mathcal{L}, \mathcal{R}] = \{Y^\mathrm{r} : \exists X^\mathrm{r} \in \mathcal{L}, \exists Z^\mathrm{r} \in \mathcal{R} \text{ such that } X^\mathrm{r} \subseteq Y^\mathrm{r} \subseteq Z^\mathrm{r}\}. \tag{5}$$

Example 7. The border of collection $\mathcal{S}$, introduced in earlier example, is equal to $[\mathcal{L}, \mathcal{R}] = [\{a_1\}, \{a_1 b_2 c_1\}]$.

The most basic operation involving borders is a border differential, defined as:

$$< \mathcal{L}, \mathcal{R} >=< \{\emptyset\}, \mathcal{R}_1 > - < \{\emptyset\}, \mathcal{R}_2 > . \tag{6}$$

As proven in [7] this operation may be reduced to a series of simpler operations. For $\mathcal{R}_1 = \{U_1, \ldots, U_m\}$:

$$< \mathcal{L}_i, \mathcal{R}_i > = < \{\emptyset\}, \{U_i^\mathrm{r}\} > - < \{\emptyset\}, \mathcal{R}_2 > . \tag{7}$$

$$< \mathcal{L}, \mathcal{R} > = < \bigcup_{i=1}^{m} \mathcal{L}_i, \bigcup_{i=1}^{m} \mathcal{R}_i > . \tag{8}$$

A direct approach to calculating the border differential would be to expand the borders and compute set differences.

Example 8. The border differential between $[\{\emptyset\}, \{a_1 b_2 c_1\}]$ and $[\{\emptyset\}, \{a_1 c_1\}]$ is equal to $[\{b_1\}, \{a_1 b_2 c_1\}]$. This is because:

$$[\{\emptyset\}, \{a_1 b_2 c_1\}] = \{a_1, b_1, b_2, c_1, a_1 b_1, a_1 b_2, a_1 c_1, b_1 c_1, b_2 c_1, a_1 b_1 c_1, a_1 b_2 c_1\}$$
$$[\{\emptyset\}, \{a_1 c_1\}] = \{a_1, c_1, a_1 c_1\}$$
$$[\{\emptyset\}, \{a_1 b_2 c_1\}] - [\{\emptyset\}, \{a_1 c_1\}] = \{b_1, b_2, a_1 b_1, a_1 b_2, b_1 c_1, b_2 c_1, a_1 b_1 c_1, a_1 b_2 c_1\}$$

6.1 Algorithm Optimizations

On the basis of optimizations proposed in [3], we now show the extensions necessary for discovering emerging patterns with occurrence counts. All of the ideas presented there for reducing the number of operations described in the context of regular EPs are also applicable for recurrent patterns. The first idea allows avoiding the expansion of borders when calculating the collection of minimal itemsets $\mathrm{Min}(\mathcal{S})$ in a border differential $\mathcal{S} = [\{\emptyset\}, \{U^{\mathrm{r}}\}] - [\{\emptyset\}, \{S_1^{\mathrm{r}}, \ldots, S_k^{\mathrm{r}}\}]$. It has been proven in [3] that $\mathrm{Min}(\mathcal{S})$ is equivalent to:

$$\mathrm{Min}(\mathcal{S}) = \mathrm{Min}(\{\bigcup\{s_1, \ldots, s_k\} : s_i \in U^{\mathrm{r}} - S_i^{\mathrm{r}}, 1 \leq i \leq k\}). \qquad (9)$$

In the case of emerging patterns with occurrence counts we need to define the left-bound union and set theoretic difference operations between multisets of items $X^{\mathrm{r}} = (X, P)$ and $Y^{\mathrm{r}} = (Y, Q)$. These operations guarantee that the resulting patterns are still minimal.

Definition 1. *The left-bound union of multisets* $X^{\mathrm{r}} \cup Y^{\mathrm{r}} = Z^{\mathrm{r}}$. $Z^{\mathrm{r}} = (Z, R)$, *where:* $Z = \{z : z \in X \vee z \in Y\}$ *and* $R = \{r(z) = max(p(z), q(z))\}$.

Definition 2. *The left-bound set theoretic difference of multisets* $X^{\mathrm{r}} - Y^{\mathrm{r}} = Z^{\mathrm{r}}$. $Z^{\mathrm{r}} = (Z, R)$, *where:* $Z = \{z : z \in X \wedge p(z) > q(z)\}$ *and* $R = \{r(z) = q(z) + 1\}$.

Example 9. For the differential: $[\{\emptyset\}, \{a_1 b_3 c_1 d_1\}] - [\{\emptyset\}, \{b_1 c_1\}, \{b_3 d_1\}, \{c_1 d_1\}]$. $U = \{a_1 b_3 c_1 d_1\}$, $S_1 = \{b_1 c_1\}$, $S_2 = \{b_3 d_1\}$, $S_3 = \{c_1 d_1\}$. $U - S_1 = \{a_1 b_2 d_1\}$, $U - S_2 = \{a_1 c_1\}$, $U - S_3 = \{a_1 b_1\}$. Calculating the Min function:

$$\mathrm{Min}([\{\emptyset\}, \{a_1 b_3 c_1 d_1\}] - [\{\emptyset\}, \{b_1 c_1\}, \{b_3 d_1\}, \{c_1 d_1\}]) =$$
$$= \mathrm{Min}(\{a_1 a_1 a_1, a_1 a_1 b_1, a_1 c_1 a_1, a_1 c_1 b_1, b_2 a_1 a_1,$$
$$b_2 a_1 b_1, b_2 c_1 b_1, d_1 a_1 a_1, d_1 a_1 b_1, d_1 c_1 a_1, d_1 c_1 b_1\}) =$$
$$= \mathrm{Min}(\{a_1, a_1 b_1, a_1 c_1, a_1 b_1 c_1, a_1 b_2, a_1 b_2, b_2 c_1, a_1 d_1, a_1 b_1 d_1, a_1 c_1 d_1, b_1 c_1 d_1\}) =$$
$$= \{a_1, b_2 c_1, b_1 c_1 d_1\} \ .$$

Similar changes are necessary when performing the border expansion in an incremental manner, which has been proposed as the second possible algorithm optimization. The union and difference operations in the following steps need to be conducted according to Definitions 1 and 2 above, see Algorithm 1.

Algorithm 1. Incremental expansion

   Input   : U^{r}, S_i^{r}
   Output: $\mathcal{L}$
1 $\mathcal{L} \longleftarrow \{\{x\} : x \in U^{\mathrm{r}} - S_1^{\mathrm{r}}\}$
2 **for** $i = 2$ **to** k **do**
3 | $\mathcal{L} \longleftarrow \mathrm{Min}\{X^{\mathrm{r}} \cup \{x\} : X^{\mathrm{r}} \in \mathcal{L}, \ x \in U^{\mathrm{r}} - S_i^{\mathrm{r}}\}$
4 **end**

Lastly, a few points need to be considered when performing the third optimization, namely avoiding generating nonminimal itemsets. Originally, the idea was to avoid expanding such itemsets during incremental processing, which are known to be minimal beforehand. This is the case when the same item is present both in an itemset in the old $\mathcal{L}$ and in the set difference $U - S_i$ (line 3 of the incremental expansion algorithm above). In case of recurrent patterns this condition is too weak to guarantee that all patterns are still going to be generated, as we have to deal with differences in the number of item occurrences. The modified conditions of itemset removal are thus as follows:

1. If an itemset X^r in the old $\mathcal{L}$ contains an item x from $T_i^r = U^r - S_i^r$ and its occurrence count is equal or greater than the one in T_i^r, then move X^r from $\mathcal{L}$ to $NewL$.
2. If the moved X^r is a singleton set $\{(x, p(x))\}$ and its occurrence count is the same in $\mathcal{L}$ and T_i^r, then remove x from T_i^r.

Example 10. Let $U^r = \{a_1 b_2\}$, $S_1^r = \{a_1\}$, $S_2^r = \{b_1\}$. Then $T_1^r = U^r - S_1^r = \{b_1\}$ and $T_2^r = U^r - S_2^r = \{a_1 b_2\}$. We initialize $\mathcal{L} = \{b_1\}$ and check it against T_2^r. While T_2^r contains $\{b_2\}$, $\{b_1\}$ may not be moved directly to $NewL$, as this would falsely result in returning $\{b_1\}$ as the only minimal itemset, instead of $\{a_1 b_1, b_2\}$. Suppose $S_1^r = \{a_1 b_1\}$, then initial $\mathcal{L} = \{b_2\}$ and this time we can see that $\{b_2\}$ does not have to be expanded, as the same item with at least equal occurrence count is present in T_2^r. Thus, $\{b_2\}$ is moved directly to $NewL$, removed from T_2^r and returned as a minimal itemset.

The final algorithm, consisting of all proposed modifications, is presented below as Algorithm 2.

Algorithm 2. Border differential

> **Input**   : $< \{\emptyset\}, \{U^r\} >, < \{\emptyset\}, \{S_1^r, \ldots, S_k^r\} >$
> **Output**: $\mathcal{L}$
> 1 $T_i^r \longleftarrow U^r - S_i^r$ **for** $1 \leq i \leq k$
> 2 **if** $\exists T_i^r = \{\emptyset\}$ **then**
> 3 $\quad$ | $\quad$ **return** $< \{\}, \{\} >$
> 4 **end**
> 5 $\mathcal{L} \longleftarrow \{\{x\} : x \in T_1^r\}$
> 6 **for** $i = 2$ **to** k **do**
> 7 $\quad$ | $\quad$ $NewL \longleftarrow \{X^r = (X, P(X)) \in \mathcal{L} : X \cap T_i \neq \emptyset \wedge \forall x \in (X \cap T_i)\ p(x) \geq t(x)\}$
> 8 $\quad$ | $\quad$ $\mathcal{L} \longleftarrow \mathcal{L} - NewL$
> 9 $\quad$ | $\quad$ $T_i^r \longleftarrow T_i^r - \{x : \{(x, p(x))\} \in NewL\}$
> 10 $\quad$ | $\quad$ **foreach** $X^r \in \mathcal{L}$ *sorted according to increasing cardinality* **do**
> 11 $\quad$ | $\quad$ $\quad$ **foreach** $x \in T_i$ **do**
> 12 $\quad$ | $\quad$ $\quad$ $\quad$ **if** $\forall Z^r \in NewL\ \mathrm{supp}_{Z^r}(X^r \cup \{x\}, 1) = 0$ **then**
> 13 $\quad$ | $\quad$ $\quad$ $\quad$ $\quad$ | $\quad$ $NewL \longleftarrow NewL \cup (X^r \cup \{x\})$
> 14 $\quad$ | $\quad$ $\quad$ $\quad$ **end**
> 15 $\quad$ | $\quad$ $\quad$ **end**
> 16 $\quad$ | $\quad$ **end**
> 17 $\quad$ | $\quad$ $\mathcal{L} \longleftarrow NewL$
> 18 **end**

6.2 Discovering occJEPs

Creating an occJEP-based classifier involves discovering all minimal occJEPs to each of the classes present in a particular decision system. We can formally define the set of patterns in a classifier $occJEP_C^\theta$ for a given occurrence threshold θ as: $occJEP_C^\theta = \bigcup_{c \in \mathcal{I}_d} occJEP_m(C_c', C_c, \theta)$, where $C_c \subseteq \mathcal{D}_L^r$ is a decision class sequence for decision item c and C_c' is a complementary sequence in a learning database $\mathcal{D}_L^r$.

To discover patterns between two dataset pairs, we first need to remove non-maximal itemsets from each them. Next, we multiply the occurrence counts of itemsets in the background dataset by the user-specified threshold. Finally, we need to iteratively call the Border-differential function and create a union of the results to find the set of all minimal jumping emerging patterns with occurrence counts from C_c' to C_c (see Algorithm 3).

Algorithm 3. Discover minimal occJEPs

    Input   : C_c', C_c, θ
    Output: $\mathcal{J}$
1 **for** $S_i^r \in \mathcal{R}$ **do**
2 | $S_i^r \longleftarrow (S_i, s(x) \cdot \theta)$
3 **end**
4 $\mathcal{J} \longleftarrow \{\emptyset\}$
5 **for** $L_i^r \in \mathcal{L}$ **do**
6 | $\mathcal{J} \longleftarrow \mathcal{J} \cup$ Border-differential($< \{\emptyset\}, \{L_i^r\} >, < \{\emptyset\}, \{S_1^r, \ldots, S_k^r\} >$)
7 **end**

Example 11. Consider a learning database $\mathcal{D}_L^r$ containing transactions of three distinct classes: $C_1, C_2, C_3 \subset \mathcal{D}_L^r$. $C_1 = \{b_2, a_1c_1\}$, $C_2 = \{a_1b_1, c_3d_1\}$ and $C_3 = \{a_3, b_1c_1d_1\}$. We need to discover occJEPs to each of the decision class sequences: $occJEP_m(C_2 \cup C_3, C_1, \theta)$, $occJEP_m(C_1 \cup C_3, C_2, \theta)$ and $occJEP_m(C_1 \cup C_2, C_3, \theta)$. Suppose $\theta = 2$. Calculating the set of all minimal patterns involves invoking the Discover-minimal-occJEPs function three times, in which the base Border-differential function is called twice each time and the resulting occJEPs are as follows: $\{a_1c_1\}$ to class 1, $\{c_3, a_1b_1\}$ to class 2 and $\{a_3, b_1c_1, b_1d_1\}$ to class 3.

7 Performing Classification

Classification of a particular transaction in the testing database $\mathcal{D}_T^r$ is performed by aggregating all minimal occJEPs, which are supported by it [9]. A scoring function is calculated and a category label is chosen by finding the class with the maximum score:

$$\text{score}(T^r, c) = \sum_{X^r} \text{supp}_{C_c}(X^r), \tag{10}$$

where $C_c \subseteq \mathcal{D}_T^r$ and $X^r \in occJEP_m(C_c', C_c)$, such that $X^r \subseteq T^r$. It is possible to normalize the score to reduce the bias induced by unequal sizes of particular decision sequences. This is performed by dividing the calculated score by a normalization factor: norm-score(T^r, c) = score(T^r, c)/base-score(c), where base-score is the median of scores of all transactions with decision item c in the learning database: base-score(c) = median$\{$score(T^r, c), for each $T^r \in C_c \subseteq \mathcal{D}_L^r\}$.

8 Experimental Results

We have used two types of data with recurrent items to assess the performance of the proposed classifier. The first is a dataset used previously in [17], which consists of images, represented by texture and color features, classified into four categories: *flower*, *food*, *mountain* and *elephant*. The data contains ca. 400 instances and 16 recurrent attributes, where each instance is an image represented by 8 types of texture and 8 types of color features, possibly occurring multiple times on a single image. The second dataset used for experiments represents the problem of text classification and has been generated on the basis of the Reuters-21578 collection of documents.

8.1 Image Dataset

The image dataset is a collection of images made available by the authors of the SIMPLIcity CBIR system [18], consisting of 1 000 photographs, which are JPEG color image files, having a resolution of 384×256 pixels. An example selection of photographs is presented on Figure 1.

Fig. 1. Example images from the SIMPLIcity test database

Feature Representation. We have used a tile-based, symbolic representation of photographs to enable using classification methods developed for transactional data in the domain of images. The images are uniformly divided into a grid of $x \times y$ tiles, where x is the number of rows and y is the number of columns, and for each of the tiles the color and texture features are calculated.

Color features are represented by a histogram calculated in the HSV color space, with the hue channel quantized to h discrete ranges, while saturation and value channels to s and v ranges respectively. In effect, the representation takes the form of a $h \times s \times v$ element vector of real values between 0 and 1. For the representation of texture we use a feature vector consisting of mean and standard deviation values calculated from the result of filtering an original image with a bank of Gabor functions. These filters are scaled and rotated versions of the base function, which is a product of a Gaussian and a sine function. By using m orientations and n different scales we get a feature vector consisting of mean (μ) and standard deviation (σ) values of each of the filtered images and thus having a size of $2 \times m \times n$ values.

In the next step we aggregate all calculated image features and employ a clustering algorithm to reduce the number of values into a chosen number of groups. In this way, we create a dictionary that consists of the most representative color and texture features of the images in the learning set. The clustering is performed using the k-Means algorithm with a histogram intersection measure for comparing color feature vectors f_c and Gabor feature distance for comparing texture feature vectors f_t. Centroids resulting from the clustering operation become the elements of the dictionary and are labeled $B_1, \ldots, B_k$ in case of color and $T_1, \ldots, T_k$ in case of texture features, where k is the feature dictionary size. These identifiers are then used to describe the images in the database by associating an appropriate label with every tile of each image. This is performed by finding the closest centroid to a feature vector calculated for a given image tile, using appropriate distance measures for each of the features. The dictionary created for the learning set is reused during the classification phase.

Figure 2 presents an example of such a symbolic image representation, showing labels of its individual tiles and the representation of the whole image as a binary database transaction and a database transaction with recurrent items. Figure 3 compares and contrasts representations of images belonging to two different categories: *flower* and *food*.

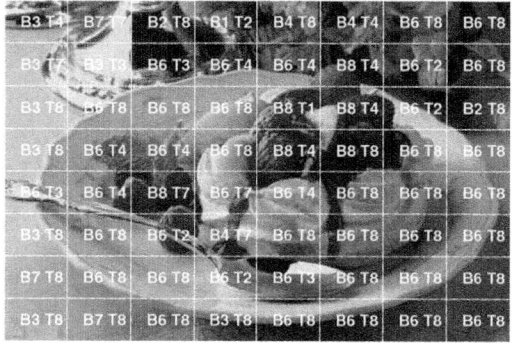

DB	Representation
$\mathcal{D}$	$B_1, B_2, B_3, B_4,$
	B_6, B_7, B_8
	$T_1, T_2, T_3, T_4,$
	T_7, T_8
$\mathcal{D}^{\mathrm{r}}$	$1 \cdot B_1,\ 2 \cdot B_2,\ 8 \cdot B_3,\ 3 \cdot B_4,$
	$41 \cdot B_6,\ 3 \cdot B_7,\ 6 \cdot B_8$
	$1 \cdot T_1,\ 5 \cdot T_2,\ 4 \cdot T_3,\ 11 \cdot T_4,$
	$5 \cdot T_7,\ 38 \cdot T_8$

Fig. 2. A symbolic representation of an image from the *food* dataset. $\mathcal{D}$ – binary transaction system, $\mathcal{D}^{\mathrm{r}}$ – transaction system with recurrent items.

Database	Representation
D_1	$T_1^{\mathrm{r}} = \{37 \cdot B_2,\ 15 \cdot B_3,\ 1 \cdot B_5,\ 11 \cdot B_7,\ 1 \cdot T_1,\ 59 \cdot T_2,\ 4 \cdot T_4\}$
	$T_2^{\mathrm{r}} = \{2 \cdot B_1,\ 8 \cdot B_2,\ 25 \cdot B_4,\ 4 \cdot B_6,\ 24 \cdot B_7,\ 1 \cdot B_8,\ 58 \cdot T_2,\ 5 \cdot T_4,\ 1 \cdot T_7\}$
	$T_3^{\mathrm{r}} = \{4 \cdot B_2,\ 36 \cdot B_3,\ 6 \cdot B_5,\ 6 \cdot B_6,\ 12 \cdot B_8,\ 1 \cdot T_1,\ 52 \cdot T_2,\ 11 \cdot T_4\}$
D_2	$T_4^{\mathrm{r}} = \{34 \cdot B_2,\ 10 \cdot B_3,\ 4 \cdot B_5,\ 3 \cdot B_7,\ 13 \cdot B_8,$
	$\qquad 1 \cdot T_1,\ 18 \cdot T_2,\ 3 \cdot T_3,\ 33 \cdot T_4,\ 9 \cdot T_6\}$
	$T_5^{\mathrm{r}} = \{4 \cdot B_1,\ 17 \cdot B_2,\ 8 \cdot B_3,\ 3 \cdot B_5,\ 12 \cdot B_6,\ 12 \cdot B_7,\ 8 \cdot B_8,$
	$\qquad 4 \cdot T_1,\ 21 \cdot T_2,\ 38 \cdot T_4,\ 1 \cdot T_6\}$
	$T_6^{\mathrm{r}} = \{1 \cdot B_1,\ 11 \cdot B_2,\ 9 \cdot B_3,\ 2 \cdot B_4,\ 10 \cdot B_5,\ 2 \cdot B_6,\ 9 \cdot B_7,\ 20 \cdot B_8,$
	$\qquad 3 \cdot T_1,\ 9 \cdot T_2,\ 6 \cdot T_3,\ 13 \cdot T_4,\ 7 \cdot T_5,\ 4 \cdot T_6,\ 3 \cdot T_7,\ 19 \cdot T_8\}$

Fig. 3. Examples of *flower* (upper row, D_1) and *food* (bottom row, D_2) images, along with their symbolic representation (numbered from left to right)

Results. We have used the following parameter values for the experiments: images partitioned into 8×8 tiles ($x = y = 8$), the sizes of feature vectors $|f_c| = 162$ ($h = 18$, $s = 3$, $v = 3$) and $|f_t| = 48$ ($m = 6$, $n = 4$) values. The dictionary size was set at $k = 8$ values. The parameters are dataset dependent: the number of tiles should be chosen based on the resolution of analyzed images and the dictionary size reflects the diversity of the dataset.

The accuracy achieved by applying the classifier based on jumping emerging patterns with occurrence counts for several threshold values and compared with other frequently used classification methods is presented in Table 4. All experiments have been conducted as a ten-fold cross-validation using the Weka package [19], having discretized the data into 10 equal-frequency bins for all algorithms, except the occJEP method. The parameters of all used classifiers have been left at their default values. The results are not directly comparable with those presented in [17], as currently the occJEP patterns are not limited to any specific length and the seed number for random instance selection during cross-validation was different than before.

Table 4. Classification accuracy of four image datasets. The performance of the classifier based on jumping emerging patterns with occurrence counts (occJEP) compared to: regular jumping emerging patterns (JEP), C4.5 and support vector machine (SVM), each after discretization into 10 equal-frequency bins.

method	θ	accuracy (%)					
		flower/ food	*flower/ elephant*	*flower/ mountain*	*food/ elephant*	*food/ mountain*	*elephant/ mountain*
	1	89.50	84.38	90.63	-	73.00	-
	1.5	94.79	96.35	98.44	78.50	87.00	87.50
occJEP	2	**97.92**	**98.96**	**97.92**	88.00	91.00	**88.50**
	2.5	92.71	97.92	95.31	83.00	90.50	85.50
	3	89.06	97.92	95.31	74.00	87.00	80.50
JEP	-	95.83	91.67	96.35	**88.50**	**93.50**	83.50
C4.5	-	93.23	89.58	85.94	87.50	92.50	82.00
SVM	-	90.63	91.15	93.75	87.50	84.50	84.50

8.2 Text Dataset

We have used the ApteMod version of the Reuters corpus [20], which originally contains 10788 documents classified into 90 categories, to assess the performance of our classifier. As the categories are highly imbalanced (the most common class contains 3937 documents, while the least common only 1), we have presented here the results of classification of the problem reduced to differentiating between the two classes with the greatest number of documents and all other combined, i.e. the new category labels are *earn* (36.5% of all instances), *acq* (21.4%) and *other* (42.1%).

Table 5. Classification accuracy of the Reuters dataset, along with precision and recall values for each of the classes, and the number of discovered emerging patterns / C4.5 tree size

method	θ	accuracy	*earn* precision	recall	*acq* precision	recall	*other* precision	recall	patterns
	1	85.12	96.2	84.7	76.6	96.1	95.5	89.4	10029
	1.5	85.58	96.2	84.7	77.8	96.1	95.5	90.4	9276
occJEP	2	84.65	96.2	87.9	78.9	95.7	**96.6**	91.5	7274
	2.5	84.19	96.2	**87.9**	77.6	95.7	96.6	90.4	7015
	10	83.72	98.00	86.2	79.3	**97.9**	95.5	91.3	3891
JEP	-	66.98	86.8	55.0	46.2	47.1	70.7	85.3	45870
C4.5	-	73.49	92.9	65.0	67.5	52.9	69.2	88.5	51
SVM	-	**86.98**	**98.1**	85.0	**85.4**	68.6	82.8	**97.1**	-

Feature Representation. Document representation has been generated by: stemming each word in the corpus using the Porter's stemmer, ignoring words, which appear on the stoplist provided with the corpus, and finally creating a vector containing the number of occurrences of words in the particular document.

Results. We have selected the 100 most relevant attributes from the resulting data, as measured by the χ^2 statistic, and sampled randomly 215 instances for cross-validation experiments, the results of which are presented in Table 5.

9 Conclusions and Future Work

We have proposed an extension of the border-based emerging patterns mining algorithm to allow discovering jumping emerging patterns with occurrence counts. Such patterns may be used to build accurate classifiers for transactional data containing recurrent attributes. By avoiding both discretization and using all values from the attribute domain, we considerably reduce the space of items and exploit the natural order of occurrence counts. We have shown that areas that could possibly benefit from using such an approach include image and text data classification.

The presented results show that the proposed classifier may achieve equal or better performance than well-known tree-based C4.5 algorithm and support vector machines (SVMs). The approach is most promising in the area of multimedia data mining, as it allows reasoning about quantitative features of images and possibly – after further research – also about their spatial relationships.

Another advantage of using a pattern-based classifier over other algorithms is the ability to analyze the created classifier, which describes the differences between two sets of data in a way easily understandable by a human. This greater insight into problem domain is an important point in many applications, e.g. bioinformatics.

The biggest drawback of the method lies in the number of discovered patterns, which is however less than in the case of regular JEPs found in discretized data. It is thus a possible area of future work to reduce the set of discovered patterns and further limit the computational complexity without influencing the classification accuracy.

References

1. Dong, G., Li, J.: Efficient mining of emerging patterns: Discovering trends and differences. In: KDD 1999: Proceedings of the Fifth ACM SIGKDD International Conference on Knowledge Discovery and Data Mining, pp. 43–52. ACM, New York (1999)
2. Dong, G., Zhang, X., Wong, L., Li, J.: CAEP: Classification by aggregating emerging patterns. In: Arikawa, S., Furukawa, K. (eds.) DS 1999. LNCS (LNAI), vol. 1721, pp. 30–42. Springer, Heidelberg (1999)
3. Dong, G., Li, J.: Mining border descriptions of emerging patterns from dataset pairs. Knowledge and Information Systems 8(2), 178–202 (2005)

4. Li, J., Dong, G., Ramamohanarao, K.: Instance-based classification by emerging patterns. In: Zighed, D.A., Komorowski, J., Żytkow, J.M. (eds.) PKDD 2000. LNCS (LNAI), vol. 1910, pp. 191–200. Springer, Heidelberg (2000)
5. Li, J., Dong, G., Ramamohanarao, K., Wong, L.: DeEPs: A new instance-based lazy discovery and classification system. Machine Learning 54(2), 99–124 (2004)
6. Li, J., Dong, G., Ramamohanarao, K.: Making use of the most expressive jumping emerging patterns for classification. Knowledge and Information Systems 3(2), 1–29 (2001)
7. Li, J., Ramamohanarao, K., Dong, G.: The space of jumping emerging patterns and its incremental maintenance algorithms. In: ICML 2000: Proceedings of the Seventeenth International Conference on Machine Learning, pp. 551–558. Morgan Kaufmann Publishers Inc., San Francisco (2000)
8. Fan, H., Ramamohanarao, K.: An efficient single-scan algorithm for mining essential jumping emerging patterns for classification. In: Chen, M.-S., Yu, P.S., Liu, B. (eds.) PAKDD 2002. LNCS (LNAI), vol. 2336, pp. 456–462. Springer, Heidelberg (2002)
9. Fan, H., Ramamohanarao, K.: Fast discovery and the generalization of strong jumping emerging patterns for building compact and accurate classifiers. IEEE Transactions on Knowledge and Data Engineering 18(6), 721–737 (2006)
10. Terlecki, P., Walczak, K.: On the relation between rough set reducts and jumping emerging patterns. Information Sciences 177(1), 74–83 (2007)
11. Li, J., Liu, G., Wong, L.: Mining statistically important equivalence classes and delta-discriminative emerging pattern. In: Proceedings of 13th International Conference on Knowledge Discovery and Data Mining, San Jose, California, pp. 430–439 (2007)
12. Li, J., Wong, L.: Emerging patterns and gene expression data. Genome Informatics 12, 3–13 (2001)
13. Li, J., Wong, L.: Identifying good diagnostic gene groups from gene expression profiles using the concept of emerging patterns. Bioinformatics 18, 725–734 (2002)
14. Zaïane, O.R., Han, J., Zhu, H.: Mining recurrent items in multimedia with progressive resolution refinement. In: Proceedings of the 16th International Conference on Data Engineering, San Diego, CA, USA, pp. 461–470 (2000)
15. Ong, K.-L., Ng, W.-K., Lim, E.-P.: Mining multi-level rules with recurrent items using FP'-Tree. In: Proceedings of the Third International Conference on Information, Communications and Signal Processing (2001)
16. Rak, R., Kurgan, L.A., Reformat, M.: A tree-projection-based algorithm for multi-label recurrent-item associative-classification rule generation. Data and Knowledge Engineering 64(1), 171–197 (2008)
17. Kobyliński, Ł., Walczak, K.: Jumping emerging patterns with occurrence count in image classification. In: Washio, T., Suzuki, E., Ting, K.M., Inokuchi, A. (eds.) PAKDD 2008. LNCS (LNAI), vol. 5012, pp. 904–909. Springer, Heidelberg (2008)
18. Wang, J.Z., Li, J., Wiederhold, G.: SIMPLIcity: Semantics-sensitive integrated matching for picture libraries. IEEE Trans. on Patt. Anal. and Machine Intell. 23, 947–963 (2001)
19. Witten, I.H., Frank, E.: Data Mining: Practical machine learning tools and techniques, 2nd edn. Morgan Kaufmann, San Francisco (2005)
20. Lewis, D.D., Williams, K.: Reuters-21578 corpus ApteMod version

Rough Entropy Hierarchical Agglomerative Clustering in Image Segmentation

Dariusz Małyszko and Jarosław Stepaniuk

Department of Computer Science
Bialystok University of Technology
Wiejska 45A, 15-351 Bialystok, Poland
{d.malyszko,j.stepaniuk}@pb.edu.pl

Abstract. In data clustering there is a constant demand on development of new algorithmic schemes capable of robust and correct data handling. This demand has been additionally highly fueled and increased by emerging new technologies in data imagery area. Hierarchical clustering represents established data grouping technique with a wide spectrum of application, especially in image analysis branch. In the paper, a new algorithmic rough entropy framework has been applied in the hierarchical clustering setting. During cluster merges the quality of the resultant merges has been assessed on the base of the rough entropy. Incorporating rough entropy measure as the evaluation of cluster quality takes into account inherent uncertainty, vagueness and impreciseness. The experimental results suggest that hierarchies created during rough entropy based merging process are robust and of high quality, giving possible area for future research applications in real implementations.

Keywords: data clustering, hierarchical agglomerative clustering, image segmentation, rough sets, rough entropy.

1 Introduction

Data clustering routines have emerged as most prominent and important data analysis methods that are primarily applied in unsupervised learning and classification problems [2], [3], [4]. Most often data clustering presents descriptive data grouping that identifies homogenous groups of data objects on the basis of the feature attributes assigned to clustered data objects. In this context, a cluster is considered as a collection of similar objects according to predefined criteria and dissimilar to the objects belonging to other clusters. Clustering routines are considered as the most important unsupervised learning problems. Data clustering is primarily concerned with finding a structure in a collection of unlabeled data.

During last decades, data clustering has extended into several distinctive algorithm groups. In data clustering methods partitional and hierarchical clustering routines seem to be of most importance. Several other clustering routines present density-based clustering, grid-based clustering and model-based clustering [5], [6], [7].

J.F. Peters et al. (Eds.): Transactions on Rough Sets XIII, LNCS 6499, pp. 89–103, 2011.

In grid-based methods, STING and CLIQUE are the most representative ones. In grid-based setting, initially a division of the spatial area into rectangle cells is established and further data manipulation is performed on the cells rather than original data objects giving high computational demand release.

In model-based clustering routines, some predefined data model is hypothesized and assumed for each data cluster and the algorithm is aimed at finding optimal model for the given data objects. The example of model-based clustering is COBWEB algorithm.

Partitional clustering groups or partitions data objects into clusters with objects in the cluster much more similar to each other than objects in different clusters. In partitional clustering routines most often k-means and k-medoids methods are applied in practical solutions.

Hierarchical clustering creates a hierarchical decomposition of the data set based on predefined criteria. In hierarchical clustering schemes there are two different kinds of clustering: divisive and agglomerative hierarchical clustering. One additional clustering approach represents incremental clustering algorithms that create the hierarchy step by step by adjusting existing hierarchy during incremental adding new data objects to the hierarchy. Most often, incremental approaches do not require storing of similarity matrices in the memory and are not iterative procedures giving possibility of high computational efficiency but at the same time less adaptable data hierarchy.

An agglomerative approach as a bottom-up approach begins with each pattern or data object in a distinct singleton cluster, and successively merges clusters together until a stopping criterion is satisfied. In the proposed solutions, the initial clusters do not represent singletons but are defined as clusters from over-segmented image input data. Each cluster is represented by its cluster center. During merges, cluster centers are successively added to the merged cluster, giving rise to group cluster representation.

Shannon's entropy concept has become useful methodology in precise description of the information content in many data models and practical data analysis applications in diverse fields. In this context, Shannon's entropy has been applied as a mathematical tool to measure uncertainty in rough set theory, see for example [15].

The presented research is based on combining the concept of rough sets and entropy measure in the area of image segmentation in the introduced **R**ough (**E**ntropy) **E**xtended **F**ramework. The theory of rough sets has been developed during past decades, detailed description of the subject has been given in [16], [17], [19], [18]. Rough set Rough entropy framework in image segmentation has been primarily introduced in [14] in the domain of image thresholding routines. Authors proposed rough entropy measure for image thresholding into two objects: foreground and background object. This type of thresholding has been extended into multilevel thresholding for one-dimensional [13] and two-dimensional domain in [11]. Further, rough entropy has been employed in image data clustering setting in [12], [10]. This research deals with extension of standard hierarchical agglomerative clustering into domain of rough sets and rough entropy during evaluation and selection merging criterion.

The main contributions of this research consists of:

1. Application of rough entropy measure in hierarchical agglomerative merging criteria.
2. Design of a hierarchical agglomerative clustering method that incorporates cluster linkage based on group of cluster centers.
3. Rough entropy direct and inverse weighting based on cardinality of cluster approximations.
4. Cluster merge strategies based on direct merge or group merge.
5. General rough entropy framework in hierarchical agglomerative clustering.

In Section 2 hierarchical clustering methods have been described. Section 3 gives description and basic notions related to rough entropy hierarchical agglomerative clustering. Experimental Setup and experimental results are presented in Section 4. Concluding remarks and areas for future research are given in Section 5.

2   Hierarchical Clustering

2.1   Hierarchical Agglomerative Clustering

Hierarchical agglomerative clustering [2] approach defines a similarity based bottom-up clustering technique that at the beginning every data object treats as a separate cluster. In the subsequent stages, the algorithm iterates over the step that merges the two most similar clusters still available, until one arrives at a universal cluster that contains all the terms.

Input data objects represent N points in d-dimensional space and during hierarchical agglomerative clustering in each algorithm step two of the clusters are merged depending upon cluster similarity measure. In this way, a hierarchy is created where the lowest level of the hierarchy - the leaves, consists of each point in its own cluster, whereas the highest level - the root is build of all points in one cluster. The leaves and the root are considered to be trivial clusterings and all the intermediate levels carry the most meaningful information about the internal data structure of the input data. The general procedure of hierarchical agglomerative clustering has been presented in Algorithm 2.1.

2.2   Clustering Quality Measures

Standard measures taken into account during image segmentation evaluation embrace broad spectrum of different source and segmented image features. Most often the following image properties are considered as image segments homogeneity, compactness, density, statistical properties, shape exactness with a great deal of specialized frameworks such as fuzzy properties, rough properties [8], [20]. In the paper a representative standard evaluation measure has been selected and used during experiments in the form of β-index.

Algorithm 1. Hierarchical Agglomerative Clustering Algorithm

Data: Input Image
Result: Cluster Hierarchy
Create initial clusters $\mathcal{C}$
while $Card(\mathcal{C}) > k$ **do**
  | **Determine the closest pair of clusters** C_l **and** C_m
  | $C_{lm} = C_l \cup C_m$ **merge the clusters**
  | $\mathcal{C} = (\mathcal{C} - \{C_l, C_m\}) \cup \{C_{lm}\}$ **update the clustering**
end

Quantitative measure: β-index. Measure in the form of β-index denotes the ratio of the total variation and within-class variation. Define n_i as the number of pixels in the i-th ($i = 1, 2, \ldots, k$) region from segmented image. Define X_{ij} as the gray value of j-th pixel ($j = 1, \ldots, n_i$) in the region i and $\overline{X}_i$ the mean of n_i values of the i-th region. The β-index is defined in the following way

$$\beta = \frac{\sum_{i=1}^{k} \sum_{j=1}^{n_i} (X_{ij} - \overline{X})^2}{\sum_{i=1}^{k} \sum_{j=1}^{n_i} (X_{ij} - \overline{X}_i)^2} \tag{1}$$

where n is the size of the image and $\overline{X}$ represents the mean value of the image pixel attributes. This index defines the ratio of the total variation and the within-class variation. In this context, important notice is the fact that index-b value increases as the increase of k number. The value of β-index should be maximized.

Quantitative measure $wVar$: within-class variance measure. Within-variance measure presents comparatively not complicated measure calculated by summing up within-variances of all clusters.

$$wVar = \frac{1}{k} \cdot \sum_{i=1}^{k} \sum_{j=1}^{n_i} \frac{1}{n_i} (X_{ij} - \overline{X}_i)^2 \tag{2}$$

This measure values are presented in experiments carried out. Within class variance should be as low as possible and during optimalization this value should be minimized.

3   Rough Entropy Hierarchical Clustering

3.1   Rough Entropy Hierarchical Agglomerative Clustering

In hierarchical agglomerative clustering approach based on rough entropy measure as a clustering quality during merges of clusters it is required to determine rough entropy type that should be calculated and group cluster linkage strategy. The flow of the agglomerative algorithm based on rough entropy measure:

1. Algorithm k-means is applied leading to over-segmentation, for example K clusters in feature domain.

Algorithm 2. Rough Entropy Hierarchical Agglomerative Clustering Algorithm

Data: Input Image, k – number of clusters,
Result: Optimal Cluster Hierarchy
Cluster the input data with a large number of clusters K $\gg$ **k**
Remember clusters as $\mathcal{C}$
while $Card(\mathcal{C}) > k$ **do**
> **foreach** $Cluster$ C_l **do**
>> **Determine the closest cluster** C_m **to** C_l
>> **Temporarily merge clusters** C_m **and** C_l
>> **Calculate and remember rough entropy partition quality measure** C
>> **Restore cluster hierarchy to the state before the merge**
>
> **end**
> **Select for the merge clusters** C_l **and** C_m **that after the merge have the highest rough entropy partition quality** C
> $C_{lm} = C_l \cup C_m$ **merge the clusters**
> $\mathcal{C} = (\mathcal{C} - \{C_l, C_m\}) \cup \{C_{lm}\}$ **update the clustering**

end

2. For each cluster determine its closest neighbor according to linkage strategies given in Subsection 3.3.
3. For each pair of the cluster and its closest neighbor make a merge and calculate the merge quality based on standard quality measures, rough entropy measures or combination of standard and rough entropy measures. Remember the best possible merge.
4. Merge the pair of clusters with the best merge quality measure.
5. Perform steps 2-4 until predefined termination criteria are met, most often when number of regions reaches threshold value, for example k segments in image attribute domain.

The general algorithm execution depends upon cluster group representation that basically for each merged cluster remembers the cluster centers of the merged clusters. In this way, after each cluster merge, the group cluster representation of the resultant cluster contains all cluster centers from two input merged clusters.

3.2 Group Cluster Representation

In group cluster representation, in addition to storing merged data objects at the same time all centers of merged clusters are remembered. In this setting, after the merge operation of two clusters, for example C_1 with cluster centers $\{c_1, \ldots, c_k\}$ and C_2 $\{c_1, \ldots, c_p\}$, data objects from two clusters are merged into one new cluster and additionally, two new cluster centers are remembered as cluster representatives, giving cluster C_{12} with cluster centers $C_m^{12} = \{c_1^1, \ldots c_k^1, c_1^2, \ldots, c_p^2\}$.

3.3   Cluster Group Similarity Measures

In the paper, in order to handle cluster group representation applicable during linkage operation the following novel three group merging operations have been introduced. Each group cluster is represented by its cluster centers $C_{\mu 1} \ldots C_{\mu k}$ as described in the previous subsection.

Group min-linkage clustering - referred to as nearest group neighbor technique with the distance between cluster groups defined as the distance between the closest pair of cluster centers representing two cluster groups. In group min-linkage strategy, at each algorithm iteration the two cluster groups are merged with the minimum distance between their group centers.

$$d(C_1, C_2) = min\{d(c_x, c_y) : c_x \in C_m^1, c_y \in C_m^2\} \tag{3}$$

where c_x and c_y represent arbitrary cluster centers respectively from cluster groups C_m^1 and C_m^2.

Group max-linkage clustering - farthest group neighbor is the opposite of the group min-linkage strategy with distance between group clusters defined as the distance between the most distant pair of cluster centers, for one cluster center from each group cluster. In group max-linkage clustering the two group clusters for which the following distance reaches the minimum value are merged

$$d(C_1, C_2) = max\{d(c_x, c_y) : c_x \in C_m^1, c_y \in C_m^2\} \tag{4}$$

Group average-linkage clustering with the established distance between two group clusters as the average of distances between all pairs of group cluster centers, each pair is made up of one group center from each group. At each stage the two group clusters are merged for which the following distance has minimal value

$$d(C_1, C_2) = \frac{\sum_{c_x \in C_m^1} \sum_{c_y \in C_m^2} d(c_x, c_y)}{Card(C_m^1) Card(C_m^2)} \tag{5}$$

3.4   Rough Entropy Measures

Rough entropy measures have been thoroughly discussed in [10], [12]. For the simplicity and information clarity, in the paper only one of the rough entropy measures has been presented - CCD RECA measure (crisp measure, crisp threshold, difference metric). This standard $RECA$ measure incorporates computation of lower and upper approximations for the given cluster centers and considering these two set cardinalities during calculation of roughness and further rough entropy clustering measure. Rough measure general calculation routine has been given in Algorithm 3.4. In all presented algorithms, before calculations, lower and upper cluster approximations should be set to zero. For each data point x_i, distance to the closest cluster C_l is denoted as $d(x_i, C_l)$ and approximations are increased by value 1 of clusters C_m that satisfy the condition:

$$|d(x_i, C_m) - d(x_i, C_l)| \leq \epsilon_{crisp} \tag{6}$$

Algorithm 3. Crisp - Crisp Difference $RECA$ - Approximations, Roughness and Rough Entropy

---

Data: Input Image, k – number of clusters
Result: R - Roghness, RE - Rough Entropy Value
foreach *Data object* x_i **do**
    Determine the closest cluster C_l **for** x_i
    Increment Lower(C_l) and Upper(C_l) by 1.0
    foreach *Cluster Cluster* $C_m \neq C_l$ *with* $|d(x_i, C_m) - d(x_i, C_l)| \leq \epsilon_{crisp}$ **do**
        Increment Upper(C_m) by 1.0
    end
end

for $l = 1$ **to** k *(number of data clusters)* **do**
    if Upper(C_l) != 0 then R(C_l) = 1 - Lower(C_l) / Upper(C_l)
end
RE = 0
for $l = 1$ **to** k *(number of data clusters)* **do**
    if R(C_l) != 0 then RE = RE - $\frac{exp}{2}$ **· R(C_l) ·** log**(R(C_l)) ;**

end

Standard rough entropy measure notion has been extended into separate measures that take into account crisp ϵ_{crisp}, fuzzy ϵ_{fuzz} and probabilistic ϵ_{pr} threshold, difference or threshold operation. Approximation measures are crisp (increased by constant value 1.0), fuzzy (increased by fuzzy membership values μ of data objects) and probabilistic (increased by probability measure).

3.5 Rough Entropy Weighting

Standard rough entropy calculation is based as presented in Algorithm 3.4 on the following formulae

$$FuzzyRE_{NW} = -\sum_{l=1}^{k} p_l \cdot \frac{e}{2} \cdot roughness(C_l) \cdot log(roughness(C_l)) \qquad (7)$$

where e denotes the base of natural logarithm and serves as a scaling factor.

In order to provide more weight for clusters depending on their cardinality, rough entropy weighting has been introduced. Giving more weight to clusters with more objects when merges with more objects are to be promoted or the second strategy when merges with less objects are to be more promoted. In the proposed algorithmic solution, two independent weighting solutions are possible. The first weighting strategy denoted as direct weighting is designed to give weight to clusters proportionally to their cardinality

$$FuzzyRE_{DW} = -\sum_{l=1}^{k} p_l \cdot \frac{e}{2} \cdot roughness(C_l) \cdot log(roughness(C_l)) \qquad (8)$$

with weight p_l values for each cluster C_l, and the second weighting strategy referred to as inverse weighting gives more weight to clusters less numerous by applying the following formulae

$$FuzzyRE_{IW} = -\sum_{l=1}^{k}(1.0 - p_l) \cdot \frac{e}{2} \cdot roughness(C_l) \cdot log(roughness(C_l)) \quad (9)$$

where k represents the number of clusters and $p_l = \frac{Card(C_l)}{N}$. The total number of data object is given as N. This type of weighting primarily makes possible promoting earlier merges between more or less numerous clusters and created hierarchies are possibly more adequate to the internal data structure.

3.6   Linkage Strategies Example

In the following subsection, grouping results for data from Figure 1 are presented. In the current linkage stage, seven distinct clusters are in the hierarchy. In the current step, three linkage strategies have been applied according to rules given in Section 3.2. After the closest clusters have been determined for each of the seven clusters relative to single linkage, complete linkage and average linkage, clusters have been initially merged and rough entropy and β-index values have been calculated separately for each of the merges. Taking into account that different linkage strategies resulted in different pairs of cluster - closest cluster, different results of merges are presented in Table 1. The numbers in the column ID describe the consecutive merges. In the other columns are contained rough entropy values and β-index values followed by their normalized values. Total quality has been given as the result of multiplication of two normalized values of rough entropy and β-index values. Total rank has been presented in the last column Total.

From the exemplary data consequent merges are visible, in case of single linkage, the best cluster quality after merge is to obtain by merge of the second

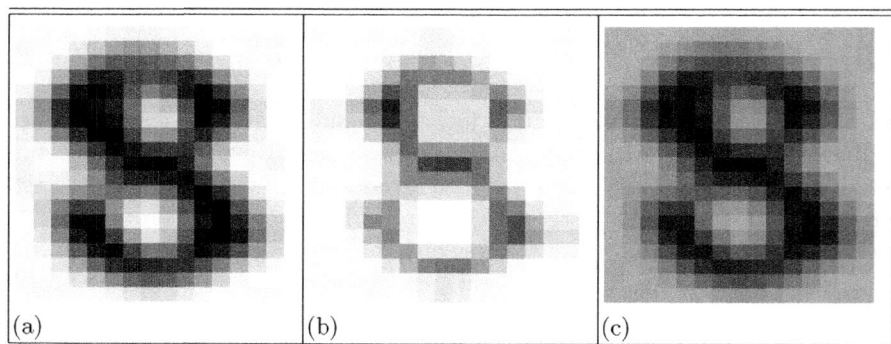

Fig. 1. 1D Eight Image as min operation windowed (a) and as max windowed channel (b), (c) max nad min features displayed as 0GB channels, stretched version

Table 1. Exemplary pre-merge stage calculations for three selected linkage strategies

ID	RE	RE Norm	β-index	β-index Norm	Total	Rank
			Group min-linkage			
1	2.19	0.84	15.66	0.41	0.35	3
2	2.17	0.65	17.12	1.00	0.65	1
3	2.09	0.00	14.65	0.00	0.00	4
4	2.21	1.00	15.79	0.46	0.46	2
5	2.21	1.00	15.79	0.46	0.46	2
6	2.09	0.00	14.65	0.00	0.00	4
7	2.19	0.84	15.66	0.41	0.35	3
			Group max-linkage			
1	2.06	0.11	16.12	0.90	0.09	3
2	2.17	0.75	17.12	1.00	0.75	2
3	2.06	0.11	16.12	0.90	0.09	3
4	2.11	0.39	07.56	0.00	0.00	4
5	2.21	1.00	15.79	0.86	0.86	1
6	2.04	0.00	09.99	0.25	0.00	4
7	2.19	0.89	15.66	0.85	0.75	2
			Group average-linkage			
1	2.19	0.84	15.66	0.89	0.75	2
2	2.17	0.65	17.12	1.00	0.65	3
3	2.09	0.00	14.65	0.81	0.00	4
4	2.21	1.00	15.79	0.90	0.90	1
5	2.21	1.00	15.79	0.90	0.90	1
6	2.18	0.72	04.29	0.00	0.00	4
7	2.19	0.84	15.66	0.89	0.75	2

cluster with the cluster that is the closest to this cluster - this data has not been included in the table. In case of complete linkage the best performance is for the cluster 5, and analogously for average linkage the best result is for 4 and 5 clusters.

4 Experimental Setup and Results

4.1 Introduction

In the experimental setting, REHA algorithm performance has been evaluated and compared with the following parameters:

- Min-linkage, average-linkage and max-linkage.
- Merge quality based upon combination of rough entropy and β-index.
- Merge enhanced by direct and inverse weighting strategy.

4.2 Image Datasets

The first group of experiments have been carried out on a text image representing image of digit eight with sizes 17×19 pixels. The input image has been preprocessed in order to make the features more distinctive two output channels have been created by applying min and max convolution operation in the window of sizes 3×3. The text image two channels are presented in Figure 1 (a) for min convoluted image and (b) for max convoluted image. In Figure 1 (c) max nad min features are displayed in 0-G-B channels (it means with all R channel values equal 0).

In the subsequent material, the image 27059 from Berkeley image database [9], [1] has been selected for experiments. The original color RGB image as given in Figure 2 (a) have been preprocessed and 2D data for bands G and B have been obtained as shown in Figure 2 (b) and 2 (c).

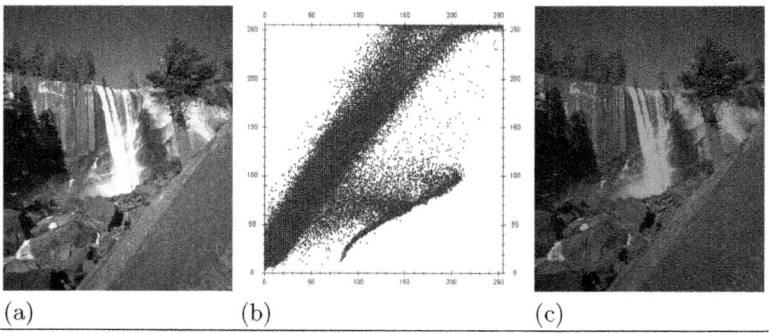

(a)   (b)   (c)

Fig. 2. Berkeley dataset image: (a) image 27059 from Berkeley image database, (b) image 27059 in green-blue attribute domain (c) image 27059 in 0GB channels

(a)

249	247	250	249	244	239	235	236	240	246	249	248	247	247	247
242	247	250	230	197	167	152	160	161	207	237	254	252	248	243
247	249	229	170	114	96	60	88	102	125	170	219	245	247	243
255	241	179	79	20	29	60	72	53	31	58	135	206	241	250
239	196	99	13	0	16	80	129	109	23	2	45	141	227	253
213	148	40	0	10	27	106	207	197	73	0	8	97	217	253
214	155	51	4	5	16	101	225	221	91	4	27	113	220	253
238	208	127	37	0	0	61	152	149	86	34	97	179	237	252
246	245	208	114	41	6	18	5	55	49	32	182	239	245	244
239	250	249	187	105	38	2	0	1	34	113	205	251	247	239
238	242	241	203	139	67	31	28	19	13	51	131	209	250	248
242	224	189	147	121	105	115	132	96	25	0	39	137	227	249
249	212	130	57	61	151	213	225	181	77	7	4	77	176	232
254	216	110	5	9	135	238	255	223	109	19	0	34	129	210
251	228	139	24	0	58	164	229	200	93	7	0	15	116	202
250	243	192	90	18	9	62	113	113	50	11	27	76	160	219
248	249	231	169	96	40	22	35	42	44	73	129	184	226	244
247	249	249	226	184	136	96	86	94	120	163	212	246	254	252

(b)

249	250	250	250	249	244	239	240	246	249	254	254	254	252	248
249	250	250	250	249	244	239	240	246	249	254	254	254	252	248
255	255	250	250	230	197	167	181	207	237	254	254	254	252	250
255	255	249	229	170	114	129	129	129	170	219	245	247	253	253
255	255	241	179	79	106	207	207	207	197	135	206	241	253	253
239	239	196	99	27	196	221	221	221	221	91	141	227	253	253
238	238	208	127	37	196	221	221	221	221	97	179	237	253	253
246	246	245	208	114	161	221	221	221	221	182	239	245	253	253
250	250	249	187	105	152	152	152	149	205	251	251	252	250
250	250	250	249	203	139	67	55	95	113	205	251	251	251	250
250	250	250	249	203	139	132	132	132	113	205	251	251	251	250
249	249	242	241	203	213	225	225	225	181	131	209	250	250	250
254	254	224	189	151	238	255	255	255	223	109	137	227	249	249
254	254	228	139	151	238	255	255	255	223	109	77	176	232	232
254	254	243	192	135	238	255	255	255	223	109	76	160	219	219
251	251	249	231	169	164	229	229	229	200	129	184	226	244	244
250	250	249	243	226	184	138	119	120	163	212	246	254	254	254
249	249	249	249	248	245	237	225	231	244	250	250	254	254	254

Fig. 3. 1D Eight Image as min operation windowed (a) and as max windowed channel (b)

4.3   Text Image Experiments

In the experiments, Text image have been over-segmented by means of k-means clustering with 12 clusters. Afterwards, REHA algorithm have been applied to these segmentations, giving cluster hierarchy, starting from 12 clusters up to 1 cluster containing all 12 input clusters.

In Table 2 data related to the consecutive merges without weighting are presented for three distinct linkage strategies: min-linkage, max-linkage and average-linkage. Starting from the first merge with ID equal 1, merges are carried out up to ID 12 merge when all data objects are combined in one cluster. Each merge operation data are accompanied by resultant β-index and rough entropy values.

In Table 3 data related to the consecutive merges with direct approach are presented for three distinct linkage strategies: min-linkage, complete-linkage and average-linkage. Starting from the first merge with ID equal 1, merges are carried out up to ID 12 merge when all data objects are combined in one cluster.

In Table 4 data related to the consecutive merges with inverse approach are presented for three distinct linkage strategies: min-linkage, complete-linkage and average-linkage. Starting from the first merge with ID equal 1, merges are carried out up to 12 merge when all data objects are combined in one cluster.

In order to assess the total hierarchy quality for all merges, values of quality indices have been summed up for each linkage and weighting strategy and results presented in Table 5.

The best performance has been attained for inverse weighting strategy in min-linkage method. Generally, the best results are obtained for min-linkage method.

Linking strategies - standard cluster representation
In this experiment, after initial over-segmentation has been defined, with one cluster center for each of the clusters, in the subsequent stages, two of the clusters have been merged by merging two relevant clusters and recalculating cluster center for the merged cluster. After merge operation is completed, the merged

Table 2. β-index values and rough entropy in merges on text image - no weighting

ID	Min-link		Max-link		Avg-link	
	β-index	RE	β-index	RE	β-index	RE
1	29.76	4.01	29.76	4.01	29.76	4.10
2	29.54	3.99	29.54	3.99	29.54	4.00
3	28.25	3.83	28.31	3.83	28.24	3.82
4	26.29	3.57	26.88	3.57	25.76	3.55
5	24.63	3.18	24.70	3.19	24.29	3.18
6	22.07	2.77	22.34	2.74	22.07	2.77
7	18.84	2.30	18.04	2.30	18.84	2.30
8	13.61	1.78	13.85	1.81	15.56	1.77
9	10.46	1.28	9.59	1.31	11.03	1.26
10	7.61	0.71	5.80	0.81	7.46	0.74
11	3.56	0.25	2.90	0.32	3.69	0.27
12	1.00	0.00	1.00	0.00	1.00	0.00

Table 3. β-index values and rough entropy in merges on text image - direct weighting

ID	Min-link		Max-link		Avg-link	
	β-index	RE	β-index	RE	β-index	RE
1	27.41	3.63	28.14	3.75	27.41	3.63
2	27.41	3.63	27.31	3.60	27.31	3.60
3	27.34	3.63	27.31	3.60	27.25	3.60
4	26.73	3.46	26.90	3.50	26.32	3.47
5	25.25	3.10	25.60	3.12	25.13	3.11
6	22.07	2.77	22.89	2.76	22.45	2.77
7	18.51	2.27	19.48	2.30	18.75	2.30
8	15.54	1.77	14.62	1.78	16.31	1.75
9	10.71	1.25	10.84	1.25	10.49	1.24
10	8.14	0.69	6.79	0.76	7.43	0.69
11	3.56	0.25	3.09	0.28	3.56	0.25
12	1.00	0.00	1.00	0.00	1.00	0.00

Table 4. β-index values and rough entropy in merges on text image - inverse weighting

ID	Min-link		Max-link		Avg-link	
	β-index	RE	β-index	RE	β-index	RE
1	37.37	4.04	29.76	4.10	29.76	4.10
2	34.95	3.87	29.54	4.00	29.54	4.00
3	34.28	3.70	28.24	3.57	28.24	3.82
4	31.77	3.30	26.88	3.57	25.76	3.55
5	25.78	2.84	24.70	3.19	24.29	3.18
6	23.04	2.40	22.34	2.74	22.07	2.77
7	21.16	1.94	18.04	2.30	18.84	2.30
8	14.71	1.50	13.85	1.81	14.33	1.78
9	11.91	1.08	9.59	1.31	10.41	1.27
10	7.52	0.65	5.80	0.81	6.70	0.77
11	3.71	0.21	2.90	0.32	3.13	0.30
12	1.00	0.00	1.00	0.00	1.00	0.00

cluster is still represented only by one cluster center. Linkage strategies give the same result because only two cluster are merged and a result the merged cluster is represented only by one cluster. On the other side, weighting strategies give different results. In Table 6 quality indices obtained in the subsequent merge stages are given together with standard β-index values and rough entropy values for three independent weighting strategies: no weighting, direct weighting and inverse weighting.

In the experiment, the best performance has been attained for no weighting and inverse weighting strategies. This result most probably depends upon optimal trade-off between the importance of the rough entropy and the cardinality of the clusters.

Table 5. Summed up β-index and rough entropy values for all merges in min-linkage, complete-linkage and average-linkage for three distinct weighting strategies: NW - no weighting, DW - direct weighting and IW - inverse weighting

	Min-link		Max-link		Avg-link	
ID	β-index	RE	β-index	RE	β-index	RE
NW	215.62	27.67	212.71	27.88	217.24	27.76
DW	213.67	26.45	213.97	26.70	213.41	26.41
IW	247.20	25.53	212.64	27.72	214.07	27.84

Table 6. β-index values and rough entropy in merges on text image with direct merging and NW - without weighting, DW - direct weighting, NI - inverse weighting

	NW - No weighting		DW - Direct weighing		IW - Inverse weighting	
ID	β-index	RE	β-index	RE	β-index	RE
1	39.31	3.93	29.41	3.17	41.85	4.09
2	35.22	3.73	26.33	2.76	36.17	3.71
3	34.11	3.54	22.24	2.38	35.08	3.53
4	30.02	3.23	20.62	2.10	30.16	3.15
5	24.06	2.74	20.14	2.02	23.07	2.65
6	22.68	2.44	17.72	1.82	21.66	2.43
7	20.26	1.92	15.64	1.62	19.83	1.95
8	12.78	1.46	12.88	1.39	11.40	1.46
9	10.98	1.15	9.91	1.07	11.16	1.14
10	7.55	0.63	6.24	0.64	8.00	0.61
11	3.96	0.21	3.97	0.21	3.96	0.21
12	1.00	0.00	1.00	0.00	1.00	0.00

4.4   Berkeley Images Experiments

In this experiment, the Berkeley 20759 image has been clustered by means of REHA algorithm. Similarly to the experiments from the previous linking strategies part, the image clusters have been merged. In the experiments, different linking strategies have been applied. In the experiments, two different cluster hierarchies have been obtained. In the experiment, three linking strategies have been tested and three weighting strategies (no weighting, direct weighting and inverse weighting), totally giving nine possible combinations. During hierarchical clustering, two distinct hierarchies that have been obtained are presented in Table 7.

From the experimental data obviously high flexibility of the combined standard and rough entropy based strategies is the primary advantage of the REHA algorithmic schemes. The standard clustering hierarchical procedures are extended and made more robust by means of incorporating the rough entropy based measures as a additional data property. These rough entropy measures are an important source of information that describes data structure. In this way, data structure is made more comprehensively put into the cluster hierarchy

Table 7. Two distinct cluster hierarchies obtained by REHA algorithm for 27059 Berkeley data image

First hierarchy				
ID	Dunn	DB	β-index	wVar	RE
1	0.54	10.11	5.38	772	0.58
2	0.75	8.65	3.88	1072	0.58
3	0.75	7.23	3.49	1197	0.59
4	0.56	6.40	2.69	1542	0.87
5	0.72	4.14	1.57	2771	0.97
Second hierarchy				
ID	Dunn	DB	β-index	wVar	RE
1	0.54	10.11	5.38	772	0.58
2	0.54	9.00	4.77	874	0.62
3	0.41	7.87	3.33	1247	0.66
4	0.56	6.40	2.69	1542	0.87
5	0.72	4.14	1.57	2771	0.97

creation procedure. At the same time, the resultant cluster hierarchies benefit considerably mainly by means of the more precise data cluster structure that adaptively describes internal data dependencies.

5   Conclusions and Future Research

In the present research, new algorithmic scheme REHA as an extension of hierarchical agglomerative clustering has been introduced in the rough entropy setting. The notion of rough entropy has been defined and incorporated as a criterion during cluster merging operation. In the REHA algorithm, group cluster representation has been introduced with storing all merged cluster center. Group cluster representation make possible taking advantage of group linkage strategies. Additionally, particular rough entropies added to total rough entropy are optionally weighed in order to give more importance to clusters with more objects, or conversely with clusters less numerous. In this way, cluster merges are promoted on the base of their cardinality. From experimental results conclusion about high usefulness and high quality of rough entropy measure in hierarchical agglomerative setting during merges evaluation may be fully reasonable. The research area of REHA clustering presents possible implementation applicability. The introduced hierarchical clustering solutions are potentially robust tool in better description of the data structure. The area of application of the REHA algorithmic schemes is not confined only to the image segmentation routines. In the straightforward extensions, the REHA clustering routines are possible to be implemented for any solutions that require hierarchical data structure.

Acknowledgments

The research is supported by the grants N N516 0692 35 and N N516 3774 36 from Ministry of Science and Higher Education of the Republic of Poland.

References

1. Berkeley Image Database,
   http://www.cs.berkeley.edu/projects/vision/grouping/segbench/
2. Duda, R.O., Hart, P.E., Stork, D.G.: Pattern Classification. John Wiley & Sons, Inc., Chichester (2001)
3. Gonzalez, R.C., Woods, R.E.: Digital image processing, 2nd edn. Prentice-Hall, NJ (2002)
4. Jahne, B.: Digital Image Processing. Springer, Berlin (1997)
5. Jain, A.K., Dubes, R.C.: Algorithms for Clustering Data. Prentice Hall, Englewood Cliffs (1988)
6. Jain, A.K., Murty, M.N., Flynn, P.J.: Data clustering: A survey. ACM Comput. Surv. 31, 264–323 (1999)
7. Jain, A.K.: Data clustering: 50 years beyond K-means. Pattern Recognition Letters 31(8), 651–666 (2009)
8. Jiang, X.: Performance evaluation of image segmentation algorithms. In: Chen, C.H., Wang, P.S.P. (eds.) Handbook of Pattern Recognition and Computer Vision, 3rd edn., pp. 525–542. World Scientific, Singapore (2005)
9. Martin, D., Fowlkes, C., Tal, D., Malik, J.: A database of human segmented natural images and its application to evaluating segmentation algorithms and measuring ecological statistics. In: ICCV 2001, vol. (2), pp. 416–423. IEEE Computer Society, Los Alamitos (2001)
10. Malyszko, D., Stepaniuk, J.: Adaptive Rough Entropy Clustering Algorithms in Image Segmentation. Fundamenta Informaticae 98(2-3), 199–231 (2010)
11. Malyszko, D., Stepaniuk, J.: Granular Multilevel Rough Entropy Thresholding in 2D Domain. In: IIS 2008, 16th International Conference Intelligent Information Systems, Zakopane, Poland, June 16-18, pp. 151–160 (2008)
12. Małyszko, D., Stepaniuk, J.: Standard and Fuzzy Rough Entropy Clustering Algorithms in Image Segmentation. In: Chan, C.-C., Grzymala-Busse, J.W., Ziarko, W.P. (eds.) RSCTC 2008. LNCS (LNAI), vol. 5306, pp. 409–418. Springer, Heidelberg (2008)
13. Malyszko, D., Stepaniuk, J.: Multilevel Rough Entropy Evolutionary Thresholding. Information Sciences 180(7), 1138–1158 (2010)
14. Pal, S.K., Shankar, B.U., Mitra, P.: Granular computing, rough entropy and object extraction. Pattern Recognition Letters 26(16), 2509–2517 (2005)
15. Sen, D., Pal, S.K.: Histogram thresholding using fuzzy and rough measures of association error. Trans. Img. Proc. 18(4), 879–888 (2009)
16. Pawlak, Z., Skowron, A.: Rudiments of rough sets. Information Sciences 177(1), 3–27 (2007); Rough sets: Some extensions. Information Sciences 177(1), 28–40 (2007); Rough sets and Boolean reasoning. Information Sciences 177(1), 41–73 (2007)
17. Pedrycz, W., Skowron, A., Kreinovich, V. (eds.): Handbook of Granular Computing. John Wiley & Sons, New York (2008)
18. Skowron, A., Stepaniuk, J.: Information granules: Towards foundations of granular computing. International Journal of Intelligent Systems 16(1), 57–86 (2001)
19. Stepaniuk, J.: Rough–Granular Computing in Knowledge Discovery and Data Mining. Springer, Heidelberg (2008)
20. Zhang, H., Fritts, J.E., Sally, A.: Image segmentation evaluation: A survey of unsupervised methods. Computer Vision and Image Understanding 110(2), 260–280 (2008)

Software Defect Prediction Based on Source Code Metrics Time Series

Łukasz Puławski

Institute of Informatics, The University of Warsaw,
Banacha 2, 02-097 Warszawa, Poland
Lukasz.Pulawski@mimuw.edu.pl

Abstract. Source code metrics have been proved to be reliable indicators of the vulnerability of the source code to defects. Typically, a source code unit with high value of a certain metric is considered to be badly structured and thus error-prone. However, analysis of source code change history shows that there are cases when source files with low values of metrics still turn out to be defective. Instead of introducing new metrics for such cases, I investigate the possibility of estimating the vulnerability of source code units to defects on the basis of the history of the values of selected well-known metrics. The experiments show that we can efficiently identify bad source code units just by looking at the history of metrics, coming from only a few revisions that precede the actual resolution of the defect.

Keywords: software defect prediction, source code metrics, classification.

1 Introduction

Software *defect*, informally known as "bug", is an error in the source code of a computer program which makes it run in an unintended or unexpected way. Typically a defect is a consequence of a mistake made by a person, who is creating or modifying the source code units. Replacing the piece of code containing a defect with an adequate fragment that resolves the problem, is called a *defect fix*.

Identification of defects in current software systems is one of the most important tasks in the quality assurance process. Typically, running a test is required to discover a defect and an even more effort-consuming analysis of defect report is needed to locate the defective source code unit and determine a way to fix it. The earlier the defect is identified, the lower its resolution cost.

1.1 Defect Prediction

Elementary post-factum analysis of the history of many software development processes shows that, statistically, defects appear more often in some source code units than in others. The challenge is to classify source code units during the

J.F. Peters et al. (Eds.): Transactions on Rough Sets XIII, LNCS 6499, pp. 104–120, 2011.

software creation process so that we know in advance, which units are more likely to contain a defect, and thus require more concern at an early stage of development. Technically speaking, the goal is to have a classifier, which, being given the history of system development, assigns a probability of finding a defect to each source code unit. In the simplest case, the decision may be binary: the classifier should point out only those source code units, which are highly likely to be defective. Files are typical source code units considered in the defect prediction problem. In this paper two types of files are distinguished: *defective* (or *defect-prone*) files and *defectless* (or *defect-free*) files. The former are the files that are expected to contain defects inside. The latter are the files that are expected not to contain any defect. More detailed definition for these two notions is given in section 4, where the setting for the experiments is described.

It is common knowledge that bad source design promotes introduction of defects. Well-structured, cohesive source code is less likely to be defective. Therefore, constructing a classifier which assigns a complexity measure to each source code unit, is a good approach to approximate the set of defect-prone elements of source code. Well-known tools used to estimate source code unit complexity are source code metrics ([12], [21]).

1.2   Software Metrics

Source code metrics are well-known tools for static code analysis. They measure the complexity of source code units and thus provide information about potentially ill-structured parts of the code, which may be error-prone or hard to maintain. The correlation between high (read: improper) values of source code metrics and the number of defects in the corresponding source code units has been widely analyzed and proved to be true ([12], [21]). A more detailed descriptions of metrics analysed in this paper appear in section 3.

1.3   Metrics Time Series

In a collaborative software development environment, a common source repository is usually used. Most often, it is one central source code management server (SCM, for short), which allows developers to apply their changes to the common source base in a transactional manner. Such atomic changes are called *commits* or *check-ins*. Every check-in has its unique number, which is called *revision*. Anyone performing a commit changes the structure of the code in the files that are modified, which may change metric values for these files. Thus, by measuring metrics right after each commit, we get a time series of metric values for each file that has been updated by anyone in the software development process. These time series are used as an input for the classifier, which should distinguish defective from defectless files.

1.4   Remainder of the Paper

Section 2 describes in detail the motivation for this work. Section 3 gives a rough explanation of metrics considered and a short reason for their use.

Section 4 describes the algorithm used in the experiment. The remaining sections discuss experiment data, results, and compare the approach presented herein with related work as well as give some notes on future work.

2 Motivation

There are situations when a certain source code unit is actually defective, but the corresponding metrics are still within acceptable boundaries. One could think that new metrics, which properly indicate such situations, should be introduced. A different approach is proposed in this paper. Instead of measuring static information about the source code at the moment when the defect is introduced or a defect fix is applied, we can look at the values of a few existing simple metrics in the few revisions that immediately precede the defect fix. This changes the type of data used for classification: instead of static metric values that measure a snapshot of source code taken at a certain point in time, we receive a time series of values of these metrics. The experiment shows that such input might be useful for identifying potentially defective files.

Let us introduce two notions for clarity: A *static classifier* is a classifier which takes the values of static metrics at a given point in time and, based on this, decides if the corresponding source code unit is error-prone or not. A *temporal classifier*, on the other hand, takes time series of metric values, and produces a classification result of the same type. Experiments using the latter type of classifier are discussed in this paper. Pseudo-code in algorithm 1 explain how a temporal classifier is used together with a static classifier to construct a hierarchical *global classifier*. Please note that, in the setting proposed, a temporal classifier is applied only to files which were classified by a static classifier as most probably being "defectless" (that is, the values of all metrics for these files are low).

Algorithm 1. GLOBAL-CLASSIFIER(n)

Require: rev - revision of classification
Require: F - file to classify
Ensure: "defective"/"defectless" file classification at revision rev

1: **if** STATIC-CALSSIFIER(F, rev) == "defective" **then**
2:     **return** "defective";
3: **else**
4:     **return** TEMPORAL-CLASSIFIER(F, rev);
5: **end if**

The GLOBAL-CLASSIFIER, as compared to STATIC-CLASSIFIER, can be better in terms of recall and worse in terms of precision. The question is how much we can improve classification quality? In the domain of automated defect prediction, having a high level of recall is generally more important than maintaining precision at a reasonably good level. That is because, in principle, it is better to put more effort into checking a defectless source code unit for potential errors, rather than to leave unspotted faults in the code for late phases of system development. Thus, in this paper, I mostly focus on prediction quality of the TEMPORAL-CLASSIFIER. This yields the following experiment goal and question:

Goal: To Improve defect prediction quality without adding new metrics

Question: Given that STATIC-CLASSIFIER and TEMPORAL-CLASSIFIER are based on the same set of metrics, how much can we improve prediction recall, while still keeping precision at a reasonable level?

3   Metrics Used

In order to gather necessary data, I use *Checkstyle* (a tool for static java code analysis), that is able to compute a number of well-known and widely-used metrics. The metrics are considered on the file level. Therefore, all metrics used to measure smaller source code units, such as method or expression, are generalized to file level with the maximum function. For example, if a metric m is defined on the method level, and the set of methods defined in file f is denoted by Mt, the value of this metric for the file level is defined as $m(f) = max_{x \in Mt} m(x)$.

The list given in the following section shortly describes all metrics used in the experiments. More detailed descriptions of these metrics can be found in [12], [21], [23], [33] or on the tool website - [1].

All metrics in the list are expected to be low in well-structured source code. Therefore, for each metric, there is a corresponding maximum "low"(read:proper) value given. The following conjecture is made: if a metric for a given file exceeds this value, the file is badly-designed, thus defect-prone. Therefore, each metric can be treated as a function, which takes a file and a revision and produces a classification: "defective" - if the value exceeds the "low" threshold or "defectless" - otherwise. This yields a construction of metric-based STATIC-CLASSIFIER: If for a given revision and file, all metric values do not exceed the respective maximum "low" values, the file is "defectless". Otherwise it is "defective".

3.1   Metric Descriptions

This section describes all metrics provided by the *Checkstyle* and used in the experiments. Next to the name of a metric, its short name is given in brackets, which will be used in the later sections of this paper. Every metric is given a corresponding maximum "low" value. That is, a maximal value for which file is considered to be well-structured with respect to this metric. The numbers provided in the descriptions below are taken from the standard *Checkstyle* configuration and were actually used in the experiments.

- *Boolean expression complexity (Bool)*
  This metric measures the number of conjunctions (equivalently: clauses) in boolean expressions. If such an expression contains many clauses, it is

difficult to understand, debug and maintain. The maximum "low" value of this metric is 3.

– *Class data abstraction coupling (Da)*
  This metric measures the number of instantiations of other classes within the given class. The higher data abstraction coupling, the more complex the structure of dependencies between classes. The maximum "low" value of this metric is 10.
– *Class fan-out complexity (FanOut)*
  This metric measures the number of other classes that the given class depends on. Any change in another class may potentially affect the behavior of the measured class. Therefore, fan-out complexity should be kept low. The maximum "low" value of this metric is 20.
– *Cyclomatic complexity (Cycl)*
  This metric measures the number of branching instructions in the source code. The more such instructions, the more potential control flows. The maximum "low" value of this metric is 10.
– *Npath complexity (NPath)*
  This metric is similar to cyclomatic complexity, since it measures the number of possible flows through a method. It expresses a theoretical maximum number of control flows in the case when every branching instruction could independently continue in all its branches. The maximum "low" value of this metric is 200.
– *Number of effective lines in method, class, file ($NCSS_m$, $NCSS_c$, $NCSS_f$)*
  These three metrics measure the number of effective lines of code (i.e. without comments, white lines, etc.) in the respective source code unit. When the file is large (has many lines of code) it is hard to understand and maintain. The maximum "low" value of this metric for method, class and file is 50, 1500 and 2000 respectively.

4   The Algorithm

4.1   Notation and Definitions

In this and the remaining sections I use the following symbols and definitions: f is a variable which denotes a file. Variable rev denotes a revision.

$$M = \{Bool, Da, FanOut, Cycl, NPath, NCSS_m, NCSS_c, NCSS_f\},$$

stands for a set of metrics.

$$REVS(f, rev) = (rev_1, rev_2, \ldots, rev_k),$$

denotes a strictly ascending sequence of all revisions in which file f has been modified before revision rev. $|REVS(f, rev)| = k$ denotes the length of this sequence. $DF(f, rev)$ means that there has been a defect fix introduced to file f at revision rev, while $DF(f)$ stands for a set of all revisions in which there was a fix applied to f that resolved a defect in it.

4.2   Experiment Steps

Data preprocessing. In this step, on the basis of complete data encompassing the whole software development history, we choose files which are suitable for the classifier. The files are divided into two sets: $DEFECTIVE$ and $DEFECTLESS$. For the reasons explained in section 2 all files in both sets need to satisfy $scdl$ - "static classifier defectless" condition:

$$scdl(f) := \forall_{m \in M} \forall_{rev \in REVS(f,\infty)}\ m(f, rev) = \text{``defectless''},$$

(all metric values for all revisions have been below "low" value, so the file is expected to be defectless in all revisions according to the STATIC-CLASSIFIER). Please note that in one of the experiments (as described in section 6), this condition was dropped. It is equivalent to giving it an alternative trivial definition: $scdl(f) := TRUE$. If not stated otherwise, assume that the first version is valid and all files have to satisfy it.

In the end, sets $DEFECTIVE$ and $DEFECTLESS$ are defined as follows:

$$DEFECTIVE = \{f : scdl(f) \wedge |DF(f)| \geq 3\},$$

(set of files with "good" values of metrics, which have been defect-fixed at least three times).

$$DEFECTLESS = \{f : scdl(f) \wedge |DF(f)| = 0\},$$

(set of files with "good" values of metrics, which did not have any defect found in them).

As the classifier is based on a time series of metrics, we also require that the classified files have a history of revisions which is long enough. Moreover, for files which are to be classified as "defective", we require them to have an equivalently long history of revisions which precede the actual defect fix. Thus, for each file from $DEFECTLESS$ we give another lh ("long history") condition:

$$lh(f) := |REVS(f,\infty)| \geq 3,$$

(file has at least three commits in its history).

For files from $DEFECTIVE$ - $lhdf$ ("long history before defect fix") condition:

$$lhdf(f) := \exists rev_{df} : DF(f, rev_{df}) \wedge |REVS(f, rev_{df})| \geq 2\},$$

(file has at least two revisions that precede a revision with a defect fix).

This leads us to the following two definitions:

$$DEFECTLESS_h = \{f : f \in DEFECTLESS \wedge lh(f)\}.$$

This set contains files which satisfy the following conditions:

- they are correctly classified by STATIC-CLASSIFIER as being "defectless",
- they have been modified at least three times.

$$DEFECTIVE_h = \{f : f \in DEFECTIVE \wedge lhdf(f)\}.$$

This set contains files which satisfy the following conditions:

- they are wrongly classified by STATIC-CLASSIFIER as being "defectless",
- a defect has been fixed in them at least three times,
- they have been modified at least two times before a defect fix.

Sequences of revisions. We can construct a time series of metrics for each file in these two sets:

For $f \in DEFECTIVE_h$,

$$SEQ_{defective}(f) := \bigcup_{rev \in DF(f)} \{REVS(f, rev + 1) : |REVS(f, rev + 1)| \geq 3\},$$

For $f \in DEFECTLESS_h$,

$$SEQ_{DEFECTLESS}(f) := \bigcup_{rev \in N} \{REVS(f, rev + 1) : |REVS(f, rev + 1)| \geq 3\}.$$

The set $SEQ_{defective}(f)$ contains sequences which satisfy the following conditions:

- they are at least 3 items long,
- they end with a revision in which a defect was fixed in file f.

The set $SEQ_{defectless}(f)$ contains sequences which are at least three items long.

Time series of metrics. Having a sequence of revisions $(rev_1, rev_2, \ldots, rev_k)$, metric m and a file f, we can construct a time series of values of this metric for the given file as follows: $(m(f, rev_1), m(f, rev_2), \ldots, m(f, rev_k))$. The goal is now to use these time series as an input for a classifier. The initial attempt to create an information system would be as follows: For all files f take all sequences of revisions from $SEQ_{defective}(f)$, apply all metrics from section 3 (so we would get a vector of time series) and give these instances decision "defective". Similarly, we can do it for all files f and for all sequences from $SEQ_{defectless}(f)$ and give these instances decision "defectless". By doing so, we would get a time series with decisions assigned. However, these might potentially be of different length. This can be a problem for some classifiers. The next paragraph describes how it is dealt with.

Fixed-length differential sequences. Researchers point out (see [10]) that instead of looking at metric values, we can look at changes of these metrics across subsequent revisions. The reason for this is, that we would rather like to know how significant a change is or how much impact it has on the modified file. Standard metrics cannot measure this, as they can only be applied to a snapshot of a source code at a given point in time. Thus, instead of taking a series of metric values, we can take a derived differential series. That is, instead of taking $(m(f, rev_1), m(f, rev_2), \ldots, m(f, rev_k))$, we would take $(m(f, rev_2) - m(f, rev_1), \ldots, m(f, rev_k) - m(f, rev_{k-1}))$. This trick introduces a temporal change aspect, but still does not solve the problem of variable-length series. I decided to introduce two notions: LC (Last Change) and CFA (Change From Average) to cope with the problem.

Given a series of integers $s = (i_1, i_2, i_3, i_4, \ldots i_k)$, let us define:

$$LC(s) = i_k - i_{k-1},$$

$$CFA(s) = i_{k-1} - avg(i_1, i_2, i_3, i_4, \ldots, i_{k-2}).$$

Please note that LC is well-defined for sequences at least 2 items long and CFA - for sequences at least 3 items long.

These non-trivial notions require a short explanation. Suppose that we apply LC and CFA to a time series of one metric constructed from $SEQ_{defective}$. Let us denote one such time series by $s := (m(f, rev_1), m(f, rev_2), \ldots, m(f, rev_k))$. This yields:

$$LC(s) = m(f, rev_k) - m(f, rev_{k-1}),$$

$$CFA(s) = m(f, rev_{k-1}) - AVG(m(f, rev_1), \ldots, m(f, rev_{k-2})).$$

Please note that we are aware of the fact that defect fix is applied to the file f at revision rev_k, because s was taken from $SEQ_{defective}$. In particular, it means that f definitely contains a defect at revision rev_{k-1}. We do not know if the same defect has already appeared in the file previously, but if this is the case, this effect should be "smoothened" by the average function, so the conjecture is made that $AVG(m(f, rev_1), \ldots, m(f, rev_k - 2))$ may still be a good approximation of "normal" value for metric m for file f.

Intuitively, LC would provide information on how metrics values change after the defect fix; that is, how their values change from improper to proper.

The CFA plays the role of approximating how metrics values change from expected "normal" state(average) to a situation when a defect is definitely in the file.

Information system. The preceding sections give enough information to explain how the information system for classifier is constructed.

Objects are created in the way presented by the pseudo-code of algorithms 2 and 3. The final set of objects in the information system is constructed as DEFECTIVE-OBJECTS() ∪ DEFECTLESS-OBJECTS(). The former algorithm produces objects with decision "defective", whereas the latter - with decision "defectless". The information system has numeric attributes, whose total number is twice the number of metrics, because there is one LC and one CFA attribute per every metric. Please note that one file may be represented by many objects in this information system, because we might have many sequences bound to one file in both $SEQ_{defectless}(f)$ and $SEQ_{defective}(f)$.

Algorithm 2. DEFECTIVE-OBJECTS
___
Ensure: set of objects for information system with decision "defective"

1: **for all** $f \in DEFECTIVE_h$ **do**
2:   **for all** $s \in SEQ_{defective}(f)$ **do**
3:     $O \leftarrow$ *new object in the information system*;
4:     *give O decision "defective"*;
5:     **for all** $m \in M$ **do**
6:       $sm \leftarrow$ *sequence of metric m values for s*; {as described in section 4.2}
7:       *alter O with attribute $LC(sm)$*;
8:       *alter O with attribute $CFA(sm)$*;
9:     **end for**
10:   **end for**
11: **end for**

Algorithm 3. DEFECTLESS-OBJECTS
___
Ensure: set of objects for information system with decision "defectless"

1: **for all** $f \in DEFECTLESS_h$ **do**
2:   **for all** $s \in SEQ_{defectless}(f)$ **do**
3:     $O \leftarrow$ *new object in the information system*;
4:     *give O decision "defectless"*;
5:     **for all** $m \in M$ **do**
6:       $sm \leftarrow$ *sequence of metric m values for s*; {as described in section 4.2}
7:       *alter O with attribute $LC(sm)$*;
8:       *alter O with attribute $CFA(sm)$*;
9:     **end for**
10:   **end for**
11: **end for**

---

5 The Experiment

The experiment, performed with Rough Set Exploration System (RSES, for short) 2.2.2 tool ([2], [8]), contained three phases: data discretisation, rule generation, and rule validation. You will find a description of the data, the phases and a selection of inferred decision rules in the following sections.

5.1   Data

For the experimental data, I used ten open-source projects with public access to SCM and issue tracker. Issue tracker contains a list of tasks done in the course of project development. Each task, apart from other information, has a *type* and a set of modifications in source code files, which were necessary to resolve this issue. In this experiment issue type "Bug" was distinguished from other types. If a file was modified by a developer while resolving such an issue, this commit was considered to be a defect fix. Please recall that file is considered to be defective if a defect was fixed in it at least three times.

Actual datasets. Table 5.1 presents a list of particular datasets used in the experiments. Each row of this table defines what portion of project development history was downloaded from respective JIRA and SVN servers and transformed into datasets used in the experiment. For example, for dataset Axis2, JIRA server available at [5] was used and issues AXIS2-4400 to AXIS2-4700 were downloaded from it. Corresponding SVN system ([6]) was used to fetch content of the modified files and the revisions ranged approximately from 550000 to 950000.

Dataset	JIRA	JIRA project	Issues range	SVN	approximate revisions range
Jboss	[16]	JBAS	6500 - 7700	[17]	85000-100000
Axis2	[5]	AXIS2	4400 - 4700	[6]	550000-950000
Derby	[5]	DERBY	2500 - 4500	[6]	520000-900000
Hadoop	[5]	HADOOP	5000 - 6800	[6]	730000-960000
Geronimo	[5]	GERONIMO	4500 - 5500	[6]	730000-980000
Mapreduce	[5]	MAPREDUCE	500 - 2000	[6]	600000-1000000
Myfaces	[5]	MYFACES	1400 - 2100	[6]	440000-890000
Struts1	[5]	STR	2200 - 3200	[6]	400000-750000
Struts2	[5]	WW	2500 - 3400	[6]	630000-950000
Tomahawk	[5]	TOMAHAWK	1200 - 1500	[6]	630000-990000

Discretisation. The first step of the experiment was the discretisation of data. It was performed based on cuts generated with a global method, as implemented in RSES. The discretisation algorithm is described in [25].

Decision Rules. Finally, the decision rules on the discretised attributes were computed with the use of exhaustive algorithm in RSES system. It produces a rule-based classifier with usage of rough-sets methods (see ([28], [27], [26], [24])). The general concept of constructing rule-based classifiers in RSES is based on the extension of approximation spaces, as defined in e.g. [36]. Inferred rules were tested with 10-fold cross validation method. Standard voting, was a strategy used to resolve conflicts when a new object was classified. In this strategy, each rule receives as many votes as many supporting objects it has. For details please see [8] and [7].

Table 1. Compared accuracy for three different classifiers

Project	Accuracy for classifier:		
	STATIC	TEMPIORAL	GLOBAL
axis2	0,92	0,95	0,99
derby	0,75	0,78	0,88
geronimo	0,70	0,76	0,70
hadoop	0,85	0,85	0,88
jboss	0,58	0,75	0,56
mapreduce	0,77	0,83	0,84
myfaces	0,63	0,71	0,66
struts1	0,79	0,83	0,72
struts2	0,84	0,72	0,83
tomahawk	0,64	0,68	N/A

Table 2. Precision and recall of two classifiers

Project	STATIC-CLASSIFIER		GLOBAL-CLASSIFIER	
	precision	recall	precision	recall
axis2	0,20	0,90	0,95	0,70
derby	0,35	0,69	0,78	0,99
geronimo	0,74	0,64	0,76	0,98
hadoop	0,53	0,37	0,85	0,89
jboss	0,81	0,45	0,75	0,93
mapreduce	0,58	0,61	0,83	0,96
myfaces	0,63	0,52	0,71	0,97
struts1	0,94	0,78	0,83	0,91
struts2	0,75	0,79	0,72	0,89

The selection of rules can be found in tables 3 and 4. Quality measures for the complete classifier are presented in tables 1 and 2 and discussed in the following sections.

6   Interesting Rules

Tables 3 and 4 present a selection of inferred rules, which had 100% confidence, no more than 4 clauses and a support greater than $\frac{1}{8}$ of the number of objects in the information system (12.5%). Please recall that in section 4.2, two different definitions of *scdl* function ("static classifier defectless") were given. Table 3 contains rules inferred from the information system generated with the original definition, whereas table 4 - with the trivial definition. Intuitively, the former table represents rules inferred from files with low values of all metrics - as in the GLOBAL-CLASSIFIER, whereas the latter - rules inferred from all files, as in the TEMPORAL-CLASSIFIER used independently.

Table 3. Selection of interesting rules in global classifier

Dataset	Rule	Support (%)
axis2L	$LC(Bool) \leq 0.5$ & $CFA(NCSS_c) \geq 4.5 \Rightarrow buggy$	35%
	$LC(FanOut) \geq -0.5$ & $LC(NCSS_c) \leq 0.5$ & $LC(NCSS_f) \leq 2.5 \Rightarrow buggy$	28%
	$LC(NCSS_f) \leq 2.5$ & $CFA(NCSS_c) \geq 4.5 \Rightarrow buggy$	22%
	$LC(NCSS_c) \leq 0.5$ & $CFA(NCSS_f) \geq 3.5 \Rightarrow buggy$	20%
geronimoL	$LC(NCSS_m) \geq -0.5$ & $LC(NCSS_f) \in (-0.5; 0.5)$ & $CFA(NCSS_f) \in (-0.5; 0.5) \Rightarrow buggy$	35%
hadoopL	$LC(NCSS_c) \geq 0.5$ & $CFA(FanOut) \leq 0.5$ & $CFA(NCSS_f) \geq -6.0 \Rightarrow buggy$	25%
	$LC(NCSS_c) \leq 0.5$ & $LC(NCSS_f) \geq 1.5 \Rightarrow bugfree$	25%
	$CFA(Da) \geq 0.5$ & $CFA(FanOut) \geq 0.5 \Rightarrow bugfree$	25%
	$LC(NCSS_f) \geq 1.5$ & $CFA(FanOut) \geq 0.5 \Rightarrow bugfree$	16%
	$LC(NCSS_c) \geq 0.5$ & $CFA(FanOut) \leq 0.5$ & $CFA(NCSS_c) \geq 0.5 \Rightarrow buggy$	13%
	$LC(NCSS_f) \leq -0.5$ & $CFA(FanOut) \leq 0.5 \Rightarrow buggy$	13%
jbossL	$LC(NCSS_f) \leq 2.5$ & $CFA(NCSS_f) \leq 5.5 \Rightarrow bugfree$	59%
	$LC(NCSS_f) \leq 2.5$ & $CFA(FanOut) \leq 0.5 \Rightarrow bugfree$	48%
	$LC(NPath) \leq 2.0$ & $LC(NCSS_m) \geq 0.5 \Rightarrow buggy$	13%
myfacesL	$LC(FanOut) \leq 0.5$ & $LC(NCSS_c) \geq -0.5$ & $CFA(FanOut) \geq 0.5$ & $CFA(NCSS_c) \geq 2.5 \Rightarrow bugfree$	14%
	$LC(NCSS_c) \geq -0.5$ & $LC(NCSS_f) \leq 0.5$ & $CFA(FanOut) \geq 0.5 \Rightarrow bugfree$	13%

Table 4. Selection of interesting rules in temporal classifier

Dataset	Rule	Support (%)
axis2	$LC(Bool) \leq 0.5$ & $CFA(NCSS_c) \geq 4.5 => buggy$	72%
	$LC(Cycl) \leq 0.5$ & $CFA(NCSS_c) \geq 4.5 => buggy$	59%
	$LC(Bool) \leq 0.5$ & $LC(NCSS_c) \geq 2.5 => buggy$	52%
	$CFA(Da) \geq 0.5$ & $CFA(NCSS_c) \geq 4.5 => buggy$	52%
	$LC(Cycl) \leq 0.5$ & $LC(NCSS_c) \geq 2.5 => buggy$	39%
	$LC(NCSS_c) \geq 2.5$ & $CFA(Da) \geq 0.5 => buggy$	30%
	$LC(FanOut) \geq 0.5$ & $LC(NCSS_c) \geq 2.5 => buggy$	24%
geronimo	$LC(NCSS_m) \in (-0.5; 0.5)$ & $LC(NCSS_f) \in (-0.5; 0.5)$ & $CFA(NCSS_c) \in (-0.5; 0.5)$ & $CFA(NCSS_f) \in (-0.5; 0.5) => buggy$	16%
hadoop	$LC(Da) \leq 0.5$ & $LC(NCSS_m) \leq 0.5$ & $LC(NCSS_c) \geq 0.5$ & $CFA(NCSS_c) \geq 10.5 => buggy$	17%
	$LC(FanOut) \leq 0.5$ & $LC(NCSS_m) \leq 0.5$ & $LC(NCSS_c) \geq 0.5$ & $CFA(NCSS_c) \geq 10.5 => buggy$	15%
	$CFA(NPath) \leq -367.5 => buggy$	13%
jboss	$LC(NCSS_f) \in (-0.5; 2.5)$ & $CFA(NCSS_f) \leq 7.0 => bugfree$	22%
	$LC(NCSS_f) \in (-0.5; 2.5)$ & $CFA(FanOut) \leq 0.5 => bugfree$	20%
myfaces	$LC(NPath) \leq 1.0$ & $CFA(Da) \leq 0.5$ & $CFA(NCSS_m) \geq 0.5 => buggy$	18%
	$CFA(FanOut) \in (-0.5; 0.5)$ & $CFA(NCSS_m) \geq 0.5 => buggy$	13%

7 Conclusion

This section contains a discussion of the experiments results, observations and conclusions.

7.1 Prediction Quality

The experiments show that it is possible to mine even more useful information for software defect prediction on the basis of existing simple source code metrics. This can be done if we start to look at them in a temporal rather than a static manner. Introducing a notion of metric time series by adding a time dimension, seems to increase resolution of the information provided, since we can still identify defect-prone source code units, even if the values of their static metrics imply no problem. Elementary rough-set methods allow us to generate simple decision rules with reasonable prediction quality. In particular it means that there are temporal patterns in the software development processes, which are related to software quality.

The experiment shows that the defect prediction algorithm presented in this paper based on either GLOBAL-CLASSIFIER or TEMPORAL-CLASSIFIER is typically better in terms of accuracy. Even if it turns out to be worse, the drop in prediction quality is insignificant. For details see table 1.

What is even more important (recall discussion from section 4.2) is that in almost all cases the GLOBAL-CLASSIFIER turns out to out-perform STATIC-CLASSIFIER in terms of recall, while still keeping the precision on a reasonable level. This can be read directly from table 2.

7.2 Structure of Inferred Rules

The analysis of good quality rules presented in tables 3 and 4 leads to a conclusion that not all metrics are equivalently useful. It can be observed that elementary $NCSS_?$ metrics appear in these rules more often than others. In fact, they only measure the volume of changes applied to a file, without saying anything about the structure of these changes. Metrics related to inter-class dependencies complexity, such as Da or $FanOut$, are located in the second place. What seems to be important is that metrics which measure inner-class complexity, such as $Bool$, $NPath$ or $Cycl$, virtually do not appear in these rules. This yields a conclusion that temporal patterns connected to vulnerability of software to defects are related to change of dependencies between classes, rather than to the changes of the complexity of the single class. The type of modifications within one class that makes it error-prone, is still an open issue, since in the experiments described herein it is very roughly approximated by changes of $NCSS_?$ metrics only.

8 Related Work

Many methods for defect classification focus on source code metrics. There are many widely used datasets with values of metrics, such as [32]. Some advanced data mining techniques have been used to construct classifiers for these data sets (e.g.[9], [29], [35]). The main difference between algorithms presented in these papers and the one described herein, is that the latter uses a temporal aspect by focusing on particular changes in software metrics rather than on their values measured at a certain point of time only.

The need to analyse temporal aspects of software processes instead of static metrics has been widely discussed. One of the most commonly taken approaches is to use *Process metrics* ([13], [11], [22]). Generally speaking, process metrics assign one integer value to a complete history of a file (e.g. sum of lines modified in file). Although these methods appear to be accurate, they are generally hard to apply in a local ad-hoc defect prediction. The approach proposed in this paper is different. It aims at identifying temporal patterns which may lead to software faults, in almost real time (CFA and LC can be computed for a file after every commit). It seems to be a considerable advantage, as defects tend to appear in groups, closely one after another (see [18] and [14]). Thus, if a classifier can suggest the location of potential defects at the time when they are introduced to the source code, there is good chance that the developer will identify and resolve more defects at the same time, while doing the analysis.

Application of time series analysis in defect prediction can be found in [4]. The major difference to the approach presented here is that authors generate very coarse-grained time series by measuring 4 points in time (each measure related to one phase of project development). In contrast, the approach taken in this paper creates fine-grained time series which are modified with every atomic check-in performed by a developer. This means that identification of error-prone source code units can be more detailed and can be completed at an earlier stage.

9   Future Work

I am interested in discovering spatio-temporal patterns in software development processes. By spacial aspect, I mean location of patterns in certain units of the source code. My research is mainly focused on identifying "bad smells"(see [20], [19]) of various kinds in the source code. The approach presented herein is just a basic attempt to mine useful information from temporal data, based on software metrics. My further research will focus on more sophisticated representation of spatio-temporal patterns in software development processes. The discussion from section 7 implies that there are still some open research areas which could provide a more detailed understanding of which change patterns may lead to an ill-structured and error-prone code. The ideas for improvement include hierarchical patterns (i.e. patterns built from sub-patterns) and the introduction of expert knowledge into the algorithm. One of the goals for the further research is to provide better resolution of the information mined. It means, to identify fragments of source code smaller than files (like methods or blocks) which are responsible for defects in the software.

Acknowledgments

Special thanks to professor Andrzej Skowron and professor James Peters, who significantly helped me to put this paper into its final shape.

The research has been partially supported by the grants N N516 077837, N N516 368334 from Ministry of Science and Higher Education of the Republic of Poland.

References

1. Checkstyle tool home page,
   http://checkstyle.sourceforge.net/config_metrics.html
2. Rses (rough set exploration system), http://logic.mimuw.edu.pl/~rses/
3. Alpigini, J.J., Peters, J.F., Skowron, A., Zhong, N. (eds.): RSCTC 2002. LNCS (LNAI), vol. 2475. Springer, Heidelberg (2002)
4. Amasaki, S., Yoshitomi, T., Mizuno, O., Takagi, Y., Kikuno, T.: A new challenge for applying time series metrics data to software quality estimation. Software Quality Control 13(2), 177–193 (2005)
5. Axis. Issue tracker, https://issues.apache.org/jira/
6. Axis. Scm, http://svn.apache.org/repos/asf/
7. Bazan, J.G., Nguyen, H.S., Nguyen, S.H., Synak, P., Wroblewski, J.: Rough set algorithms in classification problem. In: Polkowski, L., Tsumoto, S., Lin, T.Y. (eds.) Rough Set Methods and Applications: New Developments in Knowledge Discovery in Information Systems. Studies in Fuzziness and Soft Computing, pp. 49–88. Physica-Verlag, Heidelberg (2000)
8. Bazan, J.G., Szczuka, M.S., Wroblewski, J.: A new version of rough set exploration system. In: Alpigini et al. [3], pp. 397–404
9. Bhatt, R., Ramana, S., Peters, J.F.: Software defect classification: A comparative study of rough-neuro-fuzzy hybrid approaches with linear and non-linear svms. SCI, pp. 213–231. Springer, Heidelberg(2009)
10. Demeyer, S., Ducasse, S., Nierstrasz, O.: Finding refactorings via change metrics. In: OOPSLA 2000: Proceedings of the 15th ACM SIGPLAN Conference on Object-Oriented Programming, Systems, Languages, and Applications, pp. 166–177. ACM, New York (2000)
11. Graves, T.L., Karr, A.F., Marron, J.S., Siy, H.: Predicting fault incidence using software change history. IEEE Trans. Softw. Eng. 26(7), 653–661 (2000)
12. Halstead, M.H.: Elements of Software Science. Operating and Programming Systems Series. Elsevier Science Inc., New York (1977)
13. Hassan, A.E.: Predicting faults using the complexity of code changes. In: ICSE 2009: Proceedings of the 31st International Conference on Software Engineering, Washington, DC, USA, pp. 78–88. IEEE Computer Society, Los Alamitos (2009)
14. Hassan, A.E., Holt, R.C.: The top ten list: Dynamic fault prediction. In: ICSM 2005: Proceedings of the 21st IEEE International Conference on Software Maintenance, Washington, DC, USA, pp. 263–272. IEEE Computer Society, Los Alamitos (2005)
15. Herzner, W., Ramberger, S., Länger, T., Reumann, C., Gruber, T., Sejkora, C.: Comparing software measures with fault counts derived from unit-testing of safety-critical software. In: Winther, R., Gran, B.A., Dahll, G. (eds.) SAFECOMP 2005. LNCS, vol. 3688, pp. 81–93. Springer, Heidelberg (2005)
16. JBoss. Issue tracker, https://jira.jboss.org/browse/JBAS
17. JBoss. Scm, http://anonsvn.jboss.org/repos/jbossas/
18. Kim, S., Zimmermann, T., Whitehead Jr., E.J., Zeller, A.: Predicting faults from cached history. In: ICSE 2007: Proceedings of the 29th International Conference on Software Engineering, Washington, DC, USA, pp. 489–498. IEEE Computer Society, Los Alamitos (2007)

19. Mäntylä, M., Lassenius, C.: Subjective evaluation of software evolvability using code smells: An empirical study. Empirical Software Engineering 11, 395–431 (2006)
20. Mäntylä, M., Vanhanen, J., Lassenius, C.: A taxonomy and an initial empirical study of bad smells in code. In: ICSM 2003: Proceedings of the International Conference on Software Maintenance, Washington, DC, USA, p. 381. IEEE Computer Society, Los Alamitos (2003)
21. McCabe, T.J.: A complexity measure. IEEE Trans. Softw. Eng. 2(4), 308–320 (1976)
22. Moser, R., Pedrycz, W., Succi, G.: A comparative analysis of the efficiency of change metrics and static code attributes for defect prediction. In: ICSE 2008: Proceedings of the 30th International Conference on Software Engineering, pp. 181–190. ACM, New York (2008), doi:10.1145/1368088.1368114
23. Nejmeh, B.A.: Npath: a measure of execution path complexity and its applications. Commun. ACM 31(2), 188–200 (1988), doi:10.1145/42372.42379
24. Nguyen, H.: Approximate boolean reasoning: Foundations and applications in data mining. In: Peters, J.F., Skowron, A. (eds.) Transactions on Rough Sets V. LNCS, vol. 4100, pp. 334–506. Springer, Heidelberg (2006)
25. Nguyen, S.H., Nguyen, H.S.: Discretization methods in data mining. In: Rough Sets in Knowledge Discovery, vol. 1, pp. 451–482. Physica Verlag, Heidelberg (1998)
26. Pawlak, Z., Skowron, A.: Rough sets and boolean reasoning. Information Sciences 177(1), 41–73 (2007)
27. Pawlak, Z., Skowron, A.: Rough sets: Some extensions. Information Sciences 177(1), 28–40 (2007)
28. Pawlak, Z., Skowron, A.: Rudiments of rough sets. Information Sciences 177(1), 3–27 (2007)
29. Pelayo, L., Dick, S.: Applying novel resampling strategies to software defect prediction. In: Annual Meeting of the North American Fuzzy Information Processing Society, NAFIPS 2007 (2007)
30. Peters, J.F., Ramanna, S.: Towards a software change classification system: A rough set approach. Software Quality Journal 11, 121–147 (2003)
31. Peters, J.F., Ramanna, S.: Approximation space for software models. In: Peters, J.F., Skowron, A., Grzymała-Busse, J.W., Kostek, B.z., Świniarski, R.W., Szczuka, M.S. (eds.) Transactions on Rough Sets I. LNCS, vol. 3100, pp. 338–355. Springer, Heidelberg (2004)
32. PROMISE. Defect prediction data sets, http://promisedata.org/?cat=4
33. Ramamurthy, B., Melton, A.: A synthesis of software science measures and the cyclomatic number. IEEE Trans. Softw. Eng. 14(8), 1116–1121 (1988)
34. Ramanna, S.: Rough neural network for software change prediction. In: Alpigini, J.J., Peters, J.F., Skowron, A., Zhong, N. (eds.) RSCTC 2002. LNCS (LNAI), vol. 2475, pp. 602–686. Springer, Heidelberg (2002)
35. Ramanna, S., Bhatt, R., Biernot, P.: Software defect classification: A comparative study with rough hybrid approaches. In: Kryszkiewicz, M., Peters, J.F., Rybiński, H., Skowron, A. (eds.) RSEISP 2007. LNCS (LNAI), vol. 4585, pp. 630–638. Springer, Heidelberg (2007)
36. Skowron, A., Stepaniuk, J., Swiniarski, R.W.: Approximation spaces in rough-granular computing. Fundamenta Informaticae 100(1-4), 141–157 (2010)

Risk Assessment in Granular Environments

Marcin Szczuka

Institute of Mathematics, The University of Warsaw
Banacha 2, 02-097 Warsaw, Poland
szczuka@mimuw.edu.pl

Abstract. We discuss the problem of measuring the quality of decision support (classification) system that involves granularity. We put forward the proposal for such quality measure in the case when the underlying granular system is based on rough and fuzzy set paradigms. We introduce the notion of approximation, loss function, and empirical risk functional that are inspired by empirical risk assessment for classifiers in the field of statistical learning.

Keywords: risk measure, loss function, granularity, approximation, neighbourhood, empirical risk, classification.

1 Introduction

While constructing a decision support (classification) system for research purposes we usually rely on commonly used, convenient quality measures, such as success ratio (accuracy) on test set, coverage (support) and versatility of the classifier. While sufficient for the purposes of analysing classification methods in terms of their technical abilities, these measures sometimes fail to fit into a bigger picture.

In practical decision support applications the classifier is usually just a small gear in a larger machine. The decision whether to construct and then use such system is taken by the user on the basis of his confidence in relative "safety" of his computer-supported decision. This confidence is closely related to the users' assessment of the *risk* involved in making the decision.

The overall topics of risk assessment, risk management and decision making in presence of risk constitute a separate field of science. The ubiquity of decision-making processes that involve risk is making risk assessment a crucial element in areas such as economy, investment, medicine, engineering and many others. Numerous approaches have been developed so far, and vast literature dedicated to these issues exist (see [1], [2]). The topic or risk assessment and management is a topic of research in many fields of science, ranging from philosophy to seismology. In this article we restrict ourselves to a much narrower topic of calculating (assessing) the risk associated with the use of classifier in a decision-making process that involves granularity (granular computations). In this (granular) context the traditional quality measures and traditional risk assessment approaches need to be significantly modified to address the underlying (granular) structure of the task.

J.F. Peters et al. (Eds.): Transactions on Rough Sets XIII, LNCS 6499, pp. 121–134, 2011.
© Springer-Verlag Berlin Heidelberg 2011

We focus on one commonly used method for calculating a risk of (using) a classifier, which is known from the basics of statistical learning theory [3]. In this approach the risk is measured as a *summarised expectation* for creating a loss due to classifier error. More formally, the risk is equal to the total *loss* (integral) over the probabilistic distribution of data. Loss is expressed in terms of a specialised function which compares the answer of classifier with the desired one and returns the numerical value corresponding to the amount of "damage" resulting from misclassification. Such a measure of risk is to a large extent intuitive in many situations. It is by no means the only scheme used by humans to judge the risk, but a popular one. It is quite common to make assessment of the involved risk by hypothesising the situations in which the gain/loss can be generated in our system, and then weighting them by the likelihood of their occurrence. There are also other approaches, in particular the classical notions of risk assessment known from statistical decision theory [4], that use utility functions and Bayesian reasoning. They are, however, weakly relevant to our investigations, as they impose too many assumptions about the data source and underlying (unknown) probabilistic distribution.

We investigate the possibilities for approximating the risk in the situation when the standard numerical, statistical learning methods cannot be applied to full extent. The real life data is not always possible to be verified as representative, large enough or sufficiently compliant with assumptions of underlying analytical model. Also, the information we posses about the amount of loss and its probabilistic distribution may be expressed in granular rather than crisp, numerical way. Nevertheless, we would like to be able to provide approximate assessment of risk associated with a classification method. For this purpose we put forward some ideas regarding the approximate construction of two crucial components in measuring risk, i.e., the loss function and the summarisation method needed to estimate overall risk from the empirical, sample-dependant one.

This article is intended to pose some questions and provide suggestions in which direction we may search for answers, rather than deliver ready to use technical solutions. It is an extension of a sketch from last years SC&P and of a conference contribution [5]. The paper starts with short introduction of granular paradigm, followed by more formal introduction of risk functional, as known from statistical learning theory. Then, we discuss the possible sources of problems with such risk definition and suggest some directions, in particular an outline for a loss function approximation method. We also extend the discussion to the issue of finding the proper summarisation procedure for measuring the value of empirical risk functional. Our proposed approach utilises the approach known as *granular computing*. The granular computing paradigm in case of this paper is initially discussed as general approach, and then narrowed down to two cases. We introduce a sketch for the methods of risk calculation in case of granular systems defined by means of rough and fuzzy sets, which by no means represent the whole spectrum of granular systems. We conclude by pointing out several possible directions for further investigation.

This article is an extension of the paper [5] presented at the 6th International Conference on Rough Sets and Current Trends in Computing - RSCTC 2008. It also includes some of the ideas presented in [6] and [7].

2   Granularity and Granular Computing

The idea of granular systems and granular computing builds on general observation, that in many real-life situations we are unable to precisely discern between similar objects. Our perception of such universe is *granular*, which means, that we are only able to observe groups of objects (*granules*) with limited resolution.

The existence of granularity and the necessity of dealing with it has led to formation of the *granular computing* paradigm and research on granule-based information systems (cf. [8]). The original ideas of Lotfi Zadeh (cf. [9]) has grown over time. Currently the granular computing and the notion of granularity are becoming a unifying methodologies for many branches of soft computing. Several paradigms related to rough and fuzzy sets, interval analysis, shadowed sets as well as probabilistic reasoning can be represented within granular framework, as exemplified by the contents of the handbook [8].

Our focus is on systems that support classification and decision making in the presence of vagueness, imprecision and incompleteness of information. In this paper we only address a small portion of such systems and the granules we are using are of rather basic type. We mostly address the case when a granule corresponds to an abstraction (indescernibility) class or a simple fuzzy set, without caring of its internal structure.

3   Risk in Statistical Learning Theory

In the classical statistical learning approach, represented by seminal works of Vapnik [3,10], the risk associated with a classification method (classifier) α is defined as a functional (integral) of the *loss function* L_α calculated over an entire space with respect to probability distribution.

To put it more formally, let X^∞ be the complete (hypothetical) universe of objects from which we are drawing our finite sample $X \subset X^\infty$. Please, note that the hypothetical universe X^∞ shall not be confused with the denotation for the set of infinite sequences from the set X, that can be found in some mathematical textbooks. In the analytical model of risk we are assuming that a probability distribution P is defined for entire σ-field of measurable subsets of X^∞.

Definition 1. *The risk value for a classifier α is defined as:*

$$R(\alpha) = \int_{X^\infty} L_\alpha dP$$

where $L_\alpha = L(x, f_\alpha(x))$ is the real-valued loss function defined for every point $x \in X^\infty$ where the classifier α returns the value $f_\alpha(x)$.

The classical definition of risk, as presented above, is heavily dependent on assumptions regarding the underlying analytical model of the space of discourse. While over the years several methods have been developed within the area of statistical learning in pursuit of practical means for calculating risk, there are still some important shortcomings in this approach. Some of them are:

1. Sensitivity to scarceness of the data sample. In real life experiments we may be very far from complete knowledge of our data universe. The sample we are given may be tiny in comparison with the range of possible outcomes.
2. Incomplete definition of loss function. We expect that $L(y, f_\alpha(x))$ is integrable wherever f_α takes value. Unfortunately, in practice all we are given is the set of points from the graph of L_α. From these few points we have to extend (approximate) the function L_α.
3. Incomplete knowledge of the distribution, which is closely related to the point 1 above. Even with large data sample X we may not be certain about its representativeness.

There are also several advantages of the classical risk functional definition. Thanks to solid mathematical grounding it is possible to provide answers with provable quality. As long as we can assure sufficient compliance to assumptions of the underlying statistical methodology the task of estimating the risk is equivalent to solving a numerical optimisation problem. For a given classifier α we search for the solution to:

$$\lim_{l \to \infty} \Pr \left\{ z \in (X^\infty)^l : |R(\alpha) - R_{emp}(\alpha)| > \varepsilon \right\} = 0$$

where z is a data sample of size l, probability $\Pr$ is calculated according to distribution P (see Def. 1), $\varepsilon \geq 0$, and $R_{emp}(\alpha)$ is the *empirical risk* measured for the classifier α on (labelled) sample z. The empirical risk is usually measured as an average over values of loss function. For a labelled sample $z = \{x_1, \ldots, x_l\}$ of length l

$$R_{emp}(\alpha) = \frac{\sum_{i=1}^l L(x_i, f_\alpha(x_i))}{l}.$$

It is visible, that the ability to calculate value of loss L_α, i.e., to compare the answer of classifier with the desired one is a key element in empirical risk assessment.

4   Approximation of Loss Function and Its Integral

The formal postulates regarding the loss function may be hard to meet, or even verify in practical situations. Nevertheless, we would like to be able to asses the loss. In this section we suggest a method for approximating the loss function from the available, finite sample. In the process we will consider the influence of granularity on our ability to make valid approximations of loss function.

First, we will attempt to deal with the situation when the value of loss function L_α for a classifier α is given as a set of positive real values defined for data points from a finite sample z. Let $z \in (X^\infty)^l$ be a sample consisting of l data points, by $\mathbb{R}_+$ we denote the set of non-negative reals (including 0). A function $\hat{L}_\alpha : z \mapsto \mathbb{R}_+$

is called a sample of loss function $L_\alpha : X^\infty \mapsto \mathbb{R}_+$ if L_α is an extension of $\hat{L}_\alpha$. For any $Z \subseteq X^\infty \times \mathbb{R}_+$ we introduce two projection sets as follows:

$$\pi_1(Z) = \{x \in X^\infty : \exists y \in \mathbb{R}_+ \ (x,y) \in Z\},$$

$$\pi_2(Z) = \{y \in \mathbb{R}_+ : \exists x \in X^\infty \ (x,y) \in Z\}.$$

4.1   The Rough Set Case

We assume that we are also given a family $\mathcal{C}$ of neighbourhoods (granules), i.e, non-empty, measurable subsets of $X^\infty \times \mathbb{R}_+$. These neighbourhoods shall be defined for a particular application. Later in this section we will identify these neighbourhoods with granules defined as indiscernibility classes.

Under the assumptions presented above the lower approximation of $\hat{L}_\alpha$ relative to $\mathcal{C}$ is defined by

$$\underline{\mathcal{C}}\hat{L}_\alpha = \bigcup \{c \in \mathcal{C} : \hat{L}_\alpha(\pi_1(c) \cap z) \subseteq \pi_2(c)\}. \tag{1}$$

Note, that the definition of lower approximation given by (1) is different from the traditional one, known from rough set theory [11,12]. Also, the sample z in definition of approximations (formulæ (1),(2), and (3)) is treated as a set of its elements, i.e., a subset of $(X^\infty)^l$.

One can define the upper approximation of f relative to $\mathcal{C}$ by

$$\overline{\mathcal{C}}\hat{L}_\alpha = \bigcup \{c \in \mathcal{C} : \hat{L}_\alpha(\pi_1(c) \cap z) \cap \pi_2(c) \neq \emptyset\}. \tag{2}$$

An illustration of the upper and lower approximations of a function given by a finite sample if provided in Fig.1. Obviously, the situation presented in Fig. 1

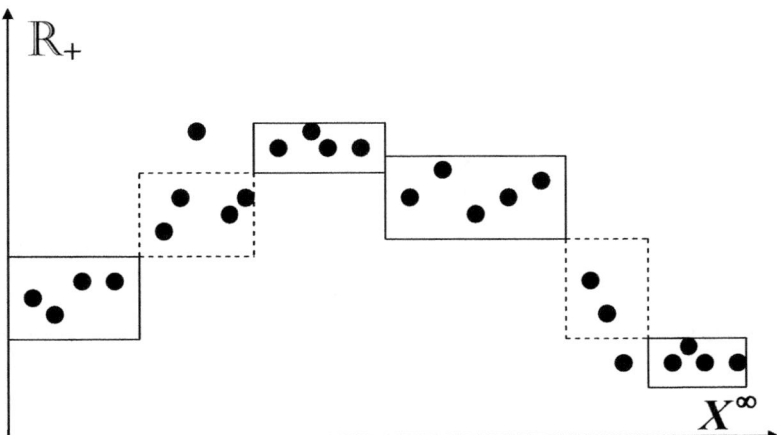

Fig. 1. Loss function approximation (neighbourhoods marked by solid lines belong to the lower approximation and with dashed lines - to the upper approximation)

is a far going simplification of actual problem. For the sake of presentation the universe X^∞ is represented as a single axis, while in the actual system it may be something much more complicated. A single point in X^∞ may represent, e.g., a patient record or a sequence of images.

Example 1. We present an illustrative example of a function $\hat{L}_\alpha : z \mapsto \mathbb{R}_+$ approximation in a simple situation where $z = \{1, 2, 4, 5, 7, 8\}$ is a sequence of $l = 6$ real numbers. Let $\hat{L}_\alpha(1) = 3$, $\hat{L}_\alpha(2) = 2$, $\hat{L}_\alpha(4) = 2$, $\hat{L}_\alpha(5) = 5$, $\hat{L}_\alpha(7) = 5$, $\hat{L}_\alpha(8) = 2$.

We consider a neighbourhood consisting of three indiscernibility classes $C_1 = [0, 3] \times [1.5, 4]$, $C_2 = [3, 6] \times [1.7, 4.5]$ and $C_3 = [6, 9] \times [3, 4]$. We compute projections of indiscernibility classes: $\pi_1(C_1) = [0, 3]$, $\pi_2(C_1) = [1.5, 4]$, $\pi_1(C_2) = [3, 6]$, $\pi_2(C_2) = [1.7, 4.5]$, $\pi_1(C_3) = [6, 9]$ and $\pi_2(C_3) = [3, 4]$.

Hence, we obtain $\hat{L}_\alpha(\pi_1(C_1) \cap z) = \hat{L}_\alpha(\{1, 2\}) = \{2, 3\} \subseteq \pi_2(C_1)$, $\hat{L}_\alpha(\pi_1(C_2) \cap z) = \hat{L}_\alpha(\{4, 5\}) = \{2, 5\} \not\subseteq \pi_2(C_2)$ but, $\hat{L}_\alpha(\pi_1(C_2) \cap z) \cap \pi_2(C_2) = \{2, 5\} \cap [1.7, 4.5] \neq \emptyset$, $\hat{L}_\alpha(\pi_1(C_3) \cap z) = \emptyset$.

We obtain the lower approximation $\underline{\mathcal{C}}\hat{L}_\alpha = C_1$ and the upper approximation $\overline{\mathcal{C}}\hat{L}_\alpha = C_1 \cup C_2$. This is illustrated by Fig. 2.

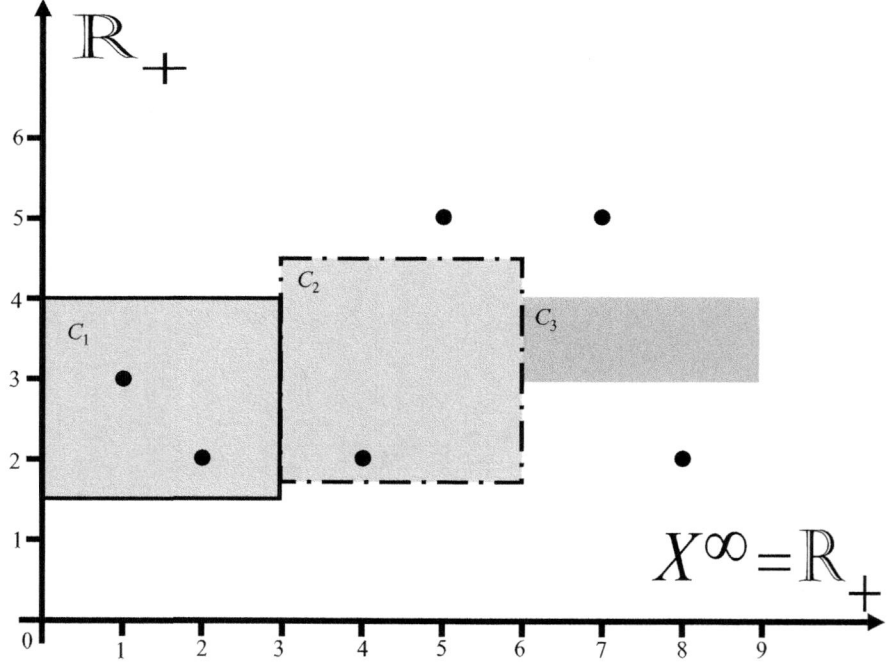

Fig. 2. Loss function approximation. Neighbourhood C_1 marked by solid lines belongs to the lower and upper approximation and C_2 marked with dashed lines - only to the upper approximation. Neighbourhood C_3 is outside of both approximations.

For the moment we have defined the approximation of loss function as a pair of sets created from the elements of neighbourhood family $\mathcal{C}$. From this approximation we would like to obtain an estimation of risk. For that purpose we need to define summarisation (integration) method analogous to Def. 1. We define an integration functional based on the idea of probabilistic version of Lebesgue-Stieltjes integral [3,13].

In order to define our integral we need to make some additional assumptions. For the universe X^∞ we assume that m is a measure[1] on a Borel σ-field of subsets of X^∞ and that $m(X^\infty) < \infty$. By m_0 we denote a measure on a σ-field of subsets of $\mathbb{R}_+$. We will also assume that $\mathcal{C}$ is a family of non-empty subsets of $X^\infty \times \mathbb{R}_+$ that are measurable relative to the product measure $\bar{m} = m \times m_0$. Finally, we assume that the value of loss function is bounded by some positive real B. Please, note that none of the above assumptions is unrealistic, and that in practical applications we are dealing with finite universes.

For the upper bound B we split the range $[0, B] \subset \mathbb{R}_+$ into $n > 0$ intervals of equal length $I_1, \ldots, I_n$, where $I_i = [\frac{(i-1)B}{n}, \frac{iB}{n}]$. This is a simplification of the most general definition, where the intervals do not have to be equal. For every interval I_i we consider the sub-family $\mathcal{C}_i \subset \mathcal{C}$ of neighbourhoods such that:

$$\mathcal{C}_i = \left\{ c \in \mathcal{C} : \forall x \in (z \cap \pi_1(c)) \quad \hat{L}_\alpha(x) > \frac{(i-1)B}{n} \right\}. \tag{3}$$

With the previous notation the estimate for empirical risk functional is given by:

$$R_{emp}(\alpha) = \sum_{i=1}^{n} \frac{B}{n} m \left(\bigcup_{c \in \mathcal{C}_i} \pi_1(c) \right) \tag{4}$$

In theoretical setting for the formula (4) above we shall derive its limit as $n \to \infty$, but in practical situation the parameter n does not have to go to infinity. It is sufficient to find n such that for every pair of points $x_1 \neq x_2$ taken from sample z if $\hat{L}_\alpha(x_1) < \hat{L}_\alpha(x_2)$ then for some integer $i \leq n$ we have $\hat{L}_\alpha(x_1) < \frac{iB}{m} < \hat{L}_\alpha(x_2)$.

4.2 The Fuzzy Set Case

If we consider another type of neighbourhoods, one defined with use of fuzzy sets, we will find ourselves in slightly different position. In fact, we may use fuzzy granules (neighbourhoods) in two ways:

1. We can define the family of neighbourhoods $\mathcal{C}$ as a family of fuzzy sets in $X^\infty \times \mathbb{R}_+$. This approach leads to more general idea of function approximation in fuzzy granular environment.
2. We restrict fuzziness to the domain of the loss function, i.e., X^∞. The values of $L_\alpha(.)$ remain crisp. That means that we have real-valued loss function that has a set of fuzzy membership values associated with each argument.

---

[1] We use m, m_0 and $\bar{m}$ to denote measures, instead of traditional $\mu, \mu_0, \bar{\mu}$, in order to avoid confusion with fuzzy membership functions used in next section.

While the former case is more general and could lead to nicer, more universal definitions of approximation (see, e.g., [14]), it is at the same time less intuitive if we want to discuss risk measures. For that reason we restrict ourselves to the latter case. The family of neighbourhoods $\mathcal{C}$ is now defined in such a way that each $c \in \mathcal{C}$ is a product of fuzzy set in X^∞ and a subset of $\mathbb{R}_+$. The family $\mathcal{C}$ directly corresponds to family of fuzzy membership functions (fuzzy sets) $\mathcal{C}_\mu$. Each $c \in \mathcal{C}$ is associated with a fuzzy membership function $\mu_c : X^\infty \mapsto [0,1]$ corresponding to the fuzzy projection of c onto X^∞. Please note that at the moment we assume nothing about the intersections of elements of $\mathcal{C}$ but, we assume that the family of fuzzy sets (neighbourhoods, granules) is finite.

Again, we start with a finite sample of points in the graph of loss function $\hat{L}_\alpha$: $z \mapsto \mathbb{R}_+$ for data points from a finite sample z. We will attempt to approximate L_α by extending its finite sample $\hat{L}_\alpha$. The lower and upper approximations given by fomulæ (1) and (2) need to accommodate for the fact that the neighbourhoods in $\mathcal{C}$ are now (semi-)fuzzy. The fact that each $c \in \mathcal{C}$ is a product of a fuzzy and a crisp set changes the definition of projections π_1, π_2. For $c \in \mathcal{C}$ we now introduce parameterised projections. For $0 \leq \lambda < 1$, we have:

$$\pi_1(c, \lambda) = \{x \in X^\infty : \exists y \in \mathbb{R}_+ \, ((x, y) \in c \wedge \mu_c(x) > \lambda)\},$$

$$\pi_2(c, \lambda) = \{y \in \mathbb{R}_+ : \exists x \in X^\infty \, ((x, y) \in c \wedge \mu_c(x) > \lambda)\}.$$

The parameter λ is used to establish a cut-off value for membership. The intention behind introduction of this parameter is that in some circumstances we may want to consider only those neigbourhoods which have sufficient level of confidence (sufficiently high membership). In terms of risk approximation, we would like to consider only these situations for which the argument of loss function is sufficiently certain. Naturally, we can make projections maximally general by putting $\lambda = 0$. The result of using projections introduced above, in particular $\pi_1(c, \lambda)$ is similar to taking an *alpha-cut* known from general fuzzy set theory (see [15]).

With previous notation and under previous assumptions we now introduce an approximation of $\hat{L}_\alpha$ w.r.t. the family of neigbourhoods $\mathcal{C}$ and a parameter (threshold) λ.

$$\mathcal{C}_\lambda \hat{L}_\alpha = \bigcup \{c \in \mathcal{C} : \hat{L}_\alpha(\pi_1(c, \lambda) \cap z) \subseteq \pi_2(c, \lambda)\} \tag{5}$$

It is important to notice, that while projections $\pi_1(c, \lambda)$ and $\pi_2(c, \lambda)$ are classical (crisp) sets, the resulting approximation $\mathcal{C}_\lambda \hat{L}_\alpha$ is of the same type as the original sample, i.e., it is a union of neighbourhoods (granules) which are products of fuzzy set in X^∞ and a subset of $\mathbb{R}_+$. Please also note that the union operator used in (5) is working dimension-wise, and that the resulting set $\mathcal{C}_\lambda \hat{L}_\alpha$ does not have to be a granule (neighbourhood), as we do not assume that this union operator is a granule aggregation (fusion) operator. The union operator will be also of paramount importance to the summation (integration) procedure that we about to introduce for the purpose of calculating the empirical risk.

It is relatively simple to introduce the empirical risk functional by rewriting the formulæ(3) and (4). We only need to adjust the previous definitions in such a way that they use the parameterised projections $\pi_1(c, \lambda)$, $\pi_2(c, \lambda)$.

With all the previously introduced notation, the empirical risk functional is introduced by first defining the building blocks (strata, neighbourhoods) as:

$$\mathcal{C}_i^\lambda = \left\{ c \in \mathcal{C} : \forall x \in (z \cap \pi_1(c, \lambda)) \quad \hat{L}_\alpha(x) > \frac{(i-1)B}{n} \right\}. \tag{6}$$

That leads to the estimate for empirical risk functional given by:

$$R_{emp}^\lambda(\alpha) = \sum_{i=1}^{n} \frac{B}{n} \, m \left(\bigcup_{c \in \mathcal{C}_i^\lambda} \pi_1(c, \lambda) \right) \tag{7}$$

The formulæ above, just as in the case of rough set risk estimates, are valid only if some assumptions can be made about the family of neighbourhoods $\mathcal{C}$. Again, these assumptions are rather reasonable, and quite possible to met if we are dealing with a finite sample z and a finite family of neighbourhoods $\mathcal{C}$. We have to assure that:

1. Elements of $\mathcal{C}$ are measurable w.r.t $\bar{m} = m \times m_0$ - the product measure on $X^\infty \times \mathbb{R}_+$. It means that the fuzzy part of each $c \in \mathcal{C}$ is such, that after application of projection it is still in the Borel σ-field. It has to be stated, that this is more complicated assumption to met in fuzzy than in rough case. However, in practical situations we normally deal with "standard" fuzzy membership functions μ_c. Such functions are either piecewise linear or defined by some typical analytical formulæ, and hence at least continuous (sometimes differentiable) - which provides for measurability.
2. The measure of X^∞ is finite. This is a fairly easy to achieve, as we are assuming that we are dealing with finite samples and finite number of neighbourhoods. Hence, we can always restrict X^∞ to a bounded sub-domain that contains all samples and neighbourhoods.
3. The value of loss function is always a non-negative real number, and it is bound from above by some value B.

As one can see, the risk estimator (7) is parameterised by the confidence level λ. In fact, the selection of proper value of λ in all steps of risk assessment in the fuzzy context is the crucial step. Depending on value of λ we may get (radically) different outcomes. This intuitively corresponds to the fact, that we can get different overview of the situation depending on how specific or how general we want to be.

The notions of function approximations and risk functional that we have introduced are heavily dependent on the data sample z and decomposition of our domain into family of neighbourhoods $\mathcal{C}$. It is not yet visible, how the ideas we present may help in construction of better decision support (classification) systems. In the following section we discuss these matters in some detail.

5   Classifiers, Neighbourhoods and Granulation

Insofar we have introduced the approximation of loss and the measure of risk. To show the potential use of these entities, we intend to investigate the process of creation and evaluation (scoring) of classifier-driven decision support system as a whole.

The crucial component in all our definitions is the family of non-empty sets (neighbourhoods) $\mathcal{C}$. This family represents the granular nature of the universe of discourse. We have to know this family before we can approximate loss or estimate empirical risk. In practical situations the family of neighbourhoods have to be constructed in close correlation with classifier construction. It is quite common, especially for rough sets approaches, to define these sets constructively by semantics of some formulas. An example of such formula could be the conditional part of decision rule or a template (in the sense of [16,17]). In case of fuzzy granules the neighbourhoods may be provided arbitrarily or imposed by the necessity to instantiate a set of linguistic rules (a knowledge base).

Usually the construction of proper neighbourhoods is a complicated search and optimisation task. The notions of approximation and empirical risk that we have introduced may be used to express requirements for this search/optimisation. For the purpose of making valid, low-risk decision by means of classifier α we would expect the family $\mathcal{C}$ to possess the following qualities:

1. *Precision.* In order to have really meaningful assessment of risk as well as good idea about the loss function we would like the elements of neighbourhood family to to be relatively large in terms of universe X^∞, but at the same time having possibly low variation. These requirements in rough set case translate to expectation that for the whole family $\mathcal{C}$ we want to minimise the *boundary region* in loss approximation, i.e., achieve possibly the lowest value of $\bar{m}(\overline{\mathcal{C}}\hat{L}_\alpha \setminus \underline{\mathcal{C}}\hat{L}_\alpha)$. The minimisation of boundary region shall be constrained by requirements regarding the "shape" of elements of $\mathcal{C}$. In fuzzy setting we would expect the family of neighbourhoods to be such, that the uncertain portion of each neighbourhood c, i.e., the part that has a low (close to 0) value of μ_c is relatively small. We should try to find such a family $\mathcal{C}$ that for $c \in \mathcal{C}$ the value of $m(\pi_1(c))$ $(m(\pi_1(c,\lambda)))$ is relatively large (for λ close to 1) while the value of $m_0(\pi_2(c))$ $(m_0(\pi_2(c,\lambda)))$ is relatively small. The neigbourhoods that fulfill these requirement correspond to ability of characterising with sufficient confidence large portions of the domain X^∞ as being relatively uniform in terms of the value of loss function. The low variation of loss function on any given neighbourhood can be understood as equivalent to the requirement that this neighbourhood is well contained in a granule defined by application of the classifier α.

2. *Relevance.* This requirement is closely connected with the previous one (precision). While attempting to precisely dissect the domain into neighbourhoods we have to keep under control the relative quality (relevance) of

neighbourhoods with respect to the data sample z. We are only interested in the neighbourhoods that contain sufficient number of elements of z. The actual threshold for this number is obviously dependant on the particular application and data set we are dealing with. It should be noted that without such threshold we are likely to produce neighbourhoods that are irrelevant to the data sample, hence potentially harmful for classifier learning process and risk assessment. The threshold for relevance of a neighbourhood is a subject to optimisation as well. If the threshold is too high, the resulting granularity of domain is to coarse and we are unable to make precise classification. On the other hand, if the threshold is too low and the resulting neighbourhoods contain too few elements from data sample, we may face the effect that is an equivalent of *overfitting* in classical classification systems.

3. *Coverage and adaptability.* One of the motivations that steer the process of creating the family of neighbourhoods and the classifier is the expectation regarding its ability to generalise and adapt the solution established on the basis of finite sample to a possibly large portion of the data domain. In terms of neighbourhoods it can be expressed as the requirement for minimisation of $m(X^\infty \setminus \bigcup_{c \in \mathcal{C}} \pi_1(c))$ $(m(X^\infty \setminus \bigcup_{c \in \mathcal{C}} \pi_1(c, \lambda))$ for adequately large λ). This minimisation has to be constrained by requirements expressed in points 1 and 2 above (precision and relevance). It shall also, to largest possible extent, provide for adaptability of our solution to classification problem. The adaptability requirement is understood as expectation that in the presence of newly collected data point the quality of loss approximation will not decrease dramatically as well as quality of our empirical risk assessment. In other words, the domain shall be covered by neighbourhoods in such a way that they are complaint with the expected behaviour of classifier on yet unseen examples. In terms of granulation of the domain we expect that the creation of neighbourhoods and learning of classifiers will be performed in some kind of feedback loop, that will make it possible to achieve possibly the highest compliance between the two. This requirement is formulated quite vaguely and obviously hard to turn into numerical optimisation criterion. It is, nevertheless, a crucial one, as we expect the system we are creating to be able to cater limited extensions of data sample with only small adjustments, definitely without the need for fundamental reconstruction of the entire system (classifier, neighbourhoods and loss function).

As discussed in points 1–3 above, the task of finding a family of neighbourhoods can be viewed as a multi-dimensional optimisation on meta-level. It is in par with the kind of procedure that has to be employed in construction of systems based on the granular computing paradigm [8,17].

The idea of granularity, impreciseness and limited expressiveness may surface in other places within the process of constructing the classifier, approximating the loss and assessing empirical risk. So far we have followed the assumption made at the beginning of Section 3, that the values of loss function are given as non-negative real numbers. In real application we may face the situation when the value of loss is given to us in less precise form. One such example is the loss function expressed in relative, qualitative terms. If the value of loss is given to us by the human expert, he/she may be unable to present us with precise, numerical values due to, e.g., imprecise or incompletely define nature of problem in discourse. We may then be confronted with situation when the loss is expressed in qualitative terms such as "big", "negligible", "prohibitive", "acceptable". Moreover, the value of loss may be expressed in relative terms, by reference to other, equally imprecise notions such as "greater than previous case" or "on the border between large and prohibitive". Such imprecise description of the loss function may in turn force us to introduce another training loop into our system, one that will learn how to convert the imprecise notions we have into concrete, numerical values of loss function.

Yet another axis for possible discussion and extensions of ideas presented in previous sections is associated with the choice of methodology for summation (integration). In order to introduce definition of risk measure (4) we have assumed that the the universe and corresponding sets are measurable in conventional sense. In other words, for finite samples we are dealing with, we have assumed that we ma rely on additivity of measure, and that the summation (integral) in our universe is a proper linear functional. That assumption may be challenged, as currently many authors (see [18]) bring examples showing that in several application domains we cannot easily make assumptions about linearity and sub-additivity of the mathematical structure of the universe. This is, in particular, the case of risk assessment. In complicated systems involving risk it may be necessary to perform a complex analysis that will lead to discovery of the actual properties of the domain. It is likely that such investigation will entail utilisation of expert's knowledge and domain theory/knowledge in addition to mathematical, analytic tools.

6   Summary and Conclusion

In this paper we have discussed the issues that accompany the assessment of risk in classification systems on the basis of the finite set of examples. We have pointed out some sources of possible problems and outlined some directions, in which we may search for solutions that match our expectations sufficiently well.

In conclusion, we would like to go back to the more general issue of weighting the risk involved in computer-supported decision making. As we have mentioned in the introduction to this paper, in the real-life situations the human user may display various patterns in his/her risk assessment and aversion. In particular, even with a well established mathematical model that measures the risk in a given situation, we are frequently forced to change it as new information arrives. This is a natural phenomenon that we have to take into account in design of our solutions from the very beginning. In human terms, we can talk of evolution of the concept of risk in a given system as time passes and new information arrives. Humans may wish to change the way of perceiving the risk if they are able to use more information, form new experiences and make more informed judgement. In terms of computer-aided decision making the inflow of new information contributes to changes in parameters of the model. The challenge is to devise such a model that is flexible and far-fetching enough to be able to adjust for even significant changes resulting form changes generated by inflow of new information. It is rather unrealistic to expect that it would be possible to devise and explicitly formulate a model, that sufficiently supports extensibility as well as adaptability, and at the same time applicable in many different situations. It is much more likely that in practical situation we may need to learn (or estimate) not only the parameters, but the general laws governing its dynamics, at the same time attempting preserve its flexibility and ability to adapt to new cases.

Acknowledgements. The author wishes to thank Professor Andrzej Skowron for stimulating discussions that led to formulation of some of the ideas presented in this paper. The author is grateful for the invitation from the Guest Editors to submit the paper to the Transactions on Rough Sets. Also, the thanks go to anonymous reviewers for their remarks and recommendations.

This research was supported by the grants N N516 368334 and N N516 077837 from the Ministry of Science and Higher Education of the Republic of Poland.

References

1. Warwick, B. (ed.): The Handbook of Risk. John Wiley & Sons, New York (2003)
2. Bostrom, A., French, S., Gottlieb, S. (eds.): Risk Assessment, Modeling and Decision Support. Risk, Governance and Society, vol. 14. Springer, Heidelberg (2008)
3. Vapnik, V.: Statisctical Learning Theory. John Wiley & Sons, New York (1998)
4. Berger, J.O.: Statistical Decision Theory and Bayesian Analysis, 2nd edn. Springer, New York (1985)
5. Szczuka, M.: Towards approximation of risk. In: Chan, C.-C., Grzymala-Busse, J.W., Ziarko, W.P. (eds.) RSCTC 2008. LNCS (LNAI), vol. 5306, pp. 320–328. Springer, Heidelberg (2008)
6. Szczuka, M.S.: Approximation of loss and risk in selected granular systems. In: Sakai, H., Chakraborty, M.K., Hassanien, A.E., Slezak, D., Zhu, W. (eds.) RSFD-GrC 2009. LNCS, vol. 5908, pp. 168–175. Springer, Heidelberg (2009)
7. Skowron, A., Szczuka, M.S.: Toward interactive computations: A rough-granular approach. In: Advances in Machine Learning II. SCI, vol. 263, pp. 23–42. Springer, Heidelberg (2010)
8. Pedrycz, W., Skowron, A., Kreinovich, V. (eds.): Handbook of Granular Computing. John Wiley & Sons, New York (2007)
9. Zadeh, L.: Fuzzy sets and information granularity. In: Gupta, M., Ragade, R., Yager, R. (eds.) Advances in Fuzzy Set Theory and Application, pp. 3–18. North-Holland Publishing Co., Amsterdam (1979)
10. Vapnik, V.: Principles of risk minimization for learning theory. In: Proceedings of NIPS, pp. 831–838 (1991)
11. Pawlak, Z.: Rough Sets. Theoretical Aspects of Reasoning about Data. Kluwer Academic Publishers, Dordrecht (1991)

12. Pawlak, Z.: Rough sets, rough functions and rough calculus. In: Pal, S.K., Skowron, A. (eds.) Rough Fuzzy Hybridization: A New Trend in and Decision Making, pp. 99–109. Springer, Singapore (1999)
13. Halmos, P.: Measure Theory. Springer, Berlin (1974)
14. Höeppner, F., Klawonn, F.: Systems of information granules. In: [8], pp. 187–203
15. Yager, R.R., Filev, D.P.: Essentials of fuzzy modeling and control. John Wiley & Sons, Chichester (1994)
16. Skowron, A., Synak, P.: Complex patterns. Fundamenta Informaticae 60(1-4), 351–366 (2004)
17. Jankowski, A., Peters, J.F., Skowron, A., Stepaniuk, J.: Optimization in discovery of compound granules. Fundamenta Informaticae 85, 249–265 (2008)
18. Kleinberg, J., Papadimitriou, C., Raghavan, P.: A microeconomic view of data mining. Data Mining and Knowledge Discovery 2, 311–324 (1998)

Core-Generating Discretization for Rough Set Feature Selection

David Tian[1], Xiao-jun Zeng[2], and John Keane[2]

[1] Department of Computing, Faculty of ACES, Sheffield Hallam University, Howard
Street, Sheffield, S1 1WB, UK
D.Tian@shu.ac.uk, dtian09@gmail.com
[2] School of Computer Science, University of Manchester, Kilburn Building,
Oxford Road, Manchester, M13 9PL, UK
john.keane@manchester.ac.uk, x.zeng@manchester.ac.uk

Abstract. Rough set feature selection (RSFS) can be used to improve
classifier performance. RSFS removes redundant attributes whilst keeping important ones that preserve the classification power of the original
dataset. The feature subsets selected by RSFS are called *reducts*. The intersection of all reducts is called *core*. However, RSFS handles discrete
attributes only. To process datasets consisting of real attributes, they are
discretized before applying RSFS. Discretization controls *core* of the discrete dataset. Moreover, *core* may critically affect the classification performance of *reducts*. This paper defines *core-generating discretization*, a type
of discretization method; analyzes the properties of core-generating discretization; models core-generating discretization using constraint satisfaction; defines core-generating approximate minimum entropy
(C-GAME) discretization; models C-GAME using constraint satisfaction
and evaluates the performance of C-GAME as a pre-processor of RSFS
using ten datasets from the UCI Machine Learning Repository.

Keywords: core-generating discretization, core-generating approximate
minimum entropy discretization, rough set feature selection, constraint
satisfaction.

1 Introduction

Rough set theory (RST) [31,32,17,33,27,34] is a mathematical method for approximate reasoning about concepts. In RST, datasets containing patterns are
called decision tables. Objects of a decision table have different properties. The
set of all objects with the same properties is called a concept. Thus, the decision
class is a concept. However, although there is incomplete knowledge of the concepts from the decision tables, but the concepts can be approximated using their
upper and lower approximations. In this situation, the concepts are replaced by
their approximations. Different subsets of attributes define different upper and
lower approximations. Objects of different concepts can be discerned using the
values of their corresponding attributes. The subsets of attributes which discern an equal number of objects as all the attributes are called *reducts* [27,31].
Reducts are said to preserve the classification power of the decision table.

J.F. Peters et al. (Eds.): Transactions on Rough Sets XIII, LNCS 6499, pp. 135–158, 2011.
© Springer-Verlag Berlin Heidelberg 2011

RSFS handles discrete attributes only. To handle datasets with continuous attributes, discretization can be performed prior to RSFS to map the continuous values of each attribute to a finite number of intervals. Discretization critically affects the classification performance and the effect of dimensionality reduction of RSFS because different discretization approaches result in different classification performances of the reducts and different sizes of reducts. Discretization methods have been widely studied [5,3] and can be categorized by three axes: global vs. local methods, supervised vs. unsupervised and static vs. dynamic [5]. Local methods such as Recursive Minimal Entropy Partitioning (RMEP) [5], use a subset of instances to discretize an attribute [19]. Global methods such as MD-heuristic [22] uses all the instances of a dataset to discretize an attribute [19]. Supervised methods such as 1RD [5] and ChiMerge [5] make use of the class labels during discretization, whereas unsupervised methods such as Equal Width Intervals (EWI) [5] and Equal Frequency Intervals (EFI) [5] do not make use of the class labels. Static methods such as RMEP and 1RD discretize each attribute independent of other attributes [8]. Dynamic methods take into account the interdependencies between attributes [8]. Some of the other approaches to apply RSFS to continuous datasets are fuzzy rough sets and tolerance-based rough sets (see Section 3).

RSFS selects reducts, minimal feature subsets that preserve the classification power of the dataset. For a given dataset, there may exist numerous reducts. *Core* is the set of all the common attributes of all the reducts, so *core* determines some of the attributes within each reduct. Therefore, *core* may critically affect the performance of RSFS to obtain an accurate classifier from the reduced dataset using the reduct. Significant attributes are the ones which are important for patterns classification. If *core* contains significant attributes, each reduct would contain the significant attributes and the performance of RSFS would be high. If, however, *core* does not contain any significant attributes, each reduct may in turn not contain any significant attributes, so the performance of RSFS would be low. Discretization determines *core* of a dataset, so *core* can be used as a criterion during discretization. However, current methods do not consider *core* while discretizing datasets. Section 2 presents the preliminary concepts concerning rough set theory, discretization problems and constraint satisfaction. Fuzzy rough feature selection and tolerance-based feature selection are discussed in section 3. Section 4 presents the definition of core-generating discretization. Section 5 models core-generating discretization using constraint satisfaction. Section 6 presents C-GAME and evaluates its performance using ten datasets from UCI Machine Learning Repository [10]. Section 7 concludes this work.

2 Preliminary Concepts

2.1 Rough Set Theory

Let $\mathbb{A} = (U, A \cup \{d\})$ be a decision table where U (the universe) is the set of all the objects of a dataset, A is a nonempty finite set of condition attributes such that $a : U \rightarrow V_a$ for all $a \in A$ where V_a is the set of values of attribute a; d is the decision attribute such that $d : U \rightarrow V_d$, where V_d is the set of values for d. For any $B \subseteq A$ there is associated an indiscernibility relation $IND(B)$ [27]:

$$IND(B) = \{(x, x') \in U^2 | \forall a \in B \ a(x) = a(x')\}. \tag{1}$$

If $(x, x') \in IND(B)$, then objects x and x' are indiscernible from each other using attributes of B. The equivalence classes $[x]_{IND(B)}$ of $IND(B)$ are defined as follows [27]:

$$[x]_{IND(B)} = \{y \in U : (x, y) \in IND(B)\}. \tag{2}$$

The indiscernibility relation generates a partition of the universe U, denoted $U/IND(B)$ consisting of all equivalence classes of $IND(B)$ [27]:

$$U/IND(B) = \{[x]_{IND(B)} : x \in U\}. \tag{3}$$

$U/IND(\{d\})$ is the partition of the universe using the indiscernibility relation on the decision attribute. The elements of $U/IND(\{d\})$ are called decision classes [27] which express the classification of objects as done by an expert. There exists numerous decision classes. Let X_i denote the ith decision class, the lower and upper approximations of X_i using B are defined as follows [27]:

$$\underline{B}X_i = \bigcup \{Y \in U/IND(B) : Y \subseteq X_i\}. \tag{4}$$

$$\overline{B}X_i = \bigcup \{Y \in U/IND(B) : Y \cap X_i \neq \emptyset\}. \tag{5}$$

and

$$\underline{B}X_i \subseteq X_i \subseteq \overline{B}X_i, \tag{6}$$

where $\underline{B}X_i$ and $\overline{B}X_i$ are the lower and the upper approximations of X_i respectively. $\underline{B}X_i$ contains objects that can be certainly classified as members of X_i based on the knowledge in B. $\overline{B}X_i$ contains objects that can be classified as possible members of X_i using the knowledge in B. The positive region contains all objects of U that can be certainly classified to the decision classes using the knowledge in B. The positive region denoted $POS_B(d)$ is defined as follows [27]:

$$POS_B(d) = \bigcup_{X_i \in U/IND(d)} \underline{B}X_i. \tag{7}$$

Let $a \in A$, a is dispensable in $\mathbb{A}$, if $POS_{(A-\{a\})}(d) = POS_A(d)$; otherwise a is indispensable in $\mathbb{A}$. If a is an indispensable feature, deleting it from A will cause $\mathbb{A}$ to be inconsistent. $\mathbb{A}$ is independent if all $a \in A$ are indispensable. A subset of features $R \subseteq A$ is called a reduct, if R satisfies the following conditions: [27]

$$POS_B(d) = POS_A(d). \tag{8}$$

$$POS_{B-\{i\}}(d) \neq POS_B(d), \forall i \in B. \tag{9}$$

In other words, a reduct is a minimal feature subset preserving the positive region of A. For a decision table $\mathbb{A}$, there may exist numerous reducts. *Core* is defined to be the intersection of all the reducts of $\mathbb{A}$:

$$CORE(\mathbb{A}) = \bigcap_{i=1}^{|RED(\mathbb{A})|} B_i, \tag{10}$$

where $RED(\mathbb{A})$ is the set of all the reducts of $\mathbb{A}$; B_i is a reduct i.e. $B_i \in RED(\mathbb{A})$, $\forall i \in \{1, \ldots, |RED(\mathbb{A})|\}$. Core contains all indispensable features of A. The discernibility matrix M is defined as [27]:

$$M = (c_{ij}^d), \; where \begin{matrix} c_{ij}^d = \emptyset, & \text{in case } d(x_i) = d(x_j), \\ c_{ij}^d = c_{ij} - \{d\}, & \text{in case } d(x_i) \neq d(x_j). \end{matrix} \tag{11}$$

$$c_{ij} = \{a \in A : a(x_i) \neq a(x_j)\}, \tag{12}$$

where c_{ij} is a matrix entry. The core can be computed as the set of all singletons of the discernibility matrix of the decision table [39] (equation 13):

$$core = \{a \in A : c_{ij} = \{a\}, for \; some \; i, \; j \in U\}, \tag{13}$$

where c_{ij} is a matrix entry. The B-information function is defined as [25]:

$$Inf_B(x) = \{(a, a(x)) : a \in B, \; B \subseteq A \; for \; x \in U\}. \tag{14}$$

The generalized decision ∂_B of $\mathbb{A}$ is defined as follows [25]:

$$\partial_B : U \rightarrow 2^{V_d}, \tag{15}$$

where

$$\partial_B(x) = \{i : \exists x' \in U[(x, x') \in IND(B) \land d(x') = i]\}, \tag{16}$$

and 2^{V_d} is the power set of V_d.

2.2 Discretization Problems

In [24,23,2,22,25], discretization problems are defined as follows. Let $\mathbb{A} = (U, A \cup \{d\})$ be a decision table where U is a finite set of objects (the universe), A is a nonempty finite set of condition attributes such that $a : U \rightarrow V_a$ for all $a \in A$ and d is the decision attribute such that $d : U \rightarrow V_d$. It is assumed that $V_a = [l_a, r_a) \subset \mathbb{R}$ where l_a is the minmum value of a; r_a is greater than the maximum value of a in a dataset and r_a is not a value of the dataset; $\mathbb{R}$ is the set of real numbers. In discretization problems, $\mathbb{A}$ is often a consistent decision table. That is $\forall(o_i, o_j) \in U \times U$, if $d(o_i) \neq d(o_j)$, then $\exists a \in A$ such that $a(o_i) \neq a(o_j)$ [9].

Let P_a be a partition on V_a (for $a \in A$) into subintervals so that

$$P_a = \{[c_0^a, c_1^a), [c_1^a, c_2^a), \ldots, [c_{k_a}^a, c_{k_a+1}^a)\}, \tag{17}$$

where k is some integer, $l_a = c_0^a$, $c_0^a < c_1^a < c_2^a < \ldots < c_{k_a}^a < c_{k_a+1}^a$, $r_a = c_{k_a+1}^a$ and $V_a = [c_0^a, c_1^a) \cup [c_1^a, c_2^a) \cup, \ldots, [c_{k_a}^a, c_{k_a+1}^a)$. Any P_a is uniquely defined by the set $C_a = \{c_1^a, c_2^a, \ldots, c_{k_a}^a\}$ called the set of cuts on V_a (the set of cuts is empty if $card(P_a) = 1$) [25]. Then $P = \bigcup_{a \in A} \{a\} \times C_a$ represents any global family P of partitions. Thus, P defines a global discretization of the decision table. Any pair $(a, c) \in P$ is called a cut on V_a. Any set of cuts

$$C = \{c_1^a, \ldots, c_{k_1}^a, c_1^b, \ldots, c_{k_2}^b, \ldots\} \tag{18}$$

transforms from $\mathbb{A} = (U, A \cup \{d\})$ into a discrete decision table $\mathbb{A}^C = (U, A^C \cup \{d\})$, where $A^C = \{a^C : a \in A\}$ and $a^C(o) = i \leftrightarrow a(o) \in [c_i^a, c_{i+1}^a)$ for any $o \in U$ and $i \in \{0, \ldots, k_a\}$. The table $\mathbb{A}^C$ is called C-discretization of $\mathbb{A}$.

Two sets of cuts C', C are equivalent, if and only if $\mathbb{A}^C = \mathbb{A}^{C'}$. For any $a \in A$, $a(U) = \{a(o) : o \in U\} = \{v_1^a, v_2^a, \ldots, v_{n_a}^a\}$ denotes the set of all values of attribute a occurring in the decision table $\mathbb{A}$, where $U = \{o_1, o_2, \ldots, o_n\}$ and $n_a \leq n$. It is assumed that these values are sorted in ascending order, so that $v_1^a < v_2^a < \ldots < v_{n_a}^a$. Two cuts (a, c_1) and (a, c_2) are equivalent if and only if there exists $i \in \{1, \ldots, n_a - 1\}$ such that $c_1, c_2 \in (v_i^a, v_{i+1}^a]$. Hence, the equivalent cuts should not be distinguished and all cuts in the interval $(v_i^a, v_{i+1}^a]$ are unified using one representative cut $(a, \frac{v_i^a + v_{i+1}^a}{2})$ called the generic cut [25]. The set of all the generic cuts on a is denoted by

$$C_a = \{ \frac{v_1^a + v_2^a}{2}, \frac{v_2^a + v_3^a}{2}, \ldots, \frac{v_{n_a-1}^a + v_{n_a}^a}{2} \}. \tag{19}$$

The set of all candidate cuts of a given decision table is denoted by

$$C_{\mathcal{A}} = \bigcup_{a \in A} C_a. \tag{20}$$

Discernibility of Cuts. The discretization process is associated with a loss of information [25]. During discretization a smallest set of cuts C is obtained from a given decision table $\mathbb{A}$ such that, despite loss of information, the C-discretized table $\mathbb{A}^C$ still retains some useful properties of $\mathbb{A}$. The discernibility of cuts is one such property to be preserved during discretization.

Given a decision table $\mathbb{A} = (U, A \cup \{d\})$, an attribute a discerns a pair of objects $(o_i, o_j) \in U \times U$ if $a(o_i) \neq a(o_j)$. A cut c on $a \in A$ discerns (o_i, o_j) if $(a(x) - c)(a(y) - c) < 0$. Two objects are discernible by a set of cuts C if they are discernible by at least one cut from C [25]. Therefore, discernibility of cuts determines the discernibility of the corresponding attribute. The consistency of a set of cuts is defined as follows [25]:

A set of cuts is consistent with $\mathbb{A}$ (or $\mathbb{A}$-consistent) if and only if $\partial_A = \partial_{A^C}$ where ∂_A and ∂_{A^C} are generalized decisions of A and A^C.

A consistent set of cuts C is

1. $\mathbb{A}$-irreducible if for all $C' \subset C$, C' is not $\mathbb{A}$-consistent.
2. $\mathbb{A}$-optimal if for any $\mathbb{A}$-consistent set of cuts C', card(C)$\leq$card(C').

The discernibility of cuts can be represented in the form of a discernibility table $\mathbb{A}^* = (U^*, A^*)$ constructed from the decision table as follows [24]:

$$U^* = \{(o_i, o_j) \in U^2 : d(o_i) \neq d(o_j)\}. \tag{21}$$

$$A^* = \{c : c \in C\}, \text{where } c((o_i, o_j)) = d_c^{(a,(o_i,o_j))} \text{ and} \tag{22}$$

$$d_c^{(a,(o_i,o_j))} = \begin{cases} 1, \text{ if } c \text{ discerns } o_i \text{ and } o_j, \\ 0, \text{ otherwise,} \end{cases}$$

where C is the set of all possible cuts on A. An example decision table and the corresponding discernibility table are shown in tables 1 and 2.

Table 1. An example decision table

U	a	b	Class
o_1	0.8	2	1
o_2	1	0.5	0
o_3	1.3	3	0
o_4	1.4	1	1
o_5	1.4	2	0
o_6	1.6	3	1
o_7	1.3	1	1

Table 2. The discernibility table for attributes a and b

U^*	a				b		
	0.9	1.15	1.35	1.5	0.75	1.5	2.5
(o_1, o_2)	1	0	0	0	1	1	0
(o_1, o_3)	1	1	0	0	0	0	1
(o_1, o_5)	1	1	1	0	0	0	0
(o_4, o_2)	0	1	1	0	1	0	0
(o_4, o_3)	0	0	1	0	0	1	1
(o_4, o_5)	0	0	0	0	0	1	0
(o_6, o_2)	0	1	1	1	1	1	1
(o_6, o_3)	0	0	1	1	0	0	0
(o_6, o_5)	0	0	0	1	0	0	1
(o_7, o_2)	0	1	0	0	1	0	0
(o_7, o_3)	0	0	0	0	0	1	1
(o_7, o_5)	0	0	1	0	0	1	0

Recursive Minimal Entropy Partition Discretization (RMEP). RMEP
[5] is a well-known discretization method. RMEP discretizes one condition attribute at a time in conjunction with the decision attribute such that the class
information entropy of each attribute is minimized. The discretized attributes
have almost identical information gain as the original ones. For each attribute,
candidate cut points are generated and evaluated individually using the class
information entropy criterion:

$$E(A, T; S) = \frac{|S_1|}{|S|} Ent(S_1) + \frac{|S_2|}{|S|} Ent(S_2), \tag{23}$$

where A is an attribute, T is a cut point, S is a set of instances, S_1 and S_2
are subsets of S with A-values $\leq$ and $> T$ respectively. The cut point T_{mini}
with minimum $E(A, T_{mini}; S)$ is chosen to discretize the attribute. The process
is then applied recursively to both partitions S_1 and S_2 induced by T_{mini} until
the stopping condition, which makes use of the minimum description length
principle, is satisfied.

2.3   Constraint Satisfaction Problems

A constraint satisfaction problem (CSP) consists of a finite set of variables, each of which is associated with a finite domain, and a set of constraints that restricts the values the variables can simultaneously take [37]. The domain of a variable is a set of all possible values that can be assigned to the variable. A label is a variable-value pair, denoting the assignment of a value to a variable. A compound label is a set of labels over some variables. A constraint can be viewed as a set of compound labels over all variables. A total assignment is a compound label on all variables of a CSP. A solution to a CSP is a total assignment satisfying all constraints simultaneously. The search space of a CSP is the set of all possible total assignments. The size of the search space is the number of all possible total assignments [37]: $\prod_{x \in Z} D_x$, where Z is the set of all variables x; D_x is the domain of x. A CSP can be reduced to an equivalent but simpler CSP by removing redundant values from the domains and removing redundant compound labels from the constraints in the original CSP [37]. This process is called *constraint propagation*. No solutions exist if the domain of a variable or a constraint is reduced to the empty set [37]. A constraint satisfaction optimization problem (CSOP) is a CSP with an objective function. The solution to the CSOP is the total assignment satisfying all the constraints simultaneously whilst maximizing or minimizing the objective function. CSOPs are solved using optimization algorithms such as branch and bound (B&B) [37].

3   Rough Set Feature Selection for Continuous Datasets

Some of the recent research in rough set feature selection for handling continuous datasets are fuzzy rough sets [15,14,26] and tolerance rough sets [12,13,21,28,29,35,36] which do not require a discretization process beforehand. A fuzzy rough set [15] is a rough set which is approximated by a pair of fuzzy lower and fuzzy upper approximations. Fuzzy rough sets model the extent to which two objects are similar and are based on fuzzy similarity relations. A fuzzy similarity relation R_a (24), determines to what extent two objects are similar:

$$R_a = \{((x,y), \mu_{R_a}((x,y))|(x,y) \in U \times U, \mu_{R_a} \in [0,1]\}, \tag{24}$$

where a is an attribute. To handle discrete datasets, the membership function $\mu_{R_a} \in \{0,1\}$ is defined to take either 0 or 1 for each pair of objects x and y so that $\mu_{R_a}(x,y) = 1$ if $a(x) = a(y)$; $\mu_{R_a}(x,y) = 0$ otherwise. Tolerance rough set [12,13,21,28,29,35,36] is another extension of rough set theory. Dissimilarities between continuous features values of two objects are tolerated to a certain degree. Tolerance rough sets are defined based on similarity relations. A similarity relation $SIM_{P,\tau}$ where P is a feature subset, is defined based on the difference in attributes values between two objects (25):

$$SIM_{P,\tau} = \{(x,y)|(x,y) \in U \times U \wedge [\prod_{a \in P} SIM_a(x,y)] \geq \tau\} \tag{25}$$

where

$$SIM_a(x,y) = 1 - \frac{|a(x) - a(y)|}{|a_{max} - a_{min}|}. \tag{26}$$

τ is the similarity threshold. The lower the value of τ, the more tolerant the dissimilarity of the two objects and vice versa. Setting τ to 1 leads to $SIM_{P,\tau}$ becoming the indiscernibility relation of rough set theory.

4   Core-Generating Discretization

This section presents core-generating discretization problems. Firstly, core-generating discretization is defined. Secondly, the relation between core size and core-generating objects is analyzed. Thereafter, core-generating sets of cuts are defined. Finally, properties of core-generating sets of cuts are analysed.

4.1   Core-Generating vs. Non-core-Generating: An Axis

Definition 1. *Core-generating vs. non-core-generating is an axis by which discretization methods are categorized. A discretization method is core-generating if and only if it always generate a non-empty core within the discrete dataset; otherwise, it is non-core-generating.*

Current discretization methods are *non-core-generating* methods, because they do not guarantee to generate a non-empty core for a given dataset.

4.2   Core Size and Core-Generating Objects

The core can be computed as the set of all singletons of the discernibility matrix of the decision table (see equation 13). If the core size is 0, the discernibility matrix must contain no singletons and vice versa. Based on the definition of the discernibility matrix, the following is true for a core attribute [27]:

 - There is at least 1 pair of objects (o_i, o_j), such that o_i, o_j belong to 2 different decision classes and are discerned by the core attribute only.

Pairs of objects with this property are called *core-generating objects*, because the presence of such a pair results in the presence of a core attribute within the reducts. Pairs of objects without this property are called *non-core-generating objects*. The core size of a decision table is equal to the number of pairs of core-generating objects within the decision table. For the decision table in Table 3, there are 2 pairs of core-generating objects: (o_1, o_2) and (o_2, o_3). (o_1, o_2) generates c as a core attribute, whereas (o_2, o_3) generates d as a core attribute.

4.3   Core-Generating Sets of Cuts

Given a decision table of real values $\mathbb{A} = (U, A \cup \{d\})$, a discrete decision table with a core size > 0 can be created using a *core-generating* set of cuts.

Table 3. A decision table with core $= \{c, d\}$

U	a	b	c	d	Class
o_1	4	0	6	3	2
o_2	4	0	4	3	1
o_3	4	0	4	5	2
o_4	2	0	7	2	2
o_5	1	0	3	2	1

Definition 2. *A set of cuts C, is core-generating if and only if the discrete decision table $\mathbb{A}^C$ contains core-generating objects. C is core-generating if it satisfies the following conditions:*

1. *$\exists(o_i, o_j) \in \mathbb{A}^* : [\exists a \in A : \exists c_a \in C_a \Rightarrow (a(o_i) - c_a)(a(o_j) - c_a) < 0$ and $\forall b \in A, b \neq a, \forall c_b \in C_b \Rightarrow (b(o_i) - c_b)(b(o_j) - c_b) > 0]$, where $\mathbb{A}^*$ is the discernibility table corresponding to $\mathbb{A}$; a is a core attribute; b is any other attribute (including a core attribute); (o_i, o_j) is a core-generating pair of objects generating a; C_a and C_b are sets of cuts on a and b respectively and $C_a \subset C$ and $C_b \subset C$.*
2. *$\forall a \in A$ such that condition 1 is true, there exists exactly 1 $(o_i, o_j) \in \mathbb{A}^*$ such that condition 1 is true.*
3. *$\exists(o_{i'}, o_{j'}) \in \mathbb{A}^* : \exists d, e \in A \wedge \exists c_d \in C_d \wedge \exists c_e \in C_e : (d(o_{i'}) - c_d)(d(o_{j'}) - c_d) < 0$ and $(e(o_{i'}) - c_e)(e(o_{j'}) - c_e) < 0$, where d, e are non-core attributes; C_d and C_e are sets of cuts on d and e respectively and $C_d \subset C$ and $C_e \subset C$ and $(o_{i'}, o_{j'})$ is a non-core-generating pair of objects.*

The concept of core-generating cuts is interpreted as follows. A set of cuts generates a number of core attributes if 1) there is at least one attribute a such that at least one of its cuts discerns a pair of objects of different decision classes and all cuts on all other attributes do not discern this pair of objects; 2) one core attribute is generated by one pair of objects only and 3) for some other pairs of objects of different decision classes, there exists at least two attributes such that for each of them there exists at least one cut discerning the pair. A set of cuts is *non-core-generating* if any of the above 3 conditions is not satisfied. A pair of objects is *core-generating* if it satisfies both conditions 1 and 2. A pair of objects is *non-core-generating* if it satisfies condition 3.

The 2nd condition eliminates the case where a core attribute could be generated by more than one pair of objects. This case implies that some objects of different classes could consist of identical attributes values after discretization, so the discretization process could transform the original consistent decision table to an inconsistent one. This case is illustrated by the example decision table in Table 4. The decision table contains $\{b\}$ as core which is generated by (o_1, o_2), (o_2, o_3), (o_3, o_4) and (o_1, o_4). However, o_1 and o_3 consist of identical attribute values; o_2 and o_4 also consist of identical attribute values.

A core-generating set of cuts C is *s-core-generating* if C generates s core attributes for $s < |A|$.

Table 4. An inconsistent decision table

U	a	b	c	d	e	f	Class
o_1	3	1	2	4	0	6	1
o_2	3	2	2	4	0	6	2
o_3	3	1	2	4	0	6	3
o_4	3	2	2	4	0	6	4

Definition 3. *A core-generating set of cuts C is s-core-generating for $s < |A|$ if and only if there exists some $B \subset A$, $|B| = s$ and there exists some $B' = A - B$ such that for each $b \in B$, both of conditions 1 and 2 of definition 2 are true and for each $b' \in B'$ condition 3 of definition 2 is true, where B is a core with size s and B' is the set of non-core attributes.*

Definition 4. *Given a decision table and s where $0 < s < |A|$, s-core-generating discretization problem corresponds to computing a s-core-generating set of cuts for the decision table.*

4.4   Properties of a Core-Generating Set of Cuts

Lemma 1. *A core-generating set of cuts C is consistent.*

Proof. Given that C is consistent if and only if $\partial_A = \partial_{A^C}$ (see section 2.2), the following are true if C is core-generating:

$$[\forall(o_i, o_j) \in \mathbb{A}^*] \Rightarrow \exists c_a \in C^a \wedge C^a \subset C \wedge (a(o_i) - c_a)(a(o_j) - c_a) < 0 \qquad (27)$$

and

$$([\forall(o_i, o_j) \in \mathbb{A}^*] \Rightarrow \exists c_a \in C^a \wedge C^a \subset C \wedge (a(o_i) - c_a)(a(o_j) - c_a) < 0)$$
$$\Leftrightarrow (\forall i, j[d(o_i) \neq d(o_j)] \Rightarrow \exists a^C \in A^C \wedge a^C(o_i) \neq a^C(o_j))$$
$$\Leftrightarrow Inf_{A^C}(o_i) \neq Inf_{A^C}(o_j)$$
$$\Leftrightarrow \partial_A = \partial_{A^C},$$

where $\mathbb{A}^*$ is the discernibility table corresponding to $\mathbb{A}$; C_a is the set of cuts on a; a^C and A^C are the discretized attributes corresponding to a and A respectively; $Inf_{A^C}(o_i) = \{(a, a(o_i)) : a \in A^C \wedge o_i \in U\}$ and ∂_A and ∂_{A^C} are the generalized decision of A and A^C respectively.

Hence, a core-generating set of cuts C is consistent.

Lemma 2. *A consistent set of cuts C is not necessarily core-generating.*

Proof. If a consistent set of cuts C satisfies the following condition, C is consistent and non-core-generating, because C does not satisfy conditions 1 and 2 of definition 2.

$$[\forall(o_i, o_j) \in \mathbb{A}^*, \exists C_a, C_b \subset C \wedge \exists c_a \in C_a \wedge \exists c_b \in C_b]$$
$$\Rightarrow (a(o_i) - c_a)(a(o_j) - c_a) < 0 \wedge (b(o_i) - c_b)(b(o_j) - c_b) < 0.$$

Therefore, for a decision table $\mathbb{A}$, the set of all the core-generating sets of cuts is a subset of the set of all the consistent sets of cuts.

Lemma 3. *The set of all possible cuts on all attributes of a decision table always contain core-generating sets of cuts.*

Proof. We prove that C_A (the set of all possible cuts on A) have the following properties:

1. For each pair of objects belonging to different decision classes, $\exists C_a \subset C_A$ such that C_a contains at least 1 cut discerning the pair of objects.
2. For each pair of objects belonging to different decision classes, $\exists C_b \subset C_A$ where $b \neq a$ such that C_b contains at least 1 cut that does not discern this pair of objects.

To prove property 1, we have the following is true:

$$\forall i, j \, [d(o_i) \neq d(o_j)] \Rightarrow \exists a \in A \wedge a(o_i) \neq a(o_j)$$
$$\Leftrightarrow \exists c \in C_a \wedge (a(o_i) - c)(a(o_j - c) < 0, \tag{28}$$

where $C_a \subset C_A$ is the set of cuts on an attribute a.

To prove property 2, we have the following is true:

$$\exists c \in C_b \wedge [\exists [v_i, v_{i+1}) \subset b(U) \wedge c \notin [v_i, v_{i+1})]$$
$$\Leftrightarrow (b(o_i) - c)(b(o_{i+1}) - c) > 0, \tag{29}$$

where $U = \{o_1, o_2, \ldots, o_n\}$; $b(U) = \{b(o) : o \in U\} = \{v_1, v_2, v_3, \ldots, v_n\}$; $v_1 < v_2 < v_3 \ldots < v_n$ and $c = \frac{v_k + v_{k+1}}{2}$ for some $k \in \{1, 2, 3, \ldots, n-1\}$.

Therefore, a number of cuts which satisfy the conditions of definition 1 can be chosen to create a core-generating set of cuts.

5 Formulating S-Core-Generating Discretization Using Constraint Satisfaction

In order to generate a decision table containing a core of size s, s pairs of objects within the discernibility table $\mathbb{A}^*$ corresponding to the decision table must satisfy conditions 1 and 2 of definition 2; $(n - s)$ pairs must satisfy condition 3 of definition 2 where n is the total number of pairs within $\mathbb{A}^*$. The s-core-generating discretization problem can be modelled as a constraint satisfaction problem (CSP) that consist of constraints over Boolean variables.

5.1 The CSP Model

The variables of the CSP are b_i^a, B_a^p, B'^p and $B_a'^p$, where b_i^a represents the selection of the ith cut on attribute a; B_a^p represents whether the pair p of $\mathbb{A}^*$

generates a as a core attribute; B'^p represents whether the pair p is non-core-generating and $B_a'^p$ represents whether one cut on a that discerns the pair p, is selected. The domain of each of the variables is $\{0, 1\}$.

The constraints are the following. The 1st condition of definition 2 is expressed by the following constraint:

$$B_a^p \Leftrightarrow \sum_i b_i^a d_i^{(a,p)} \geq 1 \wedge \sum_{a' \neq a} \sum_j b_j^{a'} d_j^{(a',p)} = 0, \tag{30}$$

where a is a core attribute; $d_i^{(a,p)}$ is the entry of the discernibility table $\mathbb{A}^*$ corresponding to the pair p and the ith cut of a; a' is any other attribute. The 2nd condition of definition 2 is expressed by the following constraint:

$$\sum_{p \in P} B_a^p \leq 1, \forall a \in A, \tag{31}$$

where P is the set of all pairs of $\mathbb{A}^*$. The 3rd condition of definition 1 is expressed by the following 2 constraints:

$$B_a'^p \Leftrightarrow \sum_i b_i^a d_i^{(a,p)} > 0, \forall a \in A, \tag{32}$$

$$B'^p \Leftrightarrow \sum_{a \in A} B_a'^p > 1, \forall p \in P, \tag{33}$$

where $\Leftrightarrow$ is the reified (if and only if) constraint[1]. The following constraint enforces the restriction that a pair can either be core-generating or non-core-generating but not both:

$$\sum_{a \in A} B_a^p + B'^p = 1, \forall p \in P, \tag{34}$$

where A is the set of all attributes. The following constraint imposes the restriction that s core-generating pairs must be generated by the cuts:

$$\sum_{p \in P} \sum_{a \in A} B_a^p = s, \tag{35}$$

where s is the size of the core to be produced by discretization. The following constraint enforces the restriction that $(n - s)$ non-core-generating pairs must be produced by the cuts:

$$\sum_{p \in P} B'^p = n - s, \tag{36}$$

where n is the total number of pairs of the discernibility table $\mathbb{A}^*$.

---

[1] A reified constraint (C1⇔C2) connects 2 constraints C1 and C2 and expresses the fact that C1 is satisfied if and only if C2 is satisfied [20]. Reified constraints can be used to impose constraints or their negations.

The CSP (Model-1) modelling the s-core-generating discretization problem is the following:

$$B_a^p \Leftrightarrow \sum_i b_i^a d_i^{(a,p)} > 0 \wedge \sum_{a' \neq a} \sum_j b_j^{a'} d_j^{(a',p)} = 0,$$
$$\forall a \in A, \forall p \in P. \tag{37}$$

$$B_a'^p \Leftrightarrow \sum_i b_i^a d_i^{(a,p)} > 0, \forall a \in A, \forall p \in P. \tag{38}$$

$$B'^p \Leftrightarrow \sum_{a \in A} B_a'^p > 1, \forall p \in P. \tag{39}$$

$$\sum_{a \in A} B_a^p + B'^p = 1, \forall p \in P. \tag{40}$$

$$\sum_{p \in P} B_a^p \leq 1, \forall a \in A. \tag{41}$$

$$\sum_{p \in P} \sum_{a \in A} B_a^p = s. \tag{42}$$

$$\sum_{p \in P} B'^p = n - s. \tag{43}$$

Constraints 38, 39 and 40 can be replaced by the following constraint:

$$\sum_{a \in A} B_a^p \leq 1, \forall p \in P, \tag{44}$$

because constraint 44 enforces the restriction that each pair p can either be a core-generating pair or a non-core-generating pair, but not both. Constraint 43 becomes redundant and can be removed. An equivalent simplified model (Model-2) is the following:

$$B_a^p \Leftrightarrow \sum_{i \in \{1,2,\ldots,k\}} b_i^a d_i^{(a,p)} > 0 \wedge \sum_{a' \neq a} \sum_{j \in \{1,2,\ldots,m\}} b_j^{a'} d_j^{(a',p)} = 0, \quad (45)$$
$$\forall a \in A, \forall p \in P.$$

$$\sum_{a \in A} B_a^p \leq 1, \forall p \in P. \tag{46}$$

$$\sum_{p \in P} B_a^p \leq 1, \forall a \in A. \tag{47}$$

$$\sum_{p \in P} \sum_{a \in A} B_a^p = s. \tag{48}$$

However, constraint 46 becomes redundant because for each pair p, if $B_a^p = 1$ for some $a \in A$, then $B_{a'}^p = 0$ for any other $a' \neq a$, $a' \in A$, due to constraint 45, so constraint 46 is always satisfied. Model-3 is the following:

$$B_a^p \Leftrightarrow \sum_{i \in \{1,2,\ldots,k\}} b_i^a d_i^{(a,p)} > 0 \wedge \sum_{a' \neq a} \sum_{j \in \{1,2,\ldots,m\}} b_j^{a'} d_j^{(a',p)} = 0,$$
$$\forall a \in A, \forall p \in P. \tag{49}$$

$$\sum_{p \in P} B_a^p \leq 1, \forall a \in A. \tag{50}$$

$$\sum_{p \in P} \sum_{a \in A} B_a^p = s. \tag{51}$$

Setting the core size s to a value v where $v < |A|$ and assigning B_a^ps and b_i^as with Boolean values corresponds to finding a s-core-generating set of cuts.

6 Core-Generating Approximate Minimum Entropy Discretization

Core-generating approximate minimum entropy (C-GAME) discretization is to compute a s-core-generating set of cuts which contains as many RMEP-selected cuts as possible.

6.1   Degree of Approximation of Minimum Entropy

The degree of approximation of minimum entropy (DAME) of a set of cuts is the ratio of the number of minimum entropy cuts that C contains to the total number of cuts selected by RMEP:

$$M1(B) = \frac{|\{b : b \in B \cap C_{mini}\}|}{|C_{mini}|}, \tag{52}$$

where B is a s-core-generating set of cuts and C_{mini} is the set of cuts selected by RMEP. However, the cuts of minimum entropy have different significance in classification. In respect of this, weights expressing the importance of a cut can also be incorporated. For each attribute, the cut selected first by RMEP has the smallest entropy value and is the most important; the one selected last by RMEP has the largest entropy value and is the least important. Weights of cuts can be designed to be inversely proportional to the order in which RMEP selects cuts. The weight of the non-RMEP cuts is defined to be 0. This leads to the following modified measure for DAME:

$$M2(B) = \frac{\sum_{b \in B \cap C_{mini}} w(b)}{\sum_{c_{mini} \in C_{mini}} w(c_{mini})}, \tag{53}$$

where the weight of a cut on an attribute is defined as follows:

$$w(c_a^i) = \frac{\left(\sum_{c_a^i \in C_a} order(c_a^i)\right) - order(c_a^i)}{\sum_{c_a^i \in C_a} order(c_a^i)}, \tag{54}$$

where $order(c_a^i)$ is the order of selection of c_a^i by RMEP. DAME is maximized during finding a core-generating set of cuts.

Definition 5. *A C-GAME set of cuts is a s-core-generating set of cuts with the maximum DAME.*

Definition 6. *Given a decision table $(U, A \cap \{d\})$ and a core of size s where $0 < s < |A|$, C-GAME discretization problem corresponds to computing the s-core-generating set of cuts with the maximum DAME generating the core.*

6.2   Computing a C-GAME Set of Cuts by Solving a CSOP

In order to generate a decision table containing a core of size s, s pairs of objects within the discernibility table $\mathbb{A}^*$ corresponding to the decision table must satisfy conditions 1 and 2 of definition 2; (n-s) pairs must satisfy condition 3 of definition 2 where n is the total number of pairs within $\mathbb{A}^*$. Let Model-4 be the CSOP that models C-GAME. Model-4 consists of constraints 55, 56 and 57 and objective function 58:

$$\sum_i b_i^a d_i^{(a,p)} > 0 \wedge \sum_{a' \neq a} \sum_j b_j^{a'} d_j^{(a',p)} = 0, \tag{55}$$
$$\forall a \in Core, |Core| = s, \forall p \in P, |P| = s.$$

$$\sum_i b_i^a d_i^{(a,p)} > 0 \wedge \sum_i b_i^{a'} d_i^{(a',p)} > 0, \tag{56}$$

$$\exists a, a' \in A, \forall p \in P', |P'| = n - s.$$

$$\bigwedge_a \{\sum_i b_i^a d_i^{(a,p'')} = 0\} \wedge \sum_i b_i^b d_i^{(b,p'')} > 0 \wedge \sum_i b_i^c d_i^{(c,p'')} > 0, \tag{57}$$

$$\forall a \in Core, \exists b, c \in NonCore, p'' \in P''.$$

$$maximize : M2 = \frac{\sum_{a \in A} \sum_i b_i^a w(c_i^a)}{\sum_{c_{mini} \in C_{mini}} w(c_{mini})}, \tag{58}$$

where b_i^a is the Boolean variable with domain $\{0,1\}$ representing the selection of the ith cut on attribute a i.e. c_i^a; $Core$ and $NonCore$ are sets of core attributes and non-core attributes; P is the set of core-generating pairs; $P' \cup P''$ is the set of non-core-generating pairs such that $P' \cap P'' = \emptyset$; $d_i^{(a,p)}$ is the entry of the discernibility table $\mathbb{A}^*$ corresponding to the pair p and the ith cut on a; $w(c_i^a)$ is the weight of c_i^a; M2 is the DAME measure (53) and C_{mini} is the set of cuts selected by RMEP. Constraint (55) expresses conditions 1 and 2 of definition 2. Constraint (56) expresses condition 3 of definition 2. Constraint (57) eliminates the case that only 1 reduct will be present in the discretized data. Assigning the b_i^as to values such that the objective function (58) reaches a maximium is equivalent to computing a C-GAME set of cuts. B&B can be used to solve Model-4.

6.3 Effect of Different Cores on Classification Performances of Reducts

In order to justify that core attributes may critically affect the classification performance of reducts, different cores are generated by solving Model-4 using B&B. Then, for each discrete dataset, some reducts are found using a genetic algorithm (GA) [38]. Thereafter, classification performances of the reducts are compared. Three datasets: Australian, SPECTF and Wdbc are chosen from the UCI Repository [10]. Each of the datasets was split into training and testing sets as follows. For Australian, the training set consists of 460 instances and the testing set consists of 230 instances. For SPECTF, 80 instances were used as training instances and the remaining 269 instances as testing instances. For Wdbc, the training set consists of 190 instances and the testing set contains 379 instances.

Australian

1. The 1st attribute selected by C4.5 $A8$, is chosen as core. Thereafter, a C-GAME set of cuts is computed by solving Model-4 using B&B. Thereafter, GA is run on the discrete dataset 4 times (Table 5). The training dataset is reduced using each reduct and a decision tree is induced from the reduced dataset. Patterns of the testing dataset are classified using the induced decision tree. Let r_i $(i = 1 \ldots 4)$ denote the ratio of the reducts that have higher accuracies than C4.5 to the total number of reducts found during i^{th} GA run. For each run of GA, r_i is less than 1. The intersection of all those reducts that outperform C4.5 is {A8,A2}.

2. A new C-GAME set of cuts is computed to generate {A8,A2} as core. After running GA for four times on the discrete data (Table 5), four new sets of reducts are found. Additionally, the highest quality reducts (accuracy = 88.3%) also outperform the highest quality reducts (accuracy = 86.5%) of step 1. The intersection of all those reducts outperforming C4.5 is {A8,A2,A13}.

3. Another C-GAME set of cuts is computed to generate {A8,A2,A13} as core. GA is run four times on the discrete data (Table 5). For each run of GA, all the found reducts outperform C4.5. Moreover, the highest quality reducts (accuracy = 88.7%) outperform the highest quality reducts (accuracy = 88.3%) of step 2.

Table 5. Generating different cores for Australian

Step	Core	GA Run	No. of Reducts	Reducts Sizes	Ratio > C4.5	Reducts Accuracies
1	A8	1	31	3 - 6	25/31=0.81	80 - 85.7
		2	33	3 - 6	27/33=0.81	80 - 86.5
		3	33	3 - 7	29/33=0.88	81.3 - 86.5
		4	25	3 - 7	22/25=0.88	80.9 - 86.5
2	A8,A2	1	33	4 - 7	27/33=0.81	80 - 87.4
		2	41	4 - 7	34/41=0.83	80 - 87.4
		3	32	4 - 7	28/32=0.88	80 - 88.3
		4	40	4 - 7	34/40=0.85	80 - 88.3
3	A8,A2,A13	1	18	4 - 8	18/18=1.0	83 - 88.7
		2	17	4 - 8	17/17=1.0	83 - 88.3
		3	11	4 - 8	11/11=1.0	83 - 87.8
		4	13	4 - 8	13/13=1.0	83 - 87.8

SPECTF

1. The 1st five decision tree nodes that are traversed in a breadth first order, are chosen as core. A C-GAME set of cuts is computed to generate this core. GA is run four times on the discrete dataset (Table 6). For each GA run, r_i ($i \in 1, 2, 3, 4$) is less than 1. The intersection of all the reducts that outperform C4.5 contains A41, A40, A20, A17 and A4.

2. {A41,A40,A20,A17,A4} is used as the new core. A C-GAME set of cuts is computed to generate the new core. GA is run four times on the discrete dataset. For each run, r_i equals 1 (Table 6).

Wdbc

1. The 1st three decision tree nodes that are traversed in a breadth first traversal order: A28, A24 and A2, are used as core. A C-GAME set of cuts is computed to generate this core. GA is run on the discretized dataset four times (Table 7). For each GA run, r_i is less than 1. The intersection of all the reducts that outperform C4.5, contains A28, A24, A2 and A1.

Table 6. Generating different cores for SPECTF

Step	Core	GA Run	No. of Reducts	Reduct Size	Ratio > C4.5	Reducts Accuracies (%)
1	A41,A40,A20	1	65	6 - 7	9/65	74.4 - 81
	A17,A4	2	69	6 - 7	11/69	74.7 - 81
		3	51	6 - 7	3/51	75.5 - 79.6
		4	66	6 - 7	12/66	75.5 - 81
2	A41,A40,A20 A17,A4,A5,A44	1	7	8	7/7	78.4 - 81

2. {A28,A24,A2,A1} is used as the new core. A C-GAME set of cuts is computed to generate this core. GA is run four times on the discrete dataset (Table 7). For each run of the GA, r_i equals 1 (Table 7).

Table 7. Generating different cores for Wdbc

Step	Core	GA Run	No. of Reducts	Reduct Size	Ratio > C4.5	Reducts Accuracies (%)
1	A28,A24,A2	1	28	4 - 5	1/28	82.1 - 92.6
		2	22	4	1/22	82.6 - 92.6
		3	26	4 - 5	1/26	82.1 - 92.6
		4	25	4 - 5	1/25	82.1 - 92.6
2	A28,A24,A2 A1	1	23	5 - 6	23/23	91.8 - 93.4
		2	24	5 - 6	24/24	91.8 - 93.4
		3	24	5 - 6	24/24	91.8 - 93.4
		4	22	5 - 6	22/22	91.8 - 93.4

The above results indicate that as core varies, classification accuracies of reducts also varies.

6.4   C-GAME Discretization Algorithm

C-GAME discretization algorithm repeatedly discretizes a training dataset by solving Model-4 until most of the reducts would outperform C4.5 for a given testing dataset (Fig 1).

6.5   Performance Evaluation

Ten datasets of different dimensionalities were chosen from the UCI Repository [10] (Table 8). SPECTF, Water, Ionosphere and Wdbc datasets had been split into the training and testing datasets as shown by their providers. Each of the remaining six datasets was split into 2/3 training and 1/3 testing datasets.

Comparison of Reducts Sizes. C-GAME, RMEP, ChiMerge, Boolean Reasoning (BR) and Equal Frequency (EF) result in different reducts sizes for

Algorithm: C-GAME Discretization
Input: Training dataset D_1, Testing dataset D_2
Output: C-GAME-discrete training dataset D_1'
M, dataset dimensionality
$CORE$, a set of core attributes
$RED(i)$, the set of all the reducts found by GA during i^{th} run
$RED'(i)$, the set of reducts which have higher accuracies than C4.5 on D_2 and $RED'(i) \subseteq RED(i)$
B_j, the jth reduct $\forall j \in \{1, \ldots |RED'(i)|\}$ and $B_j \in RED'(i)$
$r_i = \frac{|RED'(i)|}{|RED(i)|}$
C, a C-GAME set of cuts generating $CORE$
1. Induce a decision tree T from D_1 using C4.5;
2. Start at the root node and select N nodes in a breadth first traversal order
3. such that $\frac{N}{M} \approx 0.1$;
4. Let $CORE$ be the set of the N selected nodes;
5. Solve Model-4 using B&B to obtain C and use C to discretize D_1 to obtain D_1';
6. Run GA on D_1' for k times ($k \geq 1$) to obtain $RED_{All} = \{RED(i)|i = 1 \ldots k\}$;
7. Compute $R = \{r_i|i = 1 \ldots k\}$;
8. **if** (for each $i \in \{1 \ldots k\}$, $r_i \approx 1$);
9. **then** return D_1';
10.**else** {**for** ($m \in \{1, \ldots, k\}$)**do**{
11. $I_m = \bigcap_{j=1}^{|RED'(i)|} B_j;$} /* end for */
12. $CORE = \bigcap_{m=1}^{k} I_m;$
13.   Go to step 5;}

Fig. 1. C-GAME Discretization Algorithm

training datasets. C4.5 selects nodes i.e. features during decision tree learning. The selected features by C4.5 forms a reduct. For each of the methods including C4.5, the average reduct size of each of the methods over the ten datasets is computed as follows

$$S_{average} = \frac{\sum_{i \in D} size_i}{|D|}, \tag{59}$$

where D is a set of datasets; i denotes the ith dataset; $size_i$ is the reduct size of dataset i. The average reduct size of each of the methods is illustrated in Table 9. C4.5 leads to the largest average reduct size among all the six methods. Chimerge leads to the smallest average reduct size among all the six methods. C-GAME leads to a medium average reduct size among all the six methods. For the six high dimensional datasets: Sonar, SPECTF, Water, Ionosphere, Wdbc and Wpbc, C-GAME and C4.5 lead to the largest or the 2nd largest reduct sizes among all the six methods (Table 10). For Australian, Cleveland and Hungarian, C-GAME leads to the 2nd smallest reduct sizes among the six methods. For these three datasets, C4.5 is in the top three largest reduct sizes among all the six methods. C-GAME leads to the 2nd smallest reduct size. For Housing, C-GAME leads to the 2nd largest reduct size and C4.5 leads to the 3rd largest reduct size.

Classification Results. In the training phase, C-GAME discretization algorithm was firstly applied; then, Johnson's algorithm that finds a smallest reduct

Table 8. Datasets description

	Dataset	Attributes	Train	Test	Classes
1.	Sonar	60	139	69	2
2.	SPECTF	44	80	269	2
3.	Water	38	260	261	2
4.	Ionosphere	34	105	246	2
5.	Wdbc	30	190	379	2
6.	Wpbc	33	129	65	2
7.	Australian	14	460	230	2
8.	Cleveland	13	202	101	2
9.	Hungarian	13	196	98	2
10.	Housing	13	338	168	2

Table 9. Comparison of average sizes of reducts

C-GAME	RMEP	ChiMerge	BR	EF	C4.5
5.7	4.8	2.1	3.9	7.5	7.6

Table 10. Reducts Sizes

Dataset	C-GAME	RMEP	ChiMerge	BR	EF	C4.5
Sonar	8	2	1	1	6	8
SPECTF	8	3	3	2	5	7
Water	5	3	1	1	6	5
Ionospshere	5	1	1	2	5	5
Wdbc	5	2	4	1	6	6
Wpbc	4	2	1	1	7	9
Australian	4	12	3	12	11	10
Cleveland	6	11	3	10	9	11
Hungarian	5	10	3	10	9	8
Housing	8	2	1	1	11	7

and C4.5 were applied in sequence (Fig 2). Thereafter, a testing dataset was firstly reduced using the Johnson-found reduct; then the testing patterns were classified using the learnt decision tree (Fig 3). This integrated application of C-GAME, Johnson's algorithm and C4.5 outperformed C4.5 on the ten datasets. Moreover, C-GAME outperformed four other discretization methods: RMEP, Boolean Reasoning (BR), ChiMerge and Equal Frequncy (EF) for the majority of the ten datasets (Table 11). The performances of the four methods are lower than that of C4.5 on most of the ten datasets (Table 11). The four discretization methods are non-core-generating. Therefore, for some datasets, reducts do not contain any significant attributes, so the decision trees induced from the reduced datasets do not contain significant attributes. In contrast, C-GAME generates

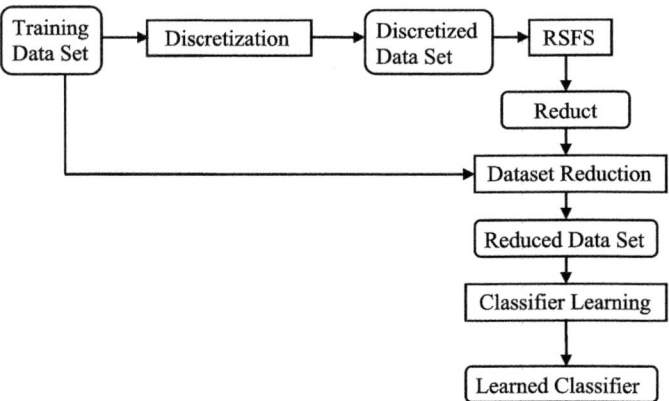

Fig. 2. Integration Framework

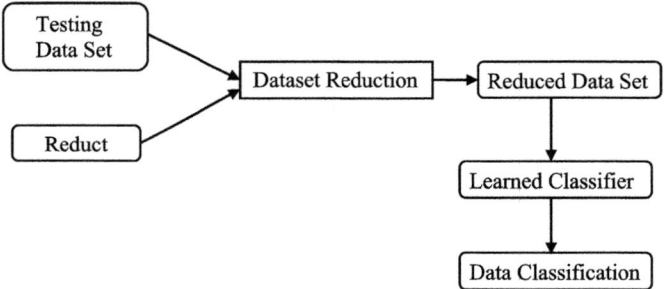

Fig. 3. Patterns Classification

a core that consists of significant attributes, so the reducts would also contain significant attributes. Consequently, decision trees would contain significant attributes. The ChiMerge integration approach has a lower accuracy than C4.5 on eight datasets. The Boolean Reasoning integration approach has a lower accuracy than that of C4.5 on six datasets. The Equal Frequency integration approach has a lower accuracy than that of C4.5 on six datasets. The RMEP integration has a worse performance than C4.5 on six datasets.

For each of the dataset, a multi-layer perceptrons (MLP) with a single hidden layer of 10 nodes and 2 output nodes was also applied in place of C4.5 (Table 12). For eight datasets, C-GAME outperforms the MLPs. For the remaining two datasets, C-GAME has the same accuracy as the MLPs. For seven datasets, C-GAME outperforms the other four discretization methods. The performances of the four discretization methods are lower than those of MLPs for most of the datasets. The results show that generating cores during discretization could lead to higher a accuracy than that of the MLP.

Table 11. Classification Accuracies with C4.5

Data set		C-GAME	C4.5	RMEP	BR	Chimerge	EF
1.	Sonar	87	84.1	50.7	72.5	72.5	84.1
2.	SPECTF	81	77.3	20.4	20.4	51.7	20.4
3.	Water	98.9	98.1	96.9	97.3	97.3	97.3
4.	Ionosphere	91.1	81.7	82.9	76.8	82.9	87.8
5.	Wdbc	92.6	91.3	69.1	69.1	88.1	69.1
6.	Wpbc	73.8	64.6	73.8	73.8	73.8	63.1
7.	Australian	83	82.6	83.9	84.3	59.6	83.9
8.	Cleveland	77.2	76.2	78.2	77.2	69.3	73.3
9.	Hungarian	85.7	79.6	79.2	85.7	68.4	77.6
10.	Housing	83.3	82.1	65.2	83.3	65.5	82.7

Table 12. Classification Accuracies with MLPs

Data sets		C-GAME	MLP	RMEP	BR	Chimerge	EF
1.	Sonar	88.4	87	56.5	71	71	78.3
2.	SPECTF	78.4	76.2	56.1	58.7	63.9	75.8
3.	Water	98.8	97.7	97.3	97.3	97.3	97.3
4.	Ionosphere	90.7	86.2	81.3	56.1	87.4	86.6
5.	Wdbc	7.9	6.9	6.4	6.3	7.9	7.9
6.	Wpbc	84.6	81.5	73.8	73.8	73.8	63.1
7.	Australian	86.5	81.7	84.3	67.4	83.9	82.1
8.	Cleveland	77.2	77.2	80.2	77.2	78.2	76.2
9.	Hungarian	83.7	80.6	83.6	79.6	81.6	83.6
10.	Housing	85.1	85.1	64.9	82.7	61.3	87.5

6.6   Ten-Fold Cross Validation

For the six datasets consisting of less than four hundred instances i.e. Sonar, SPECTF, Ionosphere, Wpbc, Cleveland and Hungarian, 10-fold cross validation experiments were performed (Table 13). C-GAME outperformed C4.5 and the other discretization methods for each dataset.

Table 13. 10-fold Cross Validation Accuracies

Data set		C-GAME	C4.5	RMEP	BR	Chimerge	EF
1.	Sonar	76.3	70.3	55.7	65.8	68.6	75.5
2.	SPECTF	88.2	84.6	80.8	71.4	81.1	81.7
3.	Ionosphere	90.9	89.2	81.7	84.9	82.3	88.3
4.	Wpbc	76.8	71.2	76.2	76.2	76.2	71.6
5.	Cleveland	81.2	79.6	80.2	77.9	75.9	79.6
6.	Hungarian	82.3	79.6	77.9	75.8	77.6	78.3

7   Conclusion

Based on the results, for a given dataset, generating a high-quality core using C-GAME discretization could lead to higher classification performance than the performance of the original classifiers - C4.5 and MLPs in this case - and the performances of non-core-generating discretization methods. In terms of dimensionality reduction, C-GAME achieves at least an average performance compared with non-core-generating discretization methods; C4.5 achieves a poor performance for dimensionality reduction compared with discretization methods. Higher dimensional datasets such as text datasets, computer networks intrusion datasets and micro array gene expression datasets could be used to further evaluate the performance of C-GAME.

Acknowledgements

This research was supported by an UK Engineering Physical Science Research Council (EPSRC) Doctoral Training Account.

References

1. Apt, K., Wallace, M.: Constraint Logic Programming using ECLiPSe. Cambridge University Press, Cambridge (2007)
2. Bazan, J.G., Nguyen, H.S., Nguyen, S.H., Synak, P., Wroblewski, J.: Rough set algorithms in classification problems. In: Polkowski, L., et al. (eds.) Rough Set Methods and Applications: New Developments in Kownledge Discovery in Information Systems, pp. 49–88. Physica-Verlag, Heidelberg (2000)
3. Chmielewski, M.R., Grzymala-Busse, J.W.: Global Discretization of Continuous Attributes as Preprocessing for Machine Learning. International Journal of Approximate Reasoning 15(4), 319–331 (1996)
4. Chai, D., Kuehlmann, A.: A Fast Pseudo-Boolean Constraint Solver. IEEE Transactions on Computer-Aided Design of Integrated Circuits and Systems 24(3), 305–317 (2005)
5. Dougherty, J., Kohavi, R., Sahami, M.: Supervised and Unsupervised Discretization of Continuous Features. In: Proceedings of the Twelfth International Conference on Machine Learning, San Francisco, CA, pp. 194–202 (1995)
6. Fayyad, M.U., Irani, B.K.: Multi-interval discretization of continuous-valued attributes for classification learning. In: Proceedings of the 13th International Joint Conference on Artificial Intelligence, pp. 1022–1027 (1993)
7. Fayyad, M.U.: On the Handling of Continuous-Valued Attributes in Decision Tree Generation. Machine Learning 8(1), 87–102 (1992)
8. Gama, J., Torgo, L., Soares, C.: Dynamic Discretization of Continuous Attributes. In: Proceedings of the Sixth Ibero-American Conference on AI, pp. 160–169 (1998)
9. Han, J., Hu, X., Lin, Y.T.: Feature Subset Selection Based on Relative Dependency between Attributes. In: Tsumoto, S., Słowiński, R., Komorowski, J., Grzymała-Busse, J.W. (eds.) RSCTC 2004. LNCS (LNAI), vol. 3066, pp. 176–185. Springer, Heidelberg (2004)

10. Hettich, S., Blake, L.C., Merz, J.C.: UCI Repository of machine learning databases, University of California, Irvine, Dept. of Information and Computer Sciences (1998), http://www.ics.uci.edu/~mlearn/MLRepository.html
11. Hentenryck, V.: Constraint Satisfaction in Logic Programming. MIT Press, Cambridge (1989)
12. Jiao, N., Miao, D.: An efficient gene selection algorithm based on tolerance rough set theory. In: Sakai, H., Chakraborty, M.K., Hassanien, A.E., Ślęzak, D., Zhu, W. (eds.) RSFDGrC 2009. LNCS, vol. 5908, pp. 176–183. Springer, Heidelberg (2009)
13. Jensen, R., Shen, Q.: Tolerance-based and Fuzzy-rough Feature Selection. In: Proceedings of the 16th IEEE International Conference on Fuzzy Systems, pp. 877–882 (2007)
14. Jensen, R., Shen, Q.: Are More Features Better? A response to Attributes Reduction Using Fuzzy Rough Sets. IEEE Transactions on Fuzzy Systems 17(6), 1456–1458 (2009)
15. Jensen, R., Shen, Q.: New Approaches to Fuzzy-Rough Feature Selection. IEEE Transactions on Fuzzy Systems 17(4), 824–838 (2009)
16. Johnson, S.D.: Approximation algorithms for combinatorial problems. Journal of Computer and System Sciences 9, 256–278 (1974)
17. Jensen, R., Shen, Q.: Semantics-preserving dimensionality reduction: Rough and fuzzy-rough-based approaches. IEEE Transactions On Knowledge and Data Engineering 16(12), 1457–1471 (2004)
18. Kohavi, R., Sahami, M.: Error-based and Entropy-based Discretization of Continous Features. In: Proceedings of the Second International Conference on Knowledge Discovery and Data Mining, Portland, Oregon, pp. 114–119 (1996)
19. Liu, H., Hussain, F., Tan, L.C., Dash, M.: Discretization: An Enabling Technique. Data Mining and Knowledge Discovery 6(4), 393–423 (2002)
20. Marriot, K., Stuckey, J.P.: Programming with Constraints: an Introduction. MIT Press, Cambridge (1998)
21. Marcus, S.: Tolerance rough sets, Cech topologies, learning processes. Bull. Polish Academy of Sciences, Technical Sciences 42(3), 471–487 (1994)
22. Nguyen, H.S., Skrowron, A.: Quantization of real values attributes, Rough set and boolean reasoning approach. In: Proceedings of the Second Joint Annual Conference on Information Sciences, Wrightsville Beach, North Carolina, pp. 34–37 (1995)
23. Nguyen, S.H., Nguyen, H.S.: Some efficient algorithms for rough set methods. In: Proceedings of the Conference of Information Processing and Management of Uncertainty in Knowledge-Based Systems IPMU 1996, Granada, Spain, pp. 1451–1456 (1996)
24. Nguyen, H.S.: Discretization Problem for Rough Sets Methods. In: Polkowski, L., Skowron, A. (eds.) RSCTC 1998. LNCS (LNAI), vol. 1424, pp. 545–552. Springer, Heidelberg (1998)
25. Nguyen, S.H.: Discretization of Real Value Attributes: Boolean Reasoning Approach. Ph.D. Thesis, Warsaw University, Warsaw, Poland (1997)
26. Parthalain, M.N., Jensen, R., Shen, Q., Zwiggelaar, R.: Fuzzy-rough approaches for mammographic risk analysis. Intelligent Data Analysis 14(2), 225–244 (2010)
27. Polkowski, L.: Rough Sets. Mathematical Foundations. Physica–Verlag, Heidelberg (2002)
28. Peters, J.F.: Tolerance near sets and image correspondence. Int. J. Bio-Inspired Computation 1(4), 239–245 (2009)
29. Peters, J.F.: Corrigenda, Addenda: Tolerance near sets and image correspondence. Int. J. Bio-Inspired Computation 2(5) (in press, 2010)

30. Quinlan, R.J.: C4.5: Programs for Machine Learning. Morgan Kaufmann, San Mateo (1993)
31. Skowron, A., Rauszer, C.: The discernibility matrices and functions in information systems. In: Slowinski, R., et al. (eds.) Intelligent Decision Support: Handbook of Applications and Advances of the Rough Set Theory, pp. 331–362. Kluwer Academic Publisher, Dordrecht (1992)
32. Shang, C., Shen, Q.: Rough Feature Selection for Neural Network Based Image Classification. International Journal of Image and Graphics 2(4), 541–555 (2002)
33. Shen, Q., Chouchoulas, A.: A rough-fuzzy approach for generating classification rules. Pattern Recognition 35(11), 341–354 (2002)
34. Swiniarski, W.R., Skowron, A.: Rough set methods in feature selection and recognition. Pattern Recognition Letters 24(6), 833–849 (2003)
35. Skowron, A., Stephaniuk, J.: Tolerance approximation spaces. Fundamenta Informaticae 27, 245–253 (1996)
36. Stepaniuk, J., Kretowski, M.: Decision systems based on tolerance rough sets. In: Proc. 4th Int. Workshop on Intelligent Information Systems, Augustow, Poland, pp. 62–73 (1995)
37. Tsang, E.P.K.: Foundations of Constraint Satisfaction. Academic Press Limited, London (1993)
38. Vinterbo, S., Ohrn, A.: Minimal approximate hitting sets and rule templates. International Journal of Approximate Reasoning 25(2), 123–143 (2000)
39. Zhong, N., Dong, J., Ohsuga, S.: Using Rough Sets with Heuristics for Feature Selection. Journal of Intelligent Information Systems 16(4), 199–214 (2001)

Perceptual Tolerance Intersection

Piotr Wasilewski[1], James F. Peters[1], and Sheela Ramanna[1,2]

[1] Computational Intelligence Laboratory,
Department of Electrical & Computer Engineering, Univ. of Manitoba,
75A Chancellor's Circle, Winnipeg, Manitoba R3T 5V6, Canada
{piotr,jfpeters}@ee.umanitoba.ca
[2] Department of Applied Computer Science
University of Winnipeg
Winnipeg, Manitoba R3B 2E9, Canada
s.ramanna@uwinnipeg.ca

Abstract. This paper elaborates on the introduction of perceptual tolerance intersection of sets as an example of a near set operation. Such operations are motivated by the need to consider similarities between digital images viewed as disjoint sets of points. The proposed approach is in keeping with work by E.C. Zeeman on tolerance spaces and visual perception and work by J.H. Poincaré on sets of similar sensations used to define representative spaces (aka *tolerance spaces*) such as visual, tactile and motile spaces. Perceptual tolerance intersection of sets is a direct consequence of recent work on near sets. The theory of perceptual set intersection has many practical applications such as a solution to the problem of how one goes about measuring the closeness of digital images. The main contribution of this article is a description-based approach to formulating perceptual set intersections between disjoint sets that resemble each other. A practical application of the proposed approach is the discovery of resemblances between sets of points in digital image regions that represent tolerance rough sets.

Keywords: Description, near sets, perceptual granules, set operations, similarity, tolerance space.

1 Introduction

This paper elaborates on the introduction of perceptual tolerance intersection of sets introduced in [1] as an example of a near set operation with a practical application in terms of the study of similarities between digital images. The proposed set operations considered in the context of tolerance spaces is directly related to work on sets of similar objects, starting with J.H. Poincaré [2] and E.C. Zeeman [3], followed by more recent studies of similarity and tolerance relations [4,5,6,7,8,9]. In general, sets are considered near each other in the case where the sets contain objects with descriptions that are similar.

The focus of this paper is on the presentation of the theory underlying perceptual tolerance intersection viewed as a near set resemblance operation. This form

J.F. Peters et al. (Eds.): Transactions on Rough Sets XIII, LNCS 6499, pp. 159–174, 2011.

of set operation takes into account the rather unusual character of near sets that can be disjoint sets that resemble each other. This resemblance between disjoint sets stems from the similarities of the descriptions of pairs of objects that are elements of the disjoint sets. This approach works well, for example, in dealing with the similarities of digital images that are viewed as disjoint sets of points. The set that is identified with the intersection of near sets is a natural outcome of the near set approach (see e.g. [10,9]). The work on near sets reported in this article is directly related to work on the relations *near* and *far* in proximity space theory (see, *e.g.*, [11,12,13]).

The paper is divided into three parts. Part one is represented by Sections 2 through 3. In Section 2, general facts about tolerance relations are presented together with Zeeman's indistinguishability of sets relation. In Section 3, a tolerance intersection of sets operation is introduced and investigated. The second part of this paper is represented by Sections 4 through 5. In Section 4, perceptual tolerance relations and perceptual tolerance intersection of sets operation are discussed. Section 5 illustrates near sets in terms of the resemblance between digital images. Section 6 provides the third part of this paper and includes a postulate on similarity measures between images which are based on perceptual tolerance relations.

2   Tolerance Relations

A relation $\tau \subseteq O \times O$ is a *tolerance on a set* O (shortly: *tolerance*, if O is understood) iff τ is reflexive and symmetric[1]. Then a pair $\langle O, \tau \rangle$ is a *tolerance space*. We denote the family of all tolerances on a set O by $Tol(O)$. Transitive tolerances are equivalence relations, i.e. $Eq(O) \subseteq Tol(O)$, where $Eq(O)$ denotes the family of all equivalences on O. An image of a set $X \subseteq O$ by a relation τ on O we denote by $\tau(X)$ (*i.e.* $\tau(X) := \{y \in O : \text{ there is } x \in X, (x, y) \in \tau\}$) with a simplifying convention where $\tau(\{x\}) = \tau(x)$. A tolerance image operator $\tau(\ )$ has some useful properties presented by the following lemma:

Lemma 1. *Let* $\langle O, \tau \rangle$ *be a tolerance space. The following conditions hold for arbitrary* $X, Y \subseteq O$:

(1) $X \subseteq \tau(X)$,   (Extensivity)
(2) $X \subseteq Y \Rightarrow \tau(X) \subseteq \tau(Y)$,   (Monotonicity)
(3) $X \subseteq Y \Rightarrow X \subseteq \tau(Y)$,
(4) $\tau(X) = \bigcup_{x \in X} \tau(x)$.

Every tolerance generates some specific coverings of a space. Two of them are mainly used. A set $A \subseteq O$ is a τ-*preclass* (or briefly *preclass* when τ is understood) if and only if for any $x, y \in A$, $(x, y) \in \tau$. The family of all preclasses

---

[1] In universal algebra or lattice theory reflexive and symmetric relations compatible with operations from a given algebra are called *tolerances*, i.e. they are generalizations of congruence relations (see e.g. [14]). We refer to such relations as *algebraic tolerances* or *algebraic tolerance relations*.

of a tolerance space is naturally ordered by set inclusion and preclasses that are maximal with respect to a set inclusion are called τ-*classes* or just *classes*, when τ is understood. The family of all classes of the space $\langle O, \tau \rangle$ is particularly interesting and is denoted by $H_\tau(O)$. The family $H_\tau(O)$ is a covering of O. However, the elements of $H_\tau(O)$ do not have to be mutually disjoint. The elements of $H_\tau(O)$ are mutually disjoint when a given tolerance τ is transitive, *i.e.*, τ is an equivalence relation. Hence, the notion of a family of tolerance classes is a natural generalization of the partition of a space.

A tolerance space $\langle O, \tau \rangle$ determines, as any relation, another family of sets, namely the family of images of elements of the space via a given tolerance relation: $\{\tau(x) : x \in O\}$. Clearly, since τ is reflexive, the family $\{\tau(x) : x \in O\}$ is a covering of a space O but it does not have to be a partition of O (analogously with the family $H_\tau(O)$, it is a partition of O when τ is transitive). However, families consisting of images of elements via tolerance relations are not natural generalizations of partitions of a space, since, for every intransitive tolerance τ on O, there is $x \in O$ and there are $a, b \in \tau(x)$ such that $(a, b) \notin \tau$. Thus one can see that images of elements with respect to a given tolerance relation are not in general tolerance classes. This holds only in the case of transitive tolerance relations, *i.e.* in the case of equivalence relations. More exactly, the following proposition holds:

Proposition 1. *Let* $\langle O, \tau \rangle$ *be a tolerance space. Then the following conditions are equivalent:*

(1) τ *is transitive,*
(2) $H_\tau(O) = \{\tau(x) : x \in O\}$,
(3) $H_\tau(O)$ *is a partition of* O,
(4) $\{\tau(x) : x \in O\}$ *is a partition of* O.

Often an image of an element x, $\tau(x)$, is called a *neighbourhood of* x while x itself is called the *centre* of $\tau(x)$. Toward the end of this section, it will become apparent that there is some reason underlying this convention. One should also note that a neighbourhood $\tau(x)$ of the element x is uniquely determined by its centre, while an element x can belong to more than one tolerance class. There is also another difference between tolerance neighbourhoods and classes that is interesting from a mathematical point of view. Families of tolerance images of elements exist for any set, finite or infinite, but this does not hold in the case of tolerance classes. If a set is finite, then, by its finiteness, every tolerance preclass is contained in some class and, in the case of an infinite set, that condition is equivalent to the Axiom of Choice in set theory [8] (in the case of algebraic tolerances on semilattices it was shown in [14] that the Axiom of Choice is equivalent to the existence of a single tolerance class). So, in general, tolerance neighbourhoods and tolerance classes are different entities.

E.C. Zeeman pointed out [3] that any tolerance relation determines in a natural way another tolerance on the subsets of the space.

Definition 1. [3] *Let $\langle O, \tau \rangle$ be a tolerance space. A relation $\sim_\tau$ on $\mathcal{P}(O)$ is defined as follows:*

$$X \sim_\tau Y \Leftrightarrow X \subseteq \tau(Y) \text{ and } Y \subseteq \tau(X)$$

X is said to be indistinguishable from Y. We refer to the relation $\sim_\tau$ as Zeeman's tolerance *or* Zeeman's indistinguishability of sets.

If a tolerance τ is treated as a formal model of similarity, then the basic intuitive interpretation given to a relation $\sim_\tau$ is that sets standing in this relation are indistinguishable with respect to a tolerance τ, as containing only mutually similar elements.

Corollary 1. *Let $\langle O, \tau \rangle$ be a tolerance space. If τ is transitive, so $\tau \in Eq(O)$, then:*

$$X \sim_\tau Y \Leftrightarrow \tau(X) = \tau(Y),$$

i.e. Zeeman's indistinguishability of sets is Z. Pawlak's upper rough equality of sets from rough set theory [15,16].

Proof. Equation $\tau(X) = \tau(Y)$ together with extensivity of the operator $\tau(\ )$ directly implies $X \sim_\tau Y$ so it is enough to prove implication $\Rightarrow$. Let $\tau \in Eq(O)$, so $\langle O, \tau \rangle$ is an approximation space while $\tau(\ )$ is an upper approximation operator [15,16]. Let $X \sim_\tau Y$, so $X \subseteq \tau(Y)$ and by monotonicity $\tau(X) \subseteq \tau(\tau(Y))$ thus $\tau(X) \subseteq \tau(Y)$ by $\tau(Y) = \tau(\tau(Y))$, one of the properties of an upper approximation operator [15,16]. Analogically for $\tau(Y) \subseteq \tau(X)$, therefore, $\tau(X) = \tau(Y)$.

One can introduce tolerance relations by defining them on the basis of pseudometric spaces. A function p on a non-empty set O is called a *pseudometric* if, and only if $p : O \times O \to [0, +\infty) \subseteq \mathbb{R}$ and the following conditions are satisfied for any $x, y, z \in O$:

(1) $p(x, x) = 0$
(2) $p(x, y) = p(y, x)$,
(3) $p(x, y) + p(y, z) \leqslant p(x, z)$.

A set O together with a pseudometric on it, is called a *pseudometric space* (denoted by $\langle O, p \rangle$). One should note that pseudometric spaces are direct generalizations of metric spaces, since there are pseudometric spaces in which for some x, y it is the case that $x \neq y$ and $p(x, y) = 0$. It is easy to see that the condition $p(x, y) = 0$ defines an equivalence relation θ_p on a space O. Thus the following holds for every $x, y \in O$: $p(x, y) = 0 \Leftrightarrow (x, y) \in \theta_p$ and this is a direct generealization of an appropriate condition for metric spaces. Using metric distance between sets, one can naturally define a metric space on the partition $O_{/\theta_p}$, where metric distance d_p is defined in the following way.

$$d_p(x_{/\theta_p}, y_{/\theta_p}) := \min\{p(a, b) : a \in x_{/\theta_p} \& b \in y_{/\theta_p}\}.$$

It follows for every $x, y \in O$ that

$$\min\{p(a, b) : a \in x_{/\theta_p} \& b \in y_{/\theta_p}\} = p(x, y),$$

because, in every pseudometric space $\langle O, p \rangle$ for all mutually different $x, y, z \in O$, it is the case that $p(x, y) = 0 \Rightarrow p(x, z) = p(y, z)$.

After the manner of E.C. Zeeman [3], every pseudometric space in a quite natural way determines tolerance relations with respect to some positive real threshold as shown by Example 1.

Example 1. Let $\langle O, p \rangle$ be pseudometric space and let $\epsilon \in (0, +\infty)$. A relation $\tau_{p,\epsilon}$ is defined for $x, y \in O$ in the following way:

$$(x, y) \in \tau_{p,\epsilon} \iff p(x, y) < \epsilon,$$

is a tolerance relation on O. Such relations we call *distance tolerance relations*.

One can show that

Proposition 2. *Let $\tau_{p,\epsilon}$ be a distance tolerance relation determined by a pseudometric space $\langle O, p \rangle$. Then, for any $x \in U$,*

$$\tau_{p,\epsilon}(x) = B_p(x, \epsilon),$$

i.e., a $\tau_{p,\epsilon}$ neighbourhood of x is just an open ball in the pseudometric space $\langle O, p \rangle$ with the centre x and radius ϵ, $B_p(x, \epsilon) := \{y \in X : p(x, y) \leq \varepsilon\}$.

Proposition 2 justifies referring to an image of the element x by any tolerance τ (not necessarily a distance tolerance) as a neighbourhood with a centre x, since in topology a named *neighbourhood of x* denotes an open ball or, as in [17], an open set containing element x.

3   Tolerance Intersection of Sets

Assuming that tolerance is a formal model of similarity, then, for any two subsets (possibly disjoint) of a tolerance space, one can ask whether the subsets contain some mutually similar elements. This motivates introducing an operation on subsets of tolerance spaces.

Definition 2. *Let $\langle O, \tau \rangle$ be a tolerance space. A tolerance intersection of sets is denoted by $\Cap_\tau$ and defined for $X, Y \subseteq O$ as follows:*

$$X \Cap_\tau Y := (X \cap \tau(Y)) \cup (Y \cap \tau(X)).$$

Let us note that disjoint sets can have a non-empty tolerance intersection as it is shown by the following example:

Example 2. Let $\langle O, \tau \rangle$ denote a tolerance space, where $O = \{a_1, a_2, b_1, b_2, c, d\}$ and $\tau := \Delta_O \cup \{(a_1, b_2), (b_2, a_1), (a_2, b_1), (b_1, a_2), (a_1, c), (c, a_1), (b_1, d), (d, b_1)\}$. Let also $A := \{a_1, a_2\}$, $B := \{b_1, b_2\}$, where Δ_O denotes the diagonal of a set O, i.e. $\Delta_O := \{(x, x) : x \in O\}$. Then, by straightforward calculations, the following equations hold:

$$\tau(A) = \{a_1, a_2, b_1, b_2, c\}, \quad \tau(B) = \{a_1, a_2, b_1, b_2, d\}.$$

Thus $A \subseteq \tau(B)$ and $B \subseteq \tau(A)$. Therefore $A \sim_\tau B$ and $A \Cap_\tau B = \{a_1, a_2, b_1, b_2\}$ but $A \cap B = \emptyset$.

Example 2 shows also that disjoint sets can be indistinguishable in Zeeman's sense. Of course, indistinguishability of disjoint sets is not equivalent to having a non-empty tolerance intersection and one can easily find a counterexample to such claim on the basis of Example 2. Let us compare tolerance intersection of sets to ordinary intersection and union of sets.

Proposition 3. *Let $\langle O, \tau \rangle$ be a tolerance space and let $X, Y \subseteq O$. Then the following conditions hold:*

1. $X \cap Y \subseteq X \Cap_\tau Y$,
2. $X \Cap_\tau Y \subseteq X \cup Y$.

Proof (1) From definition and extensivity, $Y \subseteq \tau(Y)$, we get $X \cap Y \subseteq X \cap \tau(Y)$ so $X \cap Y \subseteq (X \cap \tau(Y)) \cup (Y \cap \tau(X)) = X \Cap_\tau Y$. Thus $X \cap Y \subseteq X \Cap_\tau Y$.
(2) Since $X \cap \tau(Y) \subseteq X \cup Y$ and $Y \cap \tau(X) \subseteq X \cup Y$, thus $(X \cap \tau(Y)) \cup (Y \cap \tau(X)) \subseteq X \cup Y$ and by definition $X \Cap_\tau Y \subseteq X \cup Y$.

Lemma 2. *Let $\langle O, \tau \rangle$ be a tolerance space and let $X, Y \subseteq O$. Then*

$$X \Cap_\tau Y \subseteq \tau(X) \cap \tau(Y).$$

Considering whether a tolerance intersection of sets coincides with the ordinary intersection or union of sets, $X \Cap_\tau Y = X \cap Y$, $X \Cap_\tau Y = X \cup Y$, respectively, leads to a number of interesting observations given in Proposition 4.

Proposition 4. *Let $\langle O, \tau \rangle$ be a tolerance space and let $X, Y \subseteq O$. If $X \Cap_\tau Y = X \cap Y$, then the following conditions hold:*

1. $X \cap \tau(Y) = X \cap Y$,
2. $Y \cap \tau(X) = X \cap Y$,
3. $X \cap \tau(Y) = Y \cap \tau(X)$.

Proof. Let $X \Cap_\tau Y = X \cap Y$, so $X \cap \tau(Y) \subseteq X \cap Y$, always $X \cap Y \subseteq X \cap \tau(Y)$, thus $X \cap \tau(Y) = X \cap Y$. Analogously $Y \cap \tau(X) = X \cap Y$. 1 and 2 implies 3.

Proposition 5. *Let $\langle O, \tau \rangle$ be a tolerance space and let $X, Y \subseteq O$. Then the following condition hold:*

$$\text{If } X = \tau(X) \text{ and } Y = \tau(Y), \text{ then } X \Cap_\tau Y = X \cap Y,$$

i.e., on the family of sets closed w.r.t. the operator $\tau(\ )$ (Pawlak's definable sets in rough set theory [15,16], when τ is transitive) a tolerance intersection of sets coincides with ordinary intersection of sets.

Proposition 6. *Let $\langle O, \tau \rangle$ be a tolerance space and let $X, Y \subseteq O$. Then the following conditions are equivalent:*

1. $X \Cap_\tau Y = X \cup Y$,
2. $X \sim_\tau Y$.

i.e. *only on the families of mutually indistinguishable sets in Zeeman's sense (maximal preclasses of the tolerance $\sim_\tau$) a tolerance intersection of sets coincides with the union of sets.*

Proof. ($\Rightarrow$). If $X \mathbin{\text{ⓜ}_\tau} Y = X \cup Y$, then by lemma 2 we get $X \mathbin{\text{ⓜ}_\tau} Y \subseteq \tau(X) \cap \tau(Y)$. Thus we get that $X \cup Y \subseteq \tau(X) \cap \tau(Y)$. Thus $X \subseteq \tau(Y)$ and $Y \subseteq \tau(X)$, so $X \sim_\tau Y$.

($\Leftarrow$) Let $X \sim_\tau Y$, so $X \subseteq \tau(Y)$ and $Y \subseteq \tau(X)$. $X \subseteq \tau(Y)$ implies $X \subseteq X \cap \tau(Y)$ and so $X \subseteq X \mathbin{\text{ⓜ}_\tau} Y$. Analogically for $Y \subseteq X \mathbin{\text{ⓜ}_\tau} Y$. Thus $X \cup Y \subseteq X \mathbin{\text{ⓜ}_\tau} Y$. By proposition 3 we get $X \mathbin{\text{ⓜ}_\tau} Y \subseteq X \cup Y$. Therefore $X \mathbin{\text{ⓜ}_\tau} Y = X \cup Y$.

Proposition 7 presents some basic properties of the tolerance intersection operation.

Proposition 7. *Let $\langle O, \tau \rangle$ be a tolerance space and let $X, Y \subseteq O$. Then the following conditions hold:*

1. $X \mathbin{\text{ⓜ}_\tau} Y = Y \mathbin{\text{ⓜ}_\tau} X$,
2. $(X \cap Y) \mathbin{\text{ⓜ}_\tau} (X \cap Z) \subseteq X \cap (Y \mathbin{\text{ⓜ}_\tau} Z)$,
3. $X \cup (Y \mathbin{\text{ⓜ}_\tau} Z) \subseteq (X \cup Y) \mathbin{\text{ⓜ}_\tau} (X \cup Z)$.

Proof. (1) From the definition and commutativity of the union of sets.
(2) By monotonicity $\tau(X \cap Z) \subseteq \tau(Z)$. So $(X \cap Y) \cap \tau(X \cap Z) \subseteq (X \cap Y) \cap \tau(Z) = X \cap (Y \cap \tau(Z))$. Analogically for $(X \cap Z) \cap \tau(X \cap Y) \subseteq X \cap (Z \cap \tau(Y))$. Thus $(X \cap Y) \cap \tau(X \cap Z), (X \cap Z) \cap \tau(X \cap Y) \subseteq (X \cap (Y \cap \tau(Z))) \cup (X \cap (Z \cap \tau(Y))) = X \cap ((Y \cap \tau(Z)) \cup (Z \cap \tau(Y)))$ and so $((X \cap Y) \cap \tau(X \cap Z)) \cup ((X \cap Z) \cap \tau(X \cap Y)) \subseteq X \cap ((Y \cap \tau(Z)) \cup (Z \cap \tau(Y)))$. Therefore by the definition of a perceptual intersection one can show that $(X \cap Y) \mathbin{\text{ⓜ}_\tau} (X \cap Z) \subseteq X \cap (Y \mathbin{\text{ⓜ}_\tau} Z)$.
(3) By monotonicity $\tau(Z) \subseteq \tau(X \cup Z)$ and so $Y \cap \tau(Z) \subseteq \tau(X \cup Z)$. By extensivity $X \subseteq \tau(X \cup Z)$, thus $X \cup (Y \cap \tau(Z)) \subseteq \tau(X \cup Z)$. It also holds that $X \cup (Y \cap \tau(Z)) \subseteq X \cup Y$. Therefore $X \cup (Y \cap \tau(Z)) \subseteq (X \cup Y) \cap \tau(X \cup Z)$. Analogically one can show that $X \cup (Z \cap \tau(Y)) \subseteq (X \cup Z) \cap \tau(X \cup Y)$. Thus $[X \cup (Y \cap \tau(Z))] \cup [X \cup (Z \cap \tau(Y))] \subseteq [(X \cup Y) \cap \tau(X \cup Z)] \cup [(X \cup Z) \cap \tau(X \cup Y)]$ so $X \cup [(Y \cap \tau(Z)) \cup (Z \cap \tau(Y))] \subseteq [(X \cup Y) \cap \tau(X \cup Z)] \cup [(X \cup Z) \cap \tau(X \cup Y)]$. and $X \cup [(Y \cap \tau(Z)) \cup (Z \cap \tau(Y))] \subseteq (X \cup Y) \cap [\tau(X \cup Z) \cup \tau(X \cup Y)]$ by distributivity of set theoretical operations. Therefore by definition of a perceptual intersection it follows that $X \cup (Y \mathbin{\text{ⓜ}_\tau} Z) \subseteq (X \cup Y) \mathbin{\text{ⓜ}_\tau} (X \cup Z)$.

Let us note that inclusions opposite to inclusions 2 and 3 from Proposition 7 do not hold what is shown by the following examples.

Example 3. Let (U, τ) be a tolerance space such that $U = \{a, b, c, d, e, f\}$ and $\tau = \Delta_U \cup \{(a, e), (e, a)\}$. Let $X, Y, Z \subseteq U$ be such that $X = \{a, b, c\}$, $Y = \{a, b, d\}$ and $Z = \{e, f\}$. Since $a \in Y$ and $a \in \tau(Z)$, then $a \in Y \cap \tau(Z)$ and so $a \in Y \mathbin{\text{ⓜ}_\tau} \tau(Z)$. Since $a \in X$ thus $a \in X \cap (Y \mathbin{\text{ⓜ}_\tau} \tau(Z))$.

Since $X \cap Z = \emptyset$ so $\tau(X \cap Z) = \emptyset$ and $a \notin \tau(X \cap Z)$, thus $a \notin (X \cap Y) \cap \tau(X \cap Z)$. Since $a \notin Z$, then $a \notin X \cap Z$ and so $a \notin (X \cap Z) \cap \tau(X \cap Y)$. Therefore $a \notin (X \cap Y) \mathbin{\text{ⓜ}_\tau} (X \cap Z)$.

We have shown that $a \in X$ thus $a \in X \cap (Y \mathbin{\text{ⓜ}_\tau} \tau(Z))$ but $a \notin (X \cap Y) \mathbin{\text{ⓜ}_\tau} (X \cap Z)$ what contradicts the opposite inclusion to the inclusion (2) from Proposition 7.

Example 4. Let (U, ρ) be a tolerance space such that $U = \{a, b, c, d\}$ and $\rho = \Delta_U \cup \{(a, b), (b, a)\}$. Let $X, Y, Z \subseteq U$ be such that $X = \{b, c, d\}$, $Y = \{a, b, d\}$ and $Z = \{c, d\}$. One can note that $a \notin X$, $a \in \rho(X)$, $a \in Y$, $a \in \rho(Y)$. Thus $a \in X \cup Y$ and $a \in \rho(X \cup Z)$, so $a \in (X \cup Y) \cap \rho(X \cup Z)$. Therefore $a \in (X \cup Y) \cap_\rho (X \cup Z)$. However, $a \notin Z$ and $a \notin \rho(Z)$, thus $a \notin Y \cap \rho(Z)$ and $a \notin Z \cap \rho(Y)$. Therefore $a \notin Y \cap_\rho Z$ and since $a \notin X$, thus $a \notin X \cup (Y \cap_\rho Z)$.

We have shown that $a \in (X \cup Y) \cap_\rho (X \cup Z)$ but $a \notin X \cup (Y \cap_\rho Z)$ what contradicts the opposite inclusion to inclusion (3) from Proposition 7.

One can note that tolerances τ and ρ from examples 3 and 4, respectively, are transitive. Therefore inclusions 2 and 3 from Proposition 7 also can not be strengthened to equalities in the case of the classical Pawlak's rough set theory based on equivalence relations.

Now, keeping in mind the similarity interpretation of tolerance relations, we can introduce a tolerance intersection measure for finite subsets of a tolerance space. The family of all finite subsets of a set O is denoted by $\mathcal{P}_{fin}(O)$.

Definition 3. *Let $\langle O, \tau \rangle$ be a tolerance space and let $X, Y \in \mathcal{P}_{fin}(O)$ and at least one of them is non-empty. A* tolerance intersection measure *is denoted by* pi_τ *and defined as follows:*

$$\mathrm{pi}_\tau(X, Y) := \frac{|X \cap_\tau Y|}{|X \cup Y|}.$$

Theorem 1. *Let $\langle O, \tau \rangle$ be a tolerance space and let $X, Y \in \mathcal{P}_{fin}(O)$ and $X \neq \emptyset$ or $Y \neq \emptyset$. Then the following conditions are equivalent:*

1. *$X \sim_\tau Y$,*
2. *$X \cap_\tau Y = X \cup Y$,*
3. *$\mathrm{pi}_\tau(X, Y) = 1$.*

Proof. Because of Proposition 6 and the fact that implication $2 \Rightarrow 3$ follows directly for definition it is enough to show $3 \Rightarrow 2$. Let $\mathrm{pi}_\tau(X, Y) = 1$, thus $|X \cap_\tau Y| = |X \cup Y|$. Since $X \cap_\tau Y \subseteq X \cup Y$, then by finiteness of sets X and Y it follows that $X \cap_\tau Y = X \cup Y$.

In the light of Theorem 1, we see that a tolerance intersection measure is a measure of tolerance distinguishability of sets in Zeeman's sense.

4   Near Sets and Perceptual Tolerance Relations

Perceptual systems in near set theory [10,7,9] reflect Poincaré's idea of perception. A *perceptual system* is a pair $\langle O, \mathbb{F} \rangle$, where O is a non-empty set of *perceptual objects* and $\mathbb{F}$ is a non-empty set of real valued functions defined on O, *i.e.*, $\mathbb{F} := \{\phi \mid \phi : O \to \mathbb{R}\}$, where ϕ is called a *probe function*. Perceptual objects spring directly from the perception of physical objects derived from sets of sensations in Poincaré's view of the physical continuum [18]. A probe function

$\phi \in \mathbb{F}$ is viewed as a representation of a feature in the description of sets of sensations. So, for example, a digital image Im can be seen as a set of perceptual objects, $i.e.$, $Im \subseteq O$, where every perceptual object is described with vectors of probe function values.

A family of probe functions $\mathbb{F}$ can be infinite[2]. In applications such as image analysis, from a possibly infinite family of probe functions, we always select a finite number of probe functions, $\mathcal{B} \subseteq \mathbb{F}$ and $|\mathcal{B}| < \aleph_0$, in order to describe perceptual objects (usually pixels or pixel windows in digital images). Thus, every perceptual object $x \in O$ can be described by a vector $\phi_{\mathcal{B}}(x)$ of real values of probe functions in a space $\mathbb{R}^n$ $i.e.$

$$\phi_{\mathcal{B}}(x) = (\phi_1(x), \phi_2(x), \ \ldots \ , \phi_n(x)),$$

where $\mathcal{B} := \{\phi_1, \ \ldots \ , \phi_n\}$ for $\mathcal{B} \subseteq \mathbb{F}$.

With object descriptions, one can compare objects with respect to various metric or pseudometric distances defined on $\mathbb{R}^n$. More generally, one can introduce on the set O different topologies based on topologies determined on the space $\mathbb{R}^n$ (note that such topologies are not necessarily induced from $\mathbb{R}^n$). For example, consider a natural topology on $\mathbb{R}^n$ determined by Euclidean distance, denoted here by d. Using d one can define the distance measure on O in the following way:

$$p_{\mathcal{B}}(x, y) := d(\phi_{\mathcal{B}}(x), \phi_{\mathcal{B}}(y)) = \sqrt{\sum_{i=1}^{n}(\phi_i(x) - \phi_i(y))^2},$$

where $\mathcal{B} \subseteq \mathbb{F}$ and $\mathcal{B} := \{\phi_1, \ \ldots \ , \phi_n\}$. Notice that d is a metric on $\mathbb{R}^n$ but $p_{\mathcal{B}}$ is not necessarily a metric on O, since it is possible that there are $x, y \in O$ such that $p_{\mathcal{B}}(x, y) = 0$ but $x \neq y$, $i.e.$, two different perceptual objects can have exactly the same description over a family of probe functions. Moreover, similarly to the case of the transitivity of distance tolerances, the condition $p_{\mathcal{B}}(x, y) = 0 \Leftrightarrow x = y$ is neither implied nor excluded by the definition of $p_{\mathcal{B}}$. When the set O only consists of objects with mutually different descriptions, the function $p_{\mathcal{B}}$ is a metric on O.

From Example 1, for a perceptual system and some pseudometric one can define for a real threshold $\epsilon \in (0, +\infty)$ a distance tolerance relation

Definition 4. *Let $\langle O, \mathbb{F} \rangle$ be a perceptual system, $\langle O, p_{\mathcal{B}} \rangle$ be a pseudometric space where $\mathcal{B} := \{\phi_i(x)\}_{i=1}^{n} \subseteq \mathbb{F}$. A relation $\cong_{\mathcal{B}, \epsilon}$ is defined for any $x, y \in O$ as follows:*

$$(x, y) \in \cong_{\mathcal{B}, \epsilon} :\Leftrightarrow p_{\mathcal{B}}(x, y) < \epsilon.$$

A relation $\cong_{\mathcal{B}, \epsilon}$ is a distance tolerance relation and we call it perceptual tolerance relation.

---

[2] From a digital image analysis perspective, the potential for a countable number of probe functions has a sound interpretation, $i.e.$, the number of image probe functions is finite but unbounded, since new probe functions can be created over an indefinitely long timespan and added to the set of existing probes.

$\cong_{\mathcal{B},\epsilon}$ reflects Poincaré's idea, *i.e.*, sensations are similar if their descriptions are close enough in a space $\mathbb{R}^n$. Note that a relation $\cong_{\mathcal{B},\epsilon}$ depends not only on a choice of a threshold but also on the choice of a family of probe function. For the same threshold and for two different families of probe functions one can get two distinct perceptual tolerance relations. As a direct consequence of Proposition 2, one can infer:

Corollary 2. *Let $\langle O, \mathbb{F} \rangle$ be a perceptual system, $\langle O, p_{\mathcal{B}} \rangle$ be a pseudometric space where $\mathcal{B} := \{\phi_i(x)\}_{i=1}^{n} \subseteq \mathbb{F}$. Then for any $x \in O$ holds that*

$$\cong_{\mathcal{B},\epsilon}(x) = B_{p_{\mathcal{B}}}(x, \epsilon),$$

i.e., a $\cong_{\mathcal{B},\epsilon}$ neighbourhood of x is just an open ball in the pseudometric space $\langle O, p_{\mathcal{B}} \rangle$ with centre x and radius ϵ, where a centre x can be identified with an equivalence class $x_{/\theta_{p_{\mathcal{B}}}}$, where $(x, y) \in \theta_{p_{\mathcal{B}}} :\Leftrightarrow p_{\mathcal{B}}(x, y)$ for $x, y \in O$.

This corresponds to Poincaré's idea that sensations are identified with particular sets of sensations that are very similar. It can also be observed that when sensations $x, y \in O$ are close enough, they become indistinguishable. In a near set approach, the indistinguishability of sensations results from sensations that have the same descriptions over a selected family of probe functions $\mathcal{B} \subseteq \mathbb{F}$, *i.e.*, the pseudometric distance $p_{\mathcal{B}}$ between x and y is equal to 0. In that case, a selected family of probe functions can be interpreted as representing the sensitivity of the senses.

For apparently indistinguishable sensed objects, Poincaré's idea of similarity is based on a spatial metric distance. There is the possibility that one can have two sensations with that same descriptions and yet the sensations are spatially far apart. In that case, of course, such sensations cannot be viewed as indistinguishable. This suggests that the indistinguishability of two sensations requires that such sensations be very near each other, *i.e.*, close to each other for some spatial metric distance. Thus, a perceptual tolerance determined by some perceptual system $\langle O, \mathbb{F} \rangle$ should be based on some sensual tolerance $\cong_{\mathcal{B},\epsilon}$ defined for a selected family of probe functions $\mathcal{B} \subseteq \mathbb{F}$, some threshold $\epsilon \in (0, +\infty)$ and for a spatial tolerance $\tau_{d,\delta}$ defined for some metric space $\langle O, d \rangle$ and a threshold $\delta \in (0, +\infty)$. In that case, a perceptual tolerance is an intersection of $\cong_{\mathcal{B},\epsilon}$ and $\tau_{d,\delta}$ (notice that the intersection of two tolerances is always a tolerance).

From a near set perspective, in the light of Corollary 2 it can be also observed that similarity between perceptual objects depends on three independent factors:

- a selection of a finite family of probe functions as a basis of object descriptions,
- a selection of a pseudometric distance function for a set of perceptual objects,
- a selection of a positive real threshold.

Since probe functions represent results of perception (interaction of sensors with the environment), then the selected family of probe functions corresponds to a frame of sensors. The selected positive real threshold can represent a sensitivity of perceptual machinery interacting with the environment. Corollary 2 reflects also the fact that a process of perception (interaction of sensors with the environment) results in the first granularization, *perceptual granularization* of the set of sensations. Mathematically it is represented by a pseudometric space $\langle O, p_{_B} \rangle$ derived from a perceptual system $\langle O, \mathbb{F} \rangle$ on the basis of a finite family $\mathcal{B} \subseteq \mathbb{F}$, where the set of perceptual objects O is divided by the equivalence relation $\theta_{p_{_B}}$ into classes consisting of objects indistinguishable w.r.t. sensitivity of sensors interacting with the environment.

5   Practical Application of Near Sets in Image Analysis

Resemblance between objects such as image patches in digital image is represented by object description as it is described in the previous section. description is defined by an n-dimensional feature vector of real values, each value representing an observable feature such as average greylevel intensity or average edge orientation of pixels in subimages. Image patches (*i.e.*, collections of subimages) or features derived from image patches offer very promising approaches to content-based image retrieval (CBIR) [19]. Different types of local features can then be extracted for each subimage and used in the retrieval process. An extensive survey of CBIR systems can be found in [20]. The near set approach to CBIR has been used successfully in a number of studies (see, *e.g.*, [21,22,23].

The basic approach to digital image covers determined by a tolerance relation and then identifying interesting regions containing similar image patches in a cover an image is illustrated in Fig. 1.2 (cover of the image of Lena) and in Fig. 1.5 (cover of the image of the Mona Lisa). A basic assumption underlying the notion of an image cover is that a digital image is a set of points. In this work, a point is a set of pixels, *i.e.*, subimage. The cover of an image is determined by a tolerance relation. In this work, perceptual tolerance relations with Euclidean distances between feature vectors determine sets of similar subimages (points) that constitute the classes in an image cover. For instance, for the image of Lena in Fig. 1.1, let X denote a set of points in the digital image and let ϕ_g be a probe function denoting the average grey level of the pixels in an $n \times n, n \in \mathbb{N}$, where $\mathbb{N}$ denotes the natural numbers. Then for $\epsilon \in [0, \infty)$ one can consider a particular tolerance relation $\cong_{\phi_g, \epsilon}$ defined on the basis of an absolute value metric:

$$\cong_{\phi_g, \epsilon} = \{(x, y) \in X \times X : |\phi_g(x) - \phi_g(y)| \leq \epsilon\}.$$

Example 5. **Near Sets of Subimages in Portraits**
For example, a sample cover of the image of Lena in Fig. 1.1 is shown in Fig. 1.2. A sample tolerance class is shown in Fig. 1.3 that displays subimages where the average greylevel of the subimages along the edge of Lena's hat are similar. This class with $\epsilon = 0.01$ contains 20×20 subimages that extend along the top edge of Lena's hat and many parts of Lena's hat feather as well as in other parts of the same image. This class illustrates the scattered character (dispersion) of the subimages with similar descriptions. This dispersion phenomenon is quite common in the study of visual near sets, something that was recognized by E.C.

1.1: Lena

1.2: cover of image in Fig. 1.1

1.3: hat class in Fig. 1.2

1.4: Mona Lisa

1.5: cover of image in Fig. 1.4

1.6: head class in Fig. 1.5

Fig. 1. Sample Perceptually Near Digital Images

Zeeman [3]. Similarly, a sample cover of the image of the portrait of Mona Lisa in Fig. 1.4 is shown in Fig. 1.5. A tolerance class is shown in Fig. 1.6 that displays subimages where the average edge greylevel for a subimage along the top of forehead of the Mona Lisa portrait are similar. Again, notice the dispersion of the subimages in the same class, mostly in Mona Lisa's head and shoulders. Visually, it can be observed that the greylevel in the area of Lena's hat and feather are similar to the greylevel of parts of the forehead of the Mona Lisa. The experiment described in this example can be reproduced using the Near set Evaluation and Recognition (NEAR) system[3].

Example 6. **Near Sets of Subimages in Aircraft Images**
Another example of near sets is shown in Fig. 2. In this case, images of small aircraft are shown in Fig. 2.1 and in Fig. 2.3. A sample tolerance class is shown in Fig. 2.2 that displays a mixture of subimages (represented, for instance, by ▓ boxes with similar shades of grey), where the average edge greylevel for subimages in various parts of the aircraft as well as in the foreground and background are similar (see, *e.g.*, the tail region of the aircraft as well as the region to the left of the aircraft tail). This class again illustrates the scattered character (dispersion) of the subimages with similar descriptions (in this case, similar greylevels).

---

[3] A complete tutorial plus version 2.0 of the NEAR system are available at http://wren.ece.umanitoba.ca

2.1: aircraft 1

2.2: aircraft class in Fig. 2.1

2.3: aircraft 2

2.4: aircraft class in Fig. 2.3

Fig. 2. Sample Perceptually Near Aircraft Images

A tolerance class is shown in Fig. 2.4 that displays subimages where the average edge greylevel for a subimage in various parts of the second aircraft, foreground and background are similar. Again, notice the dispersion of the submimages in the same class, mostly in the tail and cockpit regions of the aircraft.

6 Perceptual Intersection of Sets and Perceptual Similarity Measures

On the basis of perceptual tolerance relations we can introduce perceptual intersection of sets being a particular form of tolerance intersections of sets.

Definition 5. *Let $\langle O, \mathbb{F} \rangle$ be a perceptual system and let $\langle O, \cong_{\mathcal{B}, \epsilon} \rangle$ be a perceptual tolerance space where $\mathcal{B} \subseteq \mathbb{F}$ and $\epsilon \in (0, +\infty)$. A perceptual intersection of sets based on $\langle O, \cong_{\mathcal{B}, \epsilon} \rangle$ (or shortly perceptual intersection of sets when a perceptual tolerance space is understood) is denoted by $\cap_{\mathcal{B}, \epsilon}$ and defined for $X, Y \subseteq O$ as follows:*

$$X \cap_{\mathcal{B}, \epsilon} Y := (X \cap \cong_{\mathcal{B}, \epsilon}(Y)) \cup (Y \cap \cong_{\mathcal{B}, \epsilon}(X)).$$

That $\cap_{\mathcal{B}, \epsilon}$ perceptually originated from of tolerance intersection can be seen in its similarity nature. Sets $Im_1, Im_2 \subseteq O$, where $\langle O, \mathbb{F} \rangle$ is a perceptual system, can be digital images. The perceptual intersection of Im_1 and Im_2 consists of those perceptual objects belonging to Im_1 or Im_2 which have similar 'cousins' in the other image. On the basis of perceptual intersection of sets, we can now introduce a perceptual intersection measure of the similarity of sets.

Definition 6. *Let $\langle O, \mathbb{F} \rangle$ be a perceptual system and let $\langle O, \cong_{\mathcal{B},\epsilon} \rangle$ be a perceptual tolerance space where $\mathcal{B} \subseteq \mathbb{F}$ and $\epsilon \in (0, +\infty)$. A perceptual intersection measure is denoted by $\mathrm{p}_{\widehat{\cap}_{\mathcal{B},\epsilon}}$ and defined for any $X, Y \in \mathcal{P}_{fin}(O)$, where $X \neq \emptyset$ or $Y \neq \emptyset$.*

$$\mathrm{p}_{\widehat{\cap}_{\mathcal{B},\epsilon}}(X, Y) := \frac{|X \,\widehat{\cap}_{\mathcal{B},\epsilon}\, Y|}{|X \cup Y|}.$$

Since a perceptual intersection measure is a particular form of a tolerance intersection measure, so Theorem 1 also applies to it. Additionally, one can note that when a tolerance τ is a perceptual tolerance and sets X and Y are images in some perceptual system, then Zeeman's tolerance $\sim_\tau$ becomes a perceptual indistinguishability of images. Thus a perceptual intersection measure is a measure of perceptual indistinguishability that is a kind of similarity measure between images. In sum, this paper takes into account a direct connection of a perceptual intersection measure to a perceptual form of the Zeeman's tolerance and we have formulated a postulate on similarity measures between images based on perceptual tolerance relations:

Postulate

Every similarity measure μ_ρ derived from a perceptual system $\langle O, \mathbb{F} \rangle$ on the basis of some perceptual tolerance relation ρ should fulfill the following condition for $X, Y \subseteq O$:

$$\mu_\rho(X, Y) = 1 \;\; if \; and \; only \; if \;\; X \sim_\rho Y.$$

7   Conclusion

In this work, tolerance intersection of sets and a tolerance intersection measure together with their perceptual forms derived from perceptual tolerances and perceptual systems have been investigated. The properties of the proposed set operations and measures and their connections to Zeeman's indistinguishability of sets together with their perceptual applications have been given. In addition, an indication of how to proceed in solving the digital image correspondence problem in digital image analysis is also given in terms of sample perceptual tolerance intersections relative to pairs sets extracted from digital images. Further research concerning near sets will take into account other set operations and their implications in various applications.

Acknowledgements. This research has been supported by the Natural Sciences and Engineering Research Council of Canada (NSERC) grants 185986, 194376, Canadian Arthritis Network grant SRI-BIO-05, Manitoba Centre of

Excellence Fund (MCEF) grant T277 and grants N N516 077837, N N516 368334 from the Ministry of Science and Higher Education of the Republic of Poland.

References

1. Wasilewski, P., Peters, J.F., Ramanna, S.: Perceptual tolerance intersection. In: Szczuka, M., Kryszkiewicz, M., Ramanna, S., Jensen, R., Hu, Q. (eds.) RSCTC 2010. LNCS (LNAI), vol. 6086, pp. 277–286. Springer, Heidelberg (2010)
2. Poincaré, J.: Sur certaines surfaces algébriques; troisième complément à l'analysis situs. Bulletin de la Société de France 30, 49–70 (1902)
3. Zeeman, E.: The topology of the brain and visual perception. University of Georgia Institute Conference Proceedings (1962); Published in Fort Jr., M.K. (ed.): Topology of 3-Manifolds and Related Topics, pp. 240–256. Prentice-Hall, Inc., Englewood Cliffs (1962)
4. Pogonowski, J.: Tolerance Spaces with Applications to Linguistics. University of Adam Mickiewicz Press, Poznań (1981)
5. Skowron, A., Stepaniuk, J.: Tolerance approximation spaces. Fundamenta Informaticae 27(2-3), 245–253 (1996)
6. Peters, J.F., Skowron, A., Stepaniuk, J.: Nearness of objects: Extension of approximation space model. Fundamenta Informaticae 79(3-4), 497–512 (2007)
7. Peters, J.F., Ramanna, S.: Affinities between perceptual granules: Foundations and perspectives. In: Bargiela, A., Pedrycz, W. (eds.) Human-Centric Information Processing Through Granular Modelling. SCI, vol. 182, pp. 49–66. Springer, Heidelberg (2009)
8. Wasilewski, P.: On selected similarity relations and their applications into cognitive science. PhD thesis (in Polish), Department of Logic, Cracow (2004)
9. Peters, J.F., Wasilewski, P.: Foundations of near sets. Elsevier Science 179(18), 3091–3109 (2009)
10. Peters, J.F.: Near sets. special theory about nearness of objects. Fundamenta Informaticae 75(1-4), 407–433 (2007)
11. Naimpally, S., Warrack, B.: Proximity Spaces. Cambridge Tract in Mathematics, vol. 59. Cambridge Univiversity Press, Cambridge (1970)
12. Gagrat, M., Naimpally, S.: Proximity approach to semi-metric and developable spaces. Pacific Journal of Mathematics 44(1), 93–105 (1973)
13. Császár, Á.: General Topology. Adam Hilger Ltd., Bristol (1978)
14. Grätzer, G., Wenzel, G.: Tolerances, covering systems, and the axiom of choice. Archivum Mathematicum 25(1-2), 27–34 (1989)
15. Pawlak, Z.: Rough sets. International J. Comp. Inform. Science 11, 341–356 (1981)
16. Pawlak, Z.: Rough sets. Theoretical Aspects of Reasoning About Data. Kluwer Academic Publishers, The Netherlands (1991)
17. Engelking, R.: General Topology, Revised & completed edition. Heldermann Verlag, Berlin (1989)
18. Poincaré, J.H.: Dernières pensées, trans. by J.W. Bolduc as Mathematics and Science: Last Essays. Flammarion & Kessinger Pub., Paris (1913/2009), http://docenti.lett.unisi.it/files/4/1/1/36/Dernierespenseespoinc.pdf
19. Deselares, T.: Image Retrieval, Object Recognition, and Discriminative Models. PhD thesis (2008)
20. Deselares, T., Keysers, D., Ney, H.: Features for image retrieval: an experimental comparison. Information Retrieval 11(2), 77–107 (2008)

21. Henry, C.: Near Sets: Theory and Applications. PhD thesis, Department of Electrical & Computer Engineering (2010); supervisor: J.F. Peters
22. Pal, S., Peters, J.F.: Rough Fuzzy Image Analysis. Foundations and Methodologies. CRC Press, Taylor & Francis Group (September 2010), ISBN 13: 9781439803295, ISBN 10: 1439803293
23. Henry, C., Peters, J.F.: Perception-based image classification. International Journal of Intelligent Computing and Cybernetics 3(3), 410–430 (2010), doi:10.1108/17563781011066701

Some Mathematical Structures of Generalized Rough Sets in Infinite Universes of Discourse

Wei-Zhi Wu[1] and Ju-Sheng Mi[2]

[1] School of Mathematics, Physics and Information Science
Zhejiang Ocean University, Zhoushan, Zhejiang, 316004, P.R. China
wuwz@zjou.edu.cn
[2] College of Mathematics and Information Science
Hebei Normal University, Hebei, Shijiazhuang, 050016, P.R. China
mijsh@263.net

Abstract. This paper presents a general framework for the study of mathematical structure of rough sets in infinite universes of discourse. Lower and upper approximations of a crisp set with respect to an infinite approximation space are first defined. Properties of rough approximation operators induced from various approximation spaces are examined. The relationship between a topological space and rough approximation operators is further established. By the axiomatic approach, various classes of rough approximation operators are characterized by different sets of axioms. The axiom sets of rough approximation operators guarantee the existence of certain types of crisp relations producing the same operators. The measurability structures of rough set algebras are also investigated. Finally, the connections between rough sets and Dempster-Shafer theory of evidence are also explored.

Keywords: Approximation operators, belief functions, binary relations, measurability sets, rough sets, rough set algebras, topological spaces.

1 Introduction

The theory of rough sets was originally proposed by Pawlak [20,22] as a formal tool for modelling and processing incomplete information. The basic structure of the rough set theory is an approximation space consisting of a universe of discourse and an equivalence relation imposed on it. The equivalence relation is a key notion in Pawlak's rough set model. The equivalence classes in Pawlak's rough set model provide the basis of "information granules" for database analysis discussed in Zadeh's [58,59]. Rough set theory can be viewed as a crisp-set-based granular computing method that advances research in this area [9,18,24,25,37,38,55].

However, the requirement of an equivalence relation in Pawlak's rough set model seems to be a very restrictive condition that may limit the applications of the rough set model. Thus one of the main directions of research in rough set theory is naturally the generalization of the Pawlak rough set approximations. There

J.F. Peters et al. (Eds.): Transactions on Rough Sets XIII, LNCS 6499, pp. 175–206, 2011.

are at least two approaches for the development of rough set theory, namely the constructive and axiomatic approaches. In the constructive approach, binary relations on the universe of discourse, partitions of the universe of discourse, neighborhood systems, and Boolean algebras are all the primitive notions. The lower and upper approximation operators are constructed by means of these notions [15,19,20,22,23,30,39,42,47,49,51,52,53,54,56,60].

On the other hand, the axiomatic approach, which is appropriate for studying the structures of rough set algebras, takes the lower and upper approximation operators as primitive notions. By this approach, a set of axioms is used to characterize approximation operators that are the same as the ones produced by using the constructive approach [5,17,50,51,52,53]. From this point of view, rough set theory may be interpreted as an extension theory with two additional unary operators. The lower and upper approximation operators are related respectively to the necessity (box) and possibility (diamond) operators in modal logic, and the interior and closure operators in topological space [3,4,5,11,12,13,17,40,41,43,52,56].

Another important direction for generalization of rough set theory is its relationship to the Dempster-Shafer theory of evidence which was originated by Dempster's concept of lower and upper probabilities [6] and extended by Shafer as a theory [33]. The basic representational structure in the Dempster-Shafer theory of evidence is a belief structure which consists of a family of subsets, called focal elements, with associated individual positive weights summing to one. The primitive numeric measures derived from the belief structure are a dual pair of belief and plausibility functions. There exist some natural connections between the rough set theory and the Dempster-Shafer theory of evidence [34,35,36,44,45,46,48,57]. It is demonstrated that various belief structures are associated with various rough approximation spaces such that different dual pairs of upper and lower approximation operators induced by the rough approximation spaces may be used to interpret the corresponding dual pairs of plausibility and belief functions induced by the belief structures.

In this paper, we develop a rough set model in infinite universes of discourse. We focus mainly on the study of some mathematical structures of rough sets. In the next section, we summarize existing research on generalized rough approximation operators. In Section 3, the concepts of generalized rough approximation operators in infinite universes are introduced and reviewed, and essential properties of the rough approximation operators are examined. In Section 4, we investigate the topological structure of rough sets. We will establish the relationship between rough approximation operators and topological spaces. In Section 5, we present axiomatic characterizations of rough approximation operators. We further present properties of various types of rough set algebras in Section 6. In Section 7, we explore the measurable structures of rough sets. The belief structures of rough sets are introduced in Section 8. We show that the Dempster-Shafer theory of evidence can be interpreted by rough set theory. We then conclude the paper with a summary in Section 9.

2   Related Works

From both theoretic and practical needs, many authors have generalized the notion of approximation operators by using non-equivalence binary relations which lead to various other approximation operators. For example, Slowinski and Vanderpooten [39] developed a generalized definition of rough approximations determined by a similarity binary relation. Yao [51,52,53] studied the general properties of relation based rough sets in finite universes of discourse by eliminating the transitivity, reflexivity, and symmetry axioms. Lin [15] proposed a more general framework for the study of approximation operators by using the so-called neighborhood systems from a topological space and its generalization called Frechet (V) space. With respect to a binary relation, the successor elements of a given element may be interpreted as its neighborhood. The theory of rough sets built from binary relations may therefore be related to neighborhood systems. More significant researches on this topic in finite universes of discourse were made by Yao [54], and Wu and Zhang [47]. Kondo [11] discussed rough set approximation operators in infinite universes of discourse within axiomatic definitions. We will further examine the properties of non-equivalence relation based rough set approximation operators in infinite universes of discourse in Section 3.

Topology is a branch of mathematics, whose concepts exist not only in almost all branches of mathematics, but also in many real life applications. Topological structure is an important base for knowledge extraction and processing. Therefore many works have appeared for the study of relationships between rough approximation operators and the topological structure of rough sets (see e.g. [3,4,13,14,26,27,28,29,31,43,52,61,62,63]). For example, Polkowski [26,27,28] defined the hit-or-miss topology on rough sets and proposed a scheme to approximate mathematical morphology within the general paradigm of soft computing. The main results of the relationships between rough set theory and topological spaces are that a reflexive and transitive approximation space can induce a topological space such that the lower and upper approximation operators in the approximation space are, respectively, the interior and closure operators in the topological space [29,52]. On the other hand, Kondo [12] showed that a topological space satisfying the so-called condition (comp) must be generated by a reflexive and transitive space. More recently, Qin et al. examined that Kondo 's condition (comp) can be reduced to a weak condition called (COMP). In Section 4, we will give a new sufficient and necessary condition such that there exists a one-to-one correspondence between the set of all topological spaces and the family of all approximation spaces.

Many authors explored axiomatic sets to investigate algebraic structures of relation based rough sets. For instance, Zakowski [60] presented a set of axioms on approximation operators. Comer [5] investigated axioms on approximation operators in relation to cylindric algebras. Lin and Liu [17] suggested six axioms for a pair of abstract operators on the power set of universe in the framework of topological spaces. Under these axioms, there exists an equivalence relation such that the derived lower and upper approximations are the same as the abstract operators. The similar result was also stated earlier by Wiweger [43]. A problem arisen is that all these studies are restricted to Pawlak rough set algebra defined by equivalence relations. Wybraniec-Skardowska [49] examined many axioms on various classes of approximation operators. Different constructive methods were suggested to produce such approximation operators. Thiele [40] explored axiomatic characterizations of approximation operators within modal logic for a crisp diamond and box operator represented by an arbitrary binary crisp relation. The most important axiomatic studies for crisp rough sets were made by Yao [51,52,53], Yao and Lin [56], in which various classes of crisp rough set algebras are characterized by different sets of axioms. Yang and Li [50] examined the independence of Yao's axiom sets and established the minimization of axiom sets characterizing various generalized approximation operators. In Section 5, we will give the axiomatic study on various rough approximation operators in infinite universes of discourse. Then we present in detail properties of rough set algebras in various situations in Section 6. We further show in Section 7 that, in infinite universes of discourse, the family of all definable sets in a serial rough set algebra forms a σ-algebra, conversely, in any finite universe of discourse, every σ-algebra can be induced from a rough set algebra.

It is well-known that the original concepts of belief and plausibility functions in the Dempster-Shafer theory of evidence come from the lower and upper probabilities induced by a multi-valued mapping carrying a probability measure defined over subsets of the domain of the mapping [33]. The lower and upper probabilities in the Dempster-Shafer theory of evidence have some natural correspondences with the lower and upper approximations in rough set theory. In finite universes of discourse, the study of interpretations of belief functions in rough sets is an interesting issue. Pawlak [21] is the first who showed that the probabilities of lower and upper approximations of a set in a Pawlak approximation space are respectively the belief and plausibility degrees of the set. Ruspini [32] interpreted belief functions by adopting an epistemic interpretation of modal logic $S5$. Fagin and Halpern [7] examined the interpretation of belief functions as inner measures. Lin [16], Skowron [34,35], and Skowron and Grzymala-Busse [36] investigated the relationships between the Dempster-Shafer theory of evidence and rough set theory as well as application for reasoning in information systems. However, most researches on this topic have been focused on Pawlak's rough set models and the main result is that the belief function is an inner probability. A significant research on the interpretations of Dempster-Shafer theory in the generalized rough set models using non-equivalence relations was done by Yao and Lingras [57]. It was shown that the probabilities of lower and upper approximations of a set in a serial approximation space are respectively the belief and plausibility degrees of the set. Conversely, for any belief structure, there must exist a probability approximation space such that the belief and plausibility functions can be represented as the probabilities of lower and upper approximations. Thus, it can be observed that the belief and plausibility functions in the Dempster-Shafer theory of evidence and the lower and upper approximations in rough set theory respectively capture the mechanisms of numeric and non-numeric aspects of uncertain knowledge.

It is well-known that the belief (respectively, plausibility) function is a mono-tone Choquet capacity of infinite order (respectively, alternating Choquet capacity of infinite order) [2] satisfying the sub-additive (respectively, super-additive) property at any order [33]. Sub-additivity and super-additivity at any order are therefore the essential properties of belief and plausibility functions respectively. In Section 8, we will introduce the Dempster-Shafer theory of evidence in infinite universes of discourse and make a research on the interpretations of Dempster-Shafer theory in rough set models.

3 Construction of Generalized Rough Approximation Operators

Let X be a nonempty set called the universe of discourse. The class of all subsets of X will be denoted by $\mathcal{P}(X)$. For any $A \in \mathcal{P}(X)$, we denote by $\sim A$ the complement of A.

Definition 1. *Let U and W be two nonempty universes of discourse. A subset $R \in \mathcal{P}(U \times W)$ is referred to as a binary relation from U to W. The relation R is referred to as serial if for any $x \in U$ there exists $y \in W$ such that $(x, y) \in R$. If $U = W$, $R \in \mathcal{P}(U \times U)$ is called a binary relation on U, $R \in \mathcal{P}(U \times U)$ is referred to as reflexive if $(x, x) \in R$ for all $x \in U$; R is referred to as symmetric if $(x, y) \in R$ implies $(y, x) \in R$ for all $x, y \in U$; R is referred to as transitive if for any $x, y, z \in U$, $(x, y) \in R$ and $(y, z) \in R$ imply $(x, z) \in R$; R is referred to as Euclidean if for any $x, y, z \in U$, $(x, y) \in R$ and $(x, z) \in R$ imply $(y, z) \in R$; R is referred to as a preorder if it is reflexive and transitive; R is referred to as an equivalence relation if R is reflexive, symmetric and transitive.*

Assume that R is an arbitrary binary relation from U to W. One can define a set-valued function $R_s : U \to \mathcal{P}(W)$ by:

$$R_s(x) = \{y \in W : (x, y) \in R\}, \quad x \in U. \tag{1}$$

$R_s(x)$ is called the successor neighborhood of x with respect to (w.r.t.) R [53]. Obviously, any set-valued function F from U to W defines a binary relation from U to W by setting $R = \{(x, y) \in U \times W : y \in F(x)\}$. For $A \in \mathcal{P}(W)$, let $j(A) = R_s^{-1}(A)$ be the counter-image of A under the set-valued function R_s, i.e.,

$$j(A) = \begin{cases} R_s^{-1}(A) = \{u \in U : R_s(u) = A\}, \text{ if } A \in \{R_s(x) : x \in U\}, \\ \emptyset, \qquad\qquad\qquad\qquad\qquad\qquad\qquad \text{otherwise.} \end{cases} \tag{2}$$

Then it is well-known that j satisfies the properties (J1) and (J2):

$$\text{(J1) } A \neq B \Longrightarrow j(A) \cap j(B) = \emptyset, \qquad \text{(J2) } \bigcup_{A \in \mathcal{P}(W)} j(A) = U.$$

Definition 2. *If R is an arbitrary relation from U to W, then the triple (U, W, R) is referred to as a generalized approximation space. For any set $A \subseteq W$, a pair of lower and upper approximations, $\underline{R}(A)$ and $\overline{R}(A)$, are, respectively, defined by*

$$\underline{R}(A) = \{x \in U : R_s(x) \subseteq A\}, \quad \overline{R}(A) = \{x \in U : R_s(x) \cap A \neq \emptyset\}. \quad (3)$$

The pair $(\underline{R}(A), \overline{R}(A))$ is referred to as a generalized crisp rough set, and $\underline{R}$ and $\overline{R} : \mathcal{P}(W) \to \mathcal{P}(U)$ are called the lower and upper generalized approximation operators respectively.

From the definitions of approximation operators, the following theorem can be easily derived [11,22,46,47]:

Theorem 1. *For a given approximation space (U, W, R), the lower and upper approximation operators defined in Definition 2 satisfy the following properties: for all $A, B, A_i \in \mathcal{P}(W), i \in J, J$ is an index set,*

(LD) $\underline{R}(A) =\sim \overline{R}(\sim A)$,   (UD) $\overline{R}(A) =\sim \underline{R}(\sim A)$;

(L1) $\underline{R}(W) = U$,   (U1) $\overline{R}(\emptyset) = \emptyset$;

(L2) $\underline{R}(\bigcap_{i \in J} A_i) = \bigcap_{i \in J} \underline{R}(A_i)$,   (U2) $\overline{R}(\bigcup_{i \in J} A_i) = \bigcup_{i \in J} \overline{R}(A_i)$;

(L3) $A \subseteq B \Longrightarrow \underline{R}(A) \subseteq \underline{R}(B)$, (U3) $A \subseteq B \Longrightarrow \overline{R}(A) \subseteq \overline{R}(B)$;

(L4) $\underline{R}(\bigcup_{i \in J} A_i) \supseteq \bigcup_{i \in J} \underline{R}(A_i)$,   (U4) $\overline{R}(\bigcap_{i \in J} A_i) \subseteq \bigcap_{i \in J} \overline{R}(A_i)$.

Properties (LD) and (UD) show that the rough approximation operators $\underline{R}$ and $\overline{R}$ are dual to each other. Properties with the same number may be regarded as dual properties. Properties (L2) and (U2) state that the lower rough approximation operator $\underline{R}$ is multiplicative, and the upper rough approximation operator $\overline{R}$ is additive. One may also say that $\underline{R}$ is distributive w.r.t. the intersection of sets, and $\overline{R}$ is distributive w.r.t. the union of sets. Properties (L3) and (U3) imply that $\underline{R}$ and $\overline{R}$ are monotone. Properties (L4) and (U4) show that $\underline{R}$ is not distributive w.r.t. set union, and $\overline{R}$ is not distributive w.r.t. set intersection. It can be easily checked that property (L2) implies properties (L3) and (L4), and dually, property (U2) yields properties (U3) and (U4).

By property (U2) we observe that $\overline{R}(X) = \bigcup_{x \in X} \overline{R}(\{x\})$. If we set

$$h(x) = \overline{R}(\{x\}), \quad x \in W, \quad (4)$$

then it is easy to verify that

$$h(x) = \{u \in U : x \in R_s(u)\}, \quad x \in W. \quad (5)$$

Conversely,

$$R_s(u) = \{y \in W : u \in h(y)\}, \quad u \in U. \quad (6)$$

By Eqs. (4)-(6), it is easy to see that

$$(x, y) \in R \iff x \in \overline{R}(\{y\}), \quad (x, y) \in U \times W. \quad (7)$$

Obviously,

$$\overline{R}(X) = \bigcup_{x \in X} h(x), \quad X \in \mathcal{P}(W). \tag{8}$$

Hence h is called the upper approximation distributive function [52,53]. The relationships between the inverse image j under R_s and the approximation operators can be concluded as follows:

$$\text{(JL)} \quad \underline{R}(X) = \bigcup_{Y \subseteq X} j(Y), \qquad X \subseteq W;$$

$$\text{(JU)} \quad \overline{R}(X) = \bigcup_{Y \cap X \neq \emptyset} j(Y), \qquad X \subseteq W;$$

$$\text{(LJ)} \quad j(X) = \underline{R}(X) \setminus \bigcup_{Y \subset X} \underline{R}(Y), \, X \subseteq W.$$

Analogous to Yao's study in [51,54], a serial rough set model is obtained from a serial binary relation. The property of a serial relation can be characterized by the properties of its induced rough approximation operators.

Theorem 2. *If R is an arbitrary crisp relation from U to W, and $\underline{R}$ and $\overline{R}$ are the rough approximation operators defined in Definition 2, then*

$$\begin{aligned}
R \text{ is serial} &\Longleftrightarrow \text{(L0)} \quad \underline{R}(\emptyset) = \emptyset, \\
&\Longleftrightarrow \text{(U0)} \quad \overline{R}(W) = U, \\
&\Longleftrightarrow \text{(LU0)} \quad \underline{R}(A) \subseteq \overline{R}(A), \forall A \in \mathcal{P}(W).
\end{aligned}$$

By (LU0), the pair of rough approximation operators of a serial rough set model is an interval structure. In the case of connections between other special crisp relations and rough approximation operators, we have the following theorem which may be seen as a generalization of Yao [51,52,54].

Theorem 3. *Let R be an arbitrary crisp binary relation on U, and $\underline{R}$ and $\overline{R}$ the lower and upper generalized crisp approximation operators defined in Definition 2. Then*

(1) R *is reflexive*
$\Longleftrightarrow$ (L5) $\underline{R}(A) \subseteq A, \forall A \in \mathcal{P}(U),$
$\Longleftrightarrow$ (U5) $A \subseteq \overline{R}(A), \forall A \in \mathcal{P}(U).$

(2) R *is symmetric*
$\Longleftrightarrow$ (L6) $\overline{R}(\underline{R}(A)) \subseteq A, \forall A \in \mathcal{P}(U),$
$\Longleftrightarrow$ (U6) $A \subseteq \underline{R}(\overline{R}(A)), \forall A \in \mathcal{P}(U).$

(3) R *is transitive*
$\Longleftrightarrow$ (L7) $\underline{R}(A) \subseteq \underline{R}(\underline{R}(A)), \forall A \in \mathcal{P}(U),$
$\Longleftrightarrow$ (U7) $\overline{R}(\overline{R}(A)) \subseteq \overline{R}(A), \forall A \in \mathcal{P}(U).$

(4) R *is Euclidean*
$\Longleftrightarrow$ (L8) $\overline{R}(\underline{R}(A)) \subseteq \underline{R}(A), \forall A \in \mathcal{P}(U),$
$\Longleftrightarrow$ (U8) $\overline{R}(A) \subseteq \underline{R}(\overline{R}(A)), \forall A \in \mathcal{P}(U).$

Remark 1. If R is a preorder on U, then, according to Theorem 3, the lower and upper approximation operators satisfy the following properties:
(L9) $\underline{R}(A) = \underline{R}(\underline{R}(A)), \forall A \in \mathcal{P}(U),$
(U9) $\overline{R}(\overline{R}(A)) = \overline{R}(A), \forall A \in \mathcal{P}(U).$

If R is an equivalence relation on U, then the pair (U, R) is a Pawlak approximation space and more interesting properties of lower and upper approximation operators can be derived [22].

4   Topological Spaces and Rough Approximation Operators

The relationships between topological spaces and rough approximation operators have been studied by many researchers [3,4,5,12,13,31,43,51,52,61]. In this section, we present the topological structures of rough sets in infinite universes of discourse.

4.1   Basic Concepts of Topological Spaces

Definition 3. [10] *Let U be a nonempty set. A topology on U is a family τ of sets in U satisfying the following axioms:*

(T_1) $\emptyset, U \in \tau$,

(T_2) $G_1 \cap G_2 \in \tau$ *for any* $G_1, G_2 \in \tau$,

(T_3) $\bigcup_{i \in J} G_i \in \tau$ *for any family* $\{G_i : i \in J\} \subseteq \tau$, *where J is an index set.*

In this case the pair (U, τ) is called a topological space and each set A in τ is referred to as an open set in U. The complement of an open set in the topological space (U, τ) is called a closed set in (U, τ).

Definition 4. *Let (U, τ) be a topological space and $A \in \mathcal{P}(U)$. Then the interior and closure of A are, respectively, defined by*

$int(A) = \cup\{G : G \text{ is an open set and } G \subseteq A\}$,

$cl(A) = \cap\{K : K \text{ is an closed set and } A \subseteq K\}$,

and int and cl are, respectively, called the interior operator and the closure operator of τ, and sometimes in order to distinguish, we denote them by int_τ and cl_τ.

It can be shown that $cl(A)$ is a closed set and $int(A)$ is an open set in (U, τ). A is an open set in (U, τ) if and only if $int(A) = A$, and A is a closed set in (U, τ) if and only if $cl(A) = A$. Moreover, the interior and closure operators are dual with each other, i.e.,

$$cl(\sim A) = \sim int(A), \quad \forall A \in \mathcal{P}(U), \tag{9}$$

$$int(\sim A) = \sim cl(A), \quad \forall A \in \mathcal{P}(U). \tag{10}$$

The closure operator can be also defined by axioms called the Kuratowski closure axioms [10].

Definition 5. *A mapping $cl : \mathcal{P}(U) \to \mathcal{P}(U)$ is referred to as a closure operator iff for all $A, B \in \mathcal{P}(U)$ it satisfies following axioms:*

(Cl1) $A \subseteq cl(A)$,   *extensiveness*

(Cl2) $cl(A \cup B) = cl(A) \cup cl(B)$,   *preservation of binary unions*

(Cl3) $cl(cl(A)) = cl(A)$,   *idempotency*

(Cl4) $cl(\emptyset) = \emptyset$.   *preservation of nullary unions*

Similarly, the interior operator can be defined by corresponding axioms.

Definition 6. *A mapping* $int : \mathcal{P}(U) \to \mathcal{P}(U)$ *is referred to as an interior operator iff for all* $A, B \in \mathcal{P}(U)$ *it satisfies following axioms:*

(Int1) $int(A) \subseteq A$,   *contraction*
(Int2) $int(A \cap B) = int(A) \cap int(B)$,   *preservation of binary intersections*
(Int3) $int(int(A)) = int(A)$,   *idempotency*
(Int4) $int(U) = U$.   *preservation of universal intersections*

It is easy to show that an interior operator *int* defines a topology

$$\tau_{int} = \{A \in \mathcal{P}(U) : int(A) = A\}. \tag{11}$$

So, the open sets are the fixed points of *int*. Dually, from a closure operator, one can obtain a topology on U by setting

$$\tau_{cl} = \{A \in \mathcal{P}(U) : cl(\sim A) = \sim A\}. \tag{12}$$

The results are summarized as the following

Theorem 4. [10] (1) *If an operator* $int : \mathcal{P}(U) \to \mathcal{P}(U)$ *satisfies axioms* (Int1)-(Int4), *then* τ_{int} *defined in Eq.* (11) *is a topology on* U *and*

$$int_{\tau_{int}} = int. \tag{13}$$

(2) *If an operator* $cl : \mathcal{P}(U) \to \mathcal{P}(U)$ *satisfies axioms* (Cl1)-(Cl4), *then* τ_{cl} *defined in Eq.* (12) *is a topology on* U *and*

$$cl_{\tau_{cl}} = cl. \tag{14}$$

Definition 7. [1,14] *A topology* τ *on the set* U *is called an Alexandrov topology if the intersection of arbitrarily many open sets is still open, or equivalently, the union of arbitrarily many closed sets is still closed. A topological space* (U, τ) *is said to be an Alexandrov space if* τ *is an Alexandrov topology.*

It is easy to verify that a topological space is an Alexandrov space if and only if the closure operator of the topological space distributes over arbitrary unions of subsets.

4.2   From Approximation Spaces to Topological Spaces

In this subsection we always assume that U is a nonempty universe of discourse, R a binary relation on U, and $\underline{R}$ and $\overline{R}$ the rough approximation operators defined in Definition 2.

Denote

$$\tau_R = \{A \in \mathcal{P}(U) : \underline{R}(A) = A\}. \tag{15}$$

The next theorem shows that any reflexive binary relation determines a topology.

Theorem 5. [12] *If* R *is a reflexive binary relation on* U, *then* τ_R *defined in Eq.* (15) *is a topology on* U.

By using Theorems 1, 3, Remak 1, and results in [12,20], we can obtain the following

Theorem 6. *Assume that R is a binary relation on U. Then the following are equivalent:*

(1) *R is a preorder, i.e., R is a reflexive and transitive relation;*
(2) *the upper approximation operator $\overline{R} : \mathcal{P}(U) \to \mathcal{P}(U)$ is a closure operator;*
(3) *the lower approximation operator $\underline{R} : \mathcal{P}(U) \to \mathcal{P}(U)$ is an interior operator.*

Theorem 6 shows that the lower and upper approximation operators constructed from a preorder are the interior and closure operators respectively. By Theorem 6, we can easily conclude following

Theorem 7. *Assume that R is a preorder on U. Define*

$$\tau_R = \{\underline{R}(A) : A \in \mathcal{P}(U)\}. \tag{16}$$

Then τ_R defined in Eq. (16) is a topology on U and $int_{\tau_R} = \underline{R} : \mathcal{P}(U) \to \mathcal{P}(U)$ and $cl_{\tau_R} = \overline{R} : \mathcal{P}(U) \to \mathcal{P}(U)$ are the interior and closure operators of τ_R respectively.

Remark 2. According to properties (L2) and (U2) in Theorem 1 it can be easily observed that τ_R in Theorem 7 is an Alexandrov topology. Thus Theorems 6 and 7 state that a reflexive and transitive approximation space can generate an Alexandrov space such that the family of all lower approximations of sets w.r.t. the approximation space forms the topology. And the lower and upper rough approximation operators are, respectively, the interior and closure operators of the topological space.

4.3   From Topological Spaces to Approximation Spaces

In this subsection, we show that the interior (closure, respectively) operator in a topological space can associate with a preorder such that the inducing lower (upper, respectively) rough approximation operator is exactly the interior (closure, respectively) operator if only if the given topological space is an Alexandrov space.

We now define a binary relation R_τ on U by employing the closure operator $cl : \mathcal{P}(U) \to \mathcal{P}(U)$ in a topological space (U, τ) as follows: for $(x, y) \in U \times U$,

$$(x, y) \in R_\tau \iff x \in cl(\{y\}). \tag{17}$$

By employing Eq. (17), we can conclude following

Theorem 8. *Let $cl : \mathcal{P}(U) \to \mathcal{P}(U)$ be the closure operator and $int : \mathcal{P}(U) \to \mathcal{P}(U)$ the interior operator dual to cl in a topological space (U, τ). Then there exists a preorder R_τ on U such that*

$$\overline{R_\tau}(A) = cl(A), \quad \underline{R_\tau}(A) = int(A), \quad \forall A \in \mathcal{P}(U) \tag{18}$$

iff (U, τ) *is an Alexandrov space, i.e. cl and int, respectively, satisfy following axioms* (AL2) *and* (AU2):

(AL2) $int(\bigcap\limits_{j \in J} A_j) = \bigcap\limits_{j \in J} int(A_j)$, $\quad \forall A_j \in \mathcal{P}(U)$, $j \in J$, J *is an index set,*

(AU2) $cl(\bigcup\limits_{j \in J} A_j) = \bigcup\limits_{j \in J} cl(A_j)$, $\quad \forall A_j \in \mathcal{P}(U)$, $j \in J$, J *is an index set.*

Let $\mathcal{R}$ be the set of all preorders on U and $\mathcal{T}$ the set of all Alexandrov topologies on U, in terms of Theorems 4 and 8, we can easily conclude following Theorems 9 and 10.

Theorem 9. (1) *If* $R \in \mathcal{R}$, τ_R *is defined by Eq.* (16) *and* R_{τ_R} *is defined by Eq.* (17), *then* $R_{\tau_R} = R$.
  (2) *If* $\tau \in \mathcal{T}$, R_τ *is defined by Eq.* (17), *and* τ_{R_τ} *is defined by Eq.* (16), *then* $\tau_{R_\tau} = \tau$.

Theorem 10. *There exists a one-to-one correspondence between* $\mathcal{R}$ *and* $\mathcal{T}$.

5   Axiomatic Characterizations of Rough Approximation Operators

In the axiomatic approach of rough set theory, the lower and upper approximation operators are primitive notions characterized by abstract axioms [5,11,17,50,51,52,53]. In this section, we show the axiomatic study on rough approximation operators in infinite universes of discourse. The results may be viewed as the generalized counterparts of Yao [51,52,53].

Definition 8. *Let* $L, H : \mathcal{P}(W) \to \mathcal{P}(U)$ *be two operators. They are referred to as dual operators if for all* $A \in \mathcal{P}(W)$,

$$\begin{aligned} \text{(ALD)} \quad & L(A) = \sim H(\sim A), \\ \text{(AUD)} \quad & H(A) = \sim L(\sim A). \end{aligned}$$

By the dual properties of the operators, we only need to define one operator and the other one can be computed by the dual properties. For example, one may define the operator H and regard L as an abbreviation of $\sim H \sim$. From an operator $H : \mathcal{P}(W) \to \mathcal{P}(U)$ we define a binary relation R_H from U to W as follows:

$$(x, y) \in R_H \iff x \in H(1_y), \quad (x, y) \in U \times W. \tag{19}$$

By employing Eq. (19), we can conclude following Theorem 11 via the discussion on the constructive approach which may be seen as an analogous one in [11,51].

Theorem 11. *Assume that* $L, H : \mathcal{P}(W) \to \mathcal{P}(U)$ *are dual operators. Then there exists a binary relation* R_H *from* U *to* W *such that for all* $A \in \mathcal{P}(W)$,

$$L(A) = \underline{R_H}(A), \quad and \quad H(A) = \overline{R_H}(A) \tag{20}$$

iff L satisfies axioms (AL1), (AL2), *or equivalently, H obeys axioms* (AU1), (AU2):

$$(AL1)\ L(W) = U,$$
$$(AL2)\ L(\bigcap_{i \in I} A_i) = \bigcap_{i \in I} L(A_i),\ \ \forall A_i \in \mathcal{P}(W), \forall i \in I;$$
$$(AU1)\ H(\emptyset) = \emptyset;$$
$$(AU2)\ H(\bigcup_{i \in I} A_i) = \bigcup_{i \in I} H(A_i), \forall A_i \in \mathcal{P}(W), \forall i \in I.$$

Remark 3. Axioms (AU1) and (AU2) are independent. Thus axiom set {(AUD), (AU1), (AU2)} is the basic axiom set to characterize generalized rough approximation operators.

Combining Theorems 2 and 11, we present the axiomatic characterization of serial rough set algebra as follows:

Theorem 12. *Let $L, H : \mathcal{P}(W) \longrightarrow \mathcal{P}(U)$ be a pair of dual operators. Then there exists a serial relation R_H from U to W such that $\forall A \in \mathcal{P}(W), L(A) = \underline{R_H}(A)$ and $H(A) = \overline{R_H}(A)$ if and only if L satisfies axioms* (AL1), (AL2) *and one of the following equivalent axioms, or equivalently H satisfies axioms* (AU1), (AU2) *and one of the following equivalent axioms:*

$$(AL0)\quad L(\emptyset) = \emptyset,$$
$$(AU0)\quad H(W) = U,$$
$$(ALU0)\quad L(A) \subseteq H(A),\quad \forall A \in \mathcal{P}(W).$$

Remark 4. Axioms (AU1), (AU2) and (AU0) in Theorem 12 are independent, i.e., any two of the axioms cannot derive the third one. Thus axiom set {(AUD), (AU1), (AU2), (AU0)} is the basic axiom set to characterize serial rough approximation operators.

Axiom (ALU0) states that $L(A)$ is a subset of $H(A)$. In such a case, $L, H : \mathcal{P}(P) \to \mathcal{P}(U)$ are called the lower and upper rough approximation operators and the system $(\mathcal{P}(U), \mathcal{P}(W), \cap, \cup, \sim, L, H)$ is an interval structure [53]. Similar to the axiomatic characterizations of approximation operators in finite universes of discourse [50,51,53], axiomatic characterizations of other special types of rough approximation operators in infinite universes of discourse are summarized in the following Theorems 13-17:

Theorem 13. *Let $L, H : \mathcal{P}(U) \longrightarrow \mathcal{P}(U)$ be a pair of dual operators. Then there exists a reflexive relation R_H on U such that $\forall A \in \mathcal{P}(U)$ and $L(A) = \underline{R_H}(A), H(A) = \overline{R_H}(A)$ if and only if L satisfies axioms* (AL1), (AL2) *and* (AL5), *or equivalently, H satisfies axioms* (AU1), (AU2) *and* (AU5):

$$(AL5)\ L(A) \subseteq A,\quad \forall A \in \mathcal{P}(U),$$
$$(AU5)\ A \subseteq H(A),\quad \forall A \in \mathcal{P}(U).$$

Remark 5. Axioms (AU1), (AU2) and (AU5) are independent, i.e., any two of the axioms cannot derive the third one. Thus axiom set {(AUD), (AU1), (AU2), (AU5)} is the basic axiom set to characterize reflexive rough approximation operators.

Theorem 14. *Let $L, H : \mathcal{P}(U) \longrightarrow \mathcal{P}(U)$ be a pair of dual operators. Then there exists a symmetric relation R_H on U such that $\forall A \in \mathcal{P}(U), L(A) = \underline{R_H}(A)$ and $H(A) = \overline{R_H}(A)$ if and only if L satisfies axioms (AL1), (AL2) and one of the following equivalent axioms, or equivalently H satisfies axioms (AU1), (AU2) and one of the following equivalent axioms:*

$$(AL6)' \quad y \in L(1_{U-\{x\}}) \Longleftrightarrow x \in L(1_{U-\{y\}}), \ \forall(x,y) \in U \times U,$$
$$(AU6)' \quad y \in H(1_x) \Longleftrightarrow x \in H(1_y)(x), \qquad \forall(x,y) \in U \times U,$$
$$(AL6) \quad A \subseteq L(H(A)), \qquad\qquad\qquad \forall A \in \mathcal{P}(U),$$
$$(AU6) \quad H(L(A)) \subseteq A, \qquad\qquad\qquad \forall A \in \mathcal{P}(U).$$

Remark 6. Similar to the approximation operators in finite universe of discourse in [50], we know that axiom (AU1) can be derived from axioms (AU2) and (AU6), and axioms (AU2) and (AU6) are independent. Thus axiom set $\{(AUD), (AU2), (AU6)\}$ is the basic axiom set to characterize symmetric rough approximation operators.

Theorem 15. *Let $L, H : \mathcal{P}(U) \longrightarrow \mathcal{P}(U)$ be a pair of dual operators. Then there exists a transitive relation R_H on U such that $\forall A \in \mathcal{P}(U), L(A) = \underline{R_H}(A)$ and $H(A) = \overline{R_H}(A)$ if and only if L satisfies axioms (AL1), (AL2) and (AL7), or equivalently, H satisfies axioms (AU1), (AU2) and (AU7):*

$$(AL7) \quad L(A) \subseteq L(L(A)), \quad \forall A \in \mathcal{P}(U),$$
$$(AU7) \quad H(H(A)) \subseteq H(A), \forall A \in \mathcal{P}(U).$$

Remark 7. Axioms (AU1), (AU2) and (AU7) are independent, i.e., any two of the axioms cannot derive the third one. Thus axiom set $\{(AUD), (AU1), (AU2), (AU7)\}$ is the basic axiom set to characterize transitive rough approximation operators.

Theorem 16. *Let $L, H : \mathcal{P}(U) \longrightarrow \mathcal{P}(U)$ be a pair of dual operators. Then there exists a Euclidean relation R_H on U such that $\forall A \in \mathcal{P}(U), L(A) = \underline{R_H}(A)$ and $H(A) = \overline{R_H}(A)$ if and only if L satisfies axioms (AL1), (AL2) and (AL8), or equivalently, H satisfies axioms (AU1), (AU2) and (AU8):*

$$(AL8) \quad H(L(A)) \subseteq L(A), \quad \forall A \in \mathcal{P}(U),$$
$$(AU8) \quad H(A) \subseteq L(H(A)), \quad \forall A \in \mathcal{P}(U).$$

Remark 8. Similar to the approximation operators in finite universe of discourse in [50], we know that axiom (AU1) can be derived from axioms (AU2) and (AU8), and axioms (AU2) and (AU8) are independent. Thus axiom set $\{(AUD), (AU2), (AU8)\}$ is the basic axiom set to characterize Euclidean rough approximation operators.

Theorem 17. *Let $L, H : \mathcal{P}(U) \longrightarrow \mathcal{P}(U)$ be a pair of dual operators. Then there exists an equivalence relation R_H on U such that $\forall A \in \mathcal{P}(U), L(A) = \underline{R_H}(A)$ and $H(A) = \overline{R_H}(A)$ if and only if L satisfies axioms (AL1), (AL2), (AL5), (AL6) and (AL7), or equivalently, H satisfies axioms (AU1), (AU2), (AU5), (AU6) and (AU7).*

Remark 9. Since axiom (AU1) can be induced by axioms (AU2) and (AU6), and it can be verified that axioms (AU2), (AU5), (AU6), (AU7) are independent, thus axiom set {(AUD), (AU2), (AU5), (AU6), (AU7)} is the basic axiom set to characterize Pawlak rough approximation operators. Since a reflexive and Euclidean relation is an equivalence one, by Remark 8 we state that the Pawlak rough approximation operators can be equivalently characterized by axiom set {(AUD), (AU2), (AU5), (AU8)}.

6   Rough Set Algebras in Infinite Universes of Discourse

In this section, we examine properties of approximation operators in various types of rough set algebras in infinite universes of discourse, which are the analogous results obtained by Yao in finite universes of discourse [51,53,54].

Definition 9. *Let W and U be two nonempty universes of discourse which may be infinite. A unary operator $L : \mathcal{P}(W) \to \mathcal{P}(U)$ is referred to as a rough lower approximation operator iff it satisfies axioms* (AL1) *and* (AL2) *in Theorem 11. A unary operator $H : \mathcal{P}(U) \to \mathcal{P}(U)$ is referred to as a rough upper approximation operator iff it satisfies axioms* (AU1) *and* (AU2) *in Theorem 11.*

Remark 10. According to Theorem 11, we can conclude that, for the dual rough lower and upper approximation operators $L, H : \mathcal{P}(W) \to \mathcal{P}(U)$, there exists a binary relation R_H from U to W such that

$$\underline{R_H}(A) = L(A), \quad \overline{R_H}(A) = H(A), \quad \forall A \in \mathcal{P}(W), \tag{21}$$

where $\underline{R_H}(A)$ and $\overline{R_H}(A)$ are defined in Definition 2.

Definition 10. *If $L, H : \mathcal{P}(W) \to \mathcal{P}(U)$ are dual rough lower and upper approximation operators, i.e., L satisfies axioms* (AL1), (AL2) *and* (ALD), *or equivalently, H satisfies axioms* (AU1), (AU2), *and* (AUD). *Then the system $S_L =: (\mathcal{P}(W), \mathcal{P}(U), \cap, \cup, \sim, L, H)$ is referred to as a rough set algebra (RSA). If there exists a serial relation R_H from U to W such that $L(A) = \underline{R_H}(A)$ and $H(A) = \overline{R_H}(A)$ for all $A \in \mathcal{P}(W)$, then S_L is called a serial RSA. Moreover, if $W = U$ and there exists a reflexive (respectively, a symmetric, a transitive, a Euclidean, an equivalence) relation R_H on U such that $L(A) = \underline{R_H}(A)$ and $H(A) = \overline{R_H}(A)$ for all $A \in \mathcal{P}(U)$, then S_L is called a reflexive (respectively, a symmetric, a transitive, a Euclidean, a Pawlak) RSA.*

Axioms (ALD) and (AUD) imply that operators L and H in a RSA S_L are dual with each other. It can be easily verified that axiom (AL2) implies the following axioms (AL3) and (AL4), and dually, axiom (AU2) implies the following axioms (AU3) and (AU4):

(AL3) $L(\bigcup_{j \in J} A_j) \supseteq \bigcup_{j \in J} L(A_j), \quad \forall A_j \in \mathcal{P}(W), j \in J, J$ is an index set,

(AL4) $A \subseteq B \Longrightarrow L(A) \subseteq L(B), \quad \forall A, B \in \mathcal{P}(W),$

(AU3) $H(\bigcap_{j \in J} A_j) \subseteq \bigcap_{j \in J} H(A_j), \quad \forall A_j \in \mathcal{P}(W), j \in J, J$ is an index set,

(AU4) $A \subseteq B \Longrightarrow H(A) \subseteq H(B), \forall A, B \in \mathcal{P}(W).$

In the sequel, we will denote

$$LL(A) = L(L(A)), \qquad HL(A) = H(L(A)), \ A \in \mathcal{P}(W),$$
$$HH(A) = H(H(A)), \qquad LH(A) = L(H(A)), \ A \in \mathcal{P}(W),$$
$$S_{LL} = (\mathcal{P}(W), \mathcal{P}(U), \cap, \cup, \sim, LL, HH).$$

If $W = U$, we will write $S_L = (\mathcal{P}(U), \cap, \cup, \sim, L, H)$ and $S_{LL} = (\mathcal{P}(U), \cap, \cup, \sim, LL, HH)$ for simplicity to replace $S_L = (\mathcal{P}(W), \mathcal{P}(U), \cap, \cup, \sim, L, H)$ and $S_{LL} = (\mathcal{P}(W), \mathcal{P}(U), \cap, \cup, \sim, LL, HH)$ respectively.

Theorem 18. *If $S_L = (\mathcal{P}(W), \mathcal{P}(U), \cap, \cup, \sim, L, H)$ is a RSA, then S_{LL} is also a RSA.*

Proof. It is only to prove that HH satisfies axioms (AU1), (AU2), and (AUD). For any $A \in \mathcal{P}(W)$, since H satisfies axiom (AU1), we have

$$HH(\emptyset) = H(H(\emptyset)) = H(\emptyset) = \emptyset. \tag{22}$$

Thus, operator HH obeys axiom (AU1). Since H obeys axioms (AU2) and (AUD), it is easy to verify that HH also satisfies axioms (AU2) and (AUD).

In what follows we discuss properties of approximation operators in special classes of RSAs. We will investigate the relationships between a RSA S_L and its inducing system S_{LL}.

6.1   Serial RSAs

In a serial RSA S_L, $L(A)$ is a subset of $H(A)$ for all $A \in \mathcal{P}(W)$, and L and H map $\emptyset$ and W into $\emptyset$ and U respectively. We then have the following relationships between the approximation operators:

$$LL(A) \subseteq LH(A)(\text{and } HL(A)) \subseteq HH(A), \quad A \in \mathcal{P}(W). \tag{23}$$

Thus, operators LL and HH obey axiom (ALU0), then, by Theorem 18, we can obtain the following

Theorem 19. *If S_L is a serial RSA, then S_{LL} is also a serial RSA.*

6.2   Reflexive RSAs

In a reflexive RSA S_L, L and H respectively satisfy axioms (AL5) and (AU5). It is easy to observe that LL and HH also obey axioms (AL5) and (AU5) respectively, thus by using Theorem 18 we can obtain following

Theorem 20. *If $S_L = (\mathcal{P}(U), \cap, \cup, \sim, L, H)$ is a reflexive RSA on U, then $S_{LL} = (\mathcal{P}(U), \cap, \cup, \sim, LL, HH)$ is also a reflexive RSA on U.*

In a reflexive RSA, we have the following relationships between the composed approximation operators which are stronger than those in Eq. (23):

$$LL(A) \subseteq L(A) \subseteq A \subseteq H(A) \subseteq HH(A), \quad A \in \mathcal{P}(U),$$
$$LL(A) \subseteq L(A) \subseteq LH(A) \subseteq H(A) \subseteq HH(A), \quad A \in \mathcal{P}(U), \qquad (24)$$
$$LL(A) \subseteq L(A) \subseteq HL(A) \subseteq H(A) \subseteq HH(A), \quad A \in \mathcal{P}(U).$$

Together with the monotonicity of L and H, axioms (AL5) and (AU5) imply the following properties: $\forall A, B \in \mathcal{P}(U)$,

$$
\begin{aligned}
A \subseteq L(B) &\implies L(A) \subseteq L(B), \\
H(A) \subseteq B &\implies H(A) \subseteq H(B), \\
L(A) \subseteq B &\implies L(A) \subseteq H(B), \\
A \subseteq H(B) &\implies L(A) \subseteq H(B), \\
L(A) \subseteq L(B) &\implies L(A) \subseteq B, \\
H(A) \subseteq H(B) &\implies A \subseteq H(B), \\
H(A) \subseteq L(B) &\implies A \subseteq L(B), \ H(A) \subseteq B.
\end{aligned}
$$

6.3   Symmetric RSAs

In a symmetric RSA S_L, approximation operators L and H respectively obey (AL6) and (AU6). From Theorem 14 we know that operator L in a symmetric RSA can be equivalently characterized by axioms (ALD), (AL2), and (AL6) (or equivalently, H can be characterized by axioms (AUD), (AU2), and (AU6)).

Theorem 21. *If S_L is a symmetric RSA on U, then*

$$HLH(A) = H(A), \quad LHL(A) = L(A), \ \forall A \in \mathcal{P}(U). \qquad (25)$$

Proof. Since S_L is a symmetric RSA, $A \subseteq LH(A)$ holds for all $A \in \mathcal{P}(U)$ and in terms of the monotonicity of H we then obtain

$$H(A) \subseteq HLH(A), \ \forall A \in \mathcal{P}(U). \qquad (26)$$

On the other hand, by replacing A with $H(A)$ in $HL(A) \subseteq A$, we have

$$HLH(A) \subseteq H(A), \ \forall A \in \mathcal{P}(U). \qquad (27)$$

Thus

$$HLH(A) = H(A), \ \forall A \in \mathcal{P}(U). \qquad (28)$$

Likewise, we can conclude

$$LHL(A) = L(A), \ \forall A \in \mathcal{P}(U). \qquad (29)$$

Theorem 22. *If S_L is a symmetric RSA on U, then, $\forall A, B \in \mathcal{P}(U)$,*

$$H(A) \subseteq B \iff A \subseteq L(B). \qquad (30)$$

Proof. Since S_L is a symmetric RSA, for any $A, B \in \mathcal{P}(U)$, by the monotonicity of H and the duality of L and H, we have

$$H(A) \subseteq B \Longleftrightarrow \sim L(\sim A) \subseteq B \Longleftrightarrow \sim B \subseteq L(\sim A)$$
$$\Longrightarrow H(\sim B) \subseteq HL(\sim A) \subseteq \sim A \Longrightarrow A \subseteq \sim H(\sim B) = L(B).$$

Hence

$$H(A) \subseteq B \Longrightarrow A \subseteq L(B). \tag{31}$$

On the other hand, we have

$$A \subseteq L(B) \Longleftrightarrow A \subseteq \sim H(\sim B) \Longleftrightarrow H(\sim B) \subseteq \sim A \Longrightarrow LH(\sim B) \subseteq L(\sim A)$$
$$\Longleftrightarrow \sim HL(B) \subseteq L(\sim A) = \sim H(A) \Longrightarrow H(A) \subseteq HL(B) \subseteq B.$$

Thus we conclude Eq. (30).

Theorem 23. *If S_L is a symmetric RSA on U, then S_{LL} is also a symmetric RSA on U.*

Proof. From Theorem 18 we see that S_{LL} is a RSA. By axiom (FL6), we have

$$H(A) \subseteq LHH(A), \quad \forall A \in \mathcal{P}(U). \tag{32}$$

By the monotonicity of L, it follows that

$$LH(A) \subseteq LLHH(A), \quad \forall A \in \mathcal{P}(U). \tag{33}$$

Since $A \subseteq LH(A)$, we have

$$A \subseteq LLHH(A), \quad \forall A \in \mathcal{P}(U). \tag{34}$$

Consequently, by the duality of L and H, we conclude that

$$HHLL(A) \subseteq A, \quad \forall A \in \mathcal{P}(U). \tag{35}$$

Thus operators LL and HH respectively obey axioms (AL6) and (AU6). Therefore, S_{LL} is a symmetric RSA.

6.4 Transitive RSAs

In a transitive RSA S_L, L and H respectively obey axioms (AL7) and (AU7). We then have

$$HHHH(A) \subseteq HHH(A) \subseteq HH(A), \quad \forall A \in \mathcal{P}(U). \tag{36}$$

Thus, HH obeys axiom (AU7), in terms of Theorem 18, we obtain following

Theorem 24. *If S_L is a transitive RSA on U, then S_{LL} is also a transitive RSA on U.*

In a transitive RSA, by employing the monotonicity of L and H, and in terms of axioms (AL7) and (AU7), we can conclude following properties:

$$L(A) \subseteq B \Longrightarrow L(A) \subseteq L(B), \quad A, B \in \mathcal{P}(U),$$
$$A \subseteq H(B) \Longrightarrow H(A) \subseteq H(B), \quad A, B \in \mathcal{P}(U). \tag{37}$$

6.5   Euclidean RSAs

In a Euclidean RSA S_L, L and H respectively obey axioms (AL8) and (AU8). By axiom (AL8), we have

$$HLL(A) \subseteq L(A), \quad \forall A \in \mathcal{P}(U). \tag{38}$$

Then, by the monotonicity of H and axiom (AL9), it follows that

$$HHLL(A) \subseteq HLL(A) \subseteq LL(A), \quad \forall A \in \mathcal{P}(U). \tag{39}$$

Similarly, we have

$$HH(A) \subseteq LLHH(A), \quad \forall A \in \mathcal{P}(U). \tag{40}$$

Therefore, in terms of Theorem 18 we can obtain following

Theorem 25. *Suppose that S_L is a Euclidean RSA on U, then S_{LL} is also a Euclidean RSA on U.*

It is easy to verify that approximation operators L and H in a Euclidean RSA S_L have the following properties:

$$
\begin{aligned}
&HL(A) \subseteq LL(A), \quad HH(A) \subseteq LH(A), \quad A \in \mathcal{P}(U), \\
&A \subseteq L(B) \Longrightarrow H(A) \subseteq L(B), \quad A, B \in \mathcal{P}(U), \\
&H(A) \subseteq B \Longrightarrow H(A) \subseteq L(B), \quad A, B \in \mathcal{P}(U).
\end{aligned} \tag{41}
$$

6.6   Serial and Symmetric RSAs

Operator L in a serial and symmetric RSA S_L can be characterized by axioms (ALD), (AL2), (AL0), and (AL6). It is easy to verify that

$$LL(A) \subseteq HL(A) \subseteq A \subseteq LH(A) \subseteq HH(A), \quad \forall A \in \mathcal{P}(U). \tag{42}$$

Thus, from Theorems 18 and 23 we conclude that a serial and symmetric RSA S_L produces a reflexive and symmetric RSA S_{LL}. Moreover, if S_L is a reflexive RSA, we have the following relationship: $\forall A \in \mathcal{P}(U)$,

$$LL(A) \subseteq L(A) \subseteq HL(A) \subseteq A \subseteq LH(A) \subseteq H(A) \subseteq HH(A). \tag{43}$$

In such a case, for each set $A \in \mathcal{P}(U)$, three systems S_{HL}, S_L, and S_{LL} produce a nested family of approximations, where $S_{HL} = (\mathcal{P}(U), \cap, \cup, \sim, HL, LH)$.

6.7   Serial and Transitive RSAs

Operator L in a serial and transitive RSA S_L is characterized by axioms (ALD), (AL0), (AL1), (AL2), and (AL7), then S_{LL} is also a serial and transitive RSA. It is easy to verify that

$$
\begin{aligned}
&L(A) \subseteq LL(A) \subseteq HL(A) \subseteq HH(A) \subseteq H(A), \quad A \in \mathcal{P}(U), \\
&L(A) \subseteq LL(A) \subseteq LH(A) \subseteq HH(A) \subseteq H(A), \quad A \in \mathcal{P}(U).
\end{aligned} \tag{44}
$$

Moreover, if S_L is a reflexive RSA, L and H obey axioms:

(AL9) $L(A) = LL(A), \ \forall A \in \mathcal{P}(U),$ (AU9) $H(A) = HH(A), \ \forall A \in \mathcal{P}(U).$

In such a case, two systems S_L and S_{LL} become the same one. Notice that a reflexive RSA is a serial one and thus operators L and H respectively obey axioms (AL0) and (AU0). It should be noted that axioms (AL0), (AL2), (AL5), and (AL9) of L, and (AU0), (AU2), (AU5), and (AU9) of H are the axioms of interior and closure operators of a topological space. Such an algebra is thus referred to as a topological RSA. With the topological RSA, a set A is open if $L(A) = A$, and closed if $H(A) = A$. It follows from axioms (AL9) and (AU9) that L and H respectively map any set into an open set and a closed set.

Operators L and H in a reflexive and transitive RSA S_L have relationships:

$$
\begin{array}{ll}
L(A) = LL(A) \subseteq A \subseteq HH(A) = H(A), & A \in \mathcal{P}(U), \\
L(A) = LL(A) \subseteq HL(A) \subseteq HH(A) = H(A), & A \in \mathcal{P}(U), \quad (45) \\
L(A) = LL(A) \subseteq LH(A) \subseteq HH(A) = HA, & A \in \mathcal{P}(U).
\end{array}
$$

6.8   Serial and Euclidean RSAs

Operator L in a serial and Euclidean RSA S_L is characterized by axioms (ALD), (AL0), (AL2), and (AL8), then S_{LL} is also a serial and Euclidean RSA. In terms of Eq. (41) it is easy to verify that

$$LL(A) = HL(A) \subseteq L(A) \subseteq H(A) \subseteq LH(A) = HH(A), \quad \forall A \in \mathcal{P}(U). \quad (46)$$

In such a case, two systems S_{LL} and S_{HL} become the same one and thus S_{HL} is also a serial and Euclidean RSA.

Moreover, if S_L is a reflexive rough set algebra, then

$$HL(A) \subseteq L(A) \subseteq A, \quad A \subseteq H(A) \subseteq LH(A), \quad \forall A \in \mathcal{P}(U). \quad (47)$$

From Eq. (47) we see that L and H respectively obey axioms (AL6) and (AU6), thus S_L is a symmetric RSA. On the other hand, by the monotonicity of H, axioms (AU7), (AL7), and (AL5) we have

$$HH(A) \subseteq HLH(A) \subseteq LH(A) \subseteq H(A), \quad \forall A \in \mathcal{P}(U). \quad (48)$$

That is, H obeys axiom (AU6), therefore, S_L is a transitive RSA. Thus, we conclude that a reflexive and Euclidean RSA is a Pawlak RSA. It is easy to observe

$$LL(A) = HL(A) = L(A) \subseteq A \subseteq H(A) = LH(A) = HH(A), \forall A \in \mathcal{P}(U). \quad (49)$$

In such a case, three systems S_L, S_{LL}, and S_{HL} become the same RSA.

6.9   Symmetric and Transitive RSAs

Operator L in a symmetric and transitive RSA S_L is characterized by axioms (ALD), (AL2), (AL6), and (AL7), the system S_{LL} is also a symmetric and transitive RSA. By axioms (AL6) and (AL7), we have

$$H(A) \subseteq LHH(A) \subseteq LH(A), \ \forall A \in \mathcal{P}(U). \tag{50}$$

That is, H obeys axiom (AU8), therefore, S_L is a Euclidean RSA. Hence we have following relationships:

$$HL(A) \subseteq L(A) \subseteq LL(A), \quad HH(A) \subseteq H(A) \subseteq LH(A), \ \forall A \in \mathcal{P}(U). \tag{51}$$

Moreover, if S_L is a serial rough fuzzy algebra, then

$$A \subseteq LH(A) \subseteq HH(A) \subseteq H(A), \ \forall A \in \mathcal{P}(U). \tag{52}$$

That is, S_L is a reflexive RSA. Therefore, S_L is a Pawlak RSA and three systems S_L, S_{LL}, and S_{HL} become the same RSA.

6.10   Symmetric and Euclidean RSAs

Operator L in a symmetric and Euclidean RSA S_L is characterized by axioms (ALD), (AL2), (AL6), and (AL8), the system S_{LL} is also a symmetric and Euclidean RSA. By axioms (AL6), (AU6), and (AL8) we have

$$L(A) \subseteq LHL(A) \subseteq LL(A), \quad \forall A \in \mathcal{P}(U). \tag{53}$$

That is, S_L is a transitive RSA. Moreover, if S_L is a serial RSA, then we see that S_L is a Pawlak RSA and the three systems S_L, S_{LL}, and S_{HL} become the same RSA.

6.11   Transitive and Euclidean RSAs

Operator L in a transitive and Euclidean RSA S_L is characterized by axioms (ALD), (AL2), (AL7), and (AL8), the system S_{LL} is also a transitive and Euclidean RSA. Obviously, Eq. (51) holds. Moreover, if S_L is a serial RSA, it is easy to prove

$$LL(A) = L(A) = HL(A), \quad HH(A) = H(A) = LH(A), \quad A \in \mathcal{P}(U). \tag{54}$$

Therefore, three systems S_L, S_{LL}, and S_{HL} become the same RSA. Only if S_L is a reflexive or a symmetric RSA can S_L be a Pawlak RSA.

6.12   Topological RSAs

If S_L is a serial and transitive RSA, the operator L in S_L is characterized by axioms (ALD), (AL1), (AL2), (ALU0), and (AL7), by Theorems 19 and 24, S_{LL} is also a serial and transitive RSA. It is easy to verify that

$$
\begin{aligned}
L(A) \subseteq LL(A) \subseteq HL(A) \subseteq HH(A) \subseteq H(A), \quad A \in \mathcal{P}(U), \\
L(A) \subseteq LL(A) \subseteq LH(A) \subseteq HH(A) \subseteq H(A), \quad A \in \mathcal{P}(U).
\end{aligned} \tag{55}
$$

Moreover, if S_L is a reflexive RSA, we see that L and H respectively obey following axioms (AL9) and (AU9):

$$\text{(AL9) } L(A) = LL(A), \quad \forall A \in \mathcal{P}(U),$$
$$\text{(AU9) } H(A) = HH(A), \forall A \in \mathcal{P}(U).$$

In such a case, two systems S_L and S_{LL} become the same one. Obviously, a reflexive RSA is a serial one, and thus operators L and H respectively obey axioms (AL0) and (AU0). It should be noted from Section 4 that axioms (AL0), (AL2), (AL5), and (AL9) of L, and (AU0), (AU2), (AU5), and (AU9) of H are, respectively, the axioms of interior and closure operators in a topological space. Such a rough set algebra is thus called a topological RSA. With the topological RSA S_L, a set A is open if $L(A) = A$, and closed if $H(A) = A$. It follows from axioms (AL9) and (AU9) that L and H map each set into an open set and a closed set respectively. By axioms (AL9) and (AU9), we conclude following

Theorem 26. *If S_L is a topological RSA on U, then S_{LL} is also a topological RSA on U.*

Operators L and H in a topological RSA S_L have relationships:

$$L(A) = LL(A) \subseteq A \subseteq HH(A) = H(A), \qquad A \in \mathcal{P}(U),$$
$$L(A) = LL(A) \subseteq HL(A) \subseteq HH(A) = H(A), \quad A \in \mathcal{P}(U), \qquad (56)$$
$$L(A) = LL(A) \subseteq LH(A) \subseteq HH(A) = H(A), \quad A \in \mathcal{P}(U).$$

From Theorems 20, 23, and 24, we can immediately obtain

Theorem 27. *If S_L is a Pawlak RSA on U, then S_{LL} is also a Pawlak RSA on U.*

7 Measurable Structures of Rough Set Algebras

In this section, we examine under which conditions a rough set algebra can generate a measurable space such that the class of all measurable sets is exactly the family of all definable sets in the given rough set algebra.

Definition 11. [8] *Let U be a nonempty set. A subset $\mathcal{A}$ of $\mathcal{P}(U)$ is called a σ-algebra iff*
   (A1) $U \in \mathcal{A}$,
   (A2) $\{X_n : n \in \mathbf{N}\} \subset \mathcal{A} \Longrightarrow \bigcup_{n \in \mathbf{N}} X_n \in \mathcal{A}$,
   (A3) $X \in \mathcal{A} \Longrightarrow \sim X \in \mathcal{A}$.
The sets in $\mathcal{A}$ are called measurable sets and the pair $(U, \mathcal{A})$ a measurable space.

With the definition we can see that $\emptyset \in \mathcal{A}$ and
   (A2)$'$ $\{X_n : n \in \mathbf{N}\} \subset \mathcal{A} \Longrightarrow \bigcap_{n \in \mathbf{N}} X_n \in \mathcal{A}$.
If U is a finite universe of discourse, then condition (A2) in Definition 11 can be replaced by
   (A2)$''$ $X, Y \in \mathcal{A} \Longrightarrow X \cup Y \in \mathcal{A}$.

In such a case, $\mathcal{A}$ is said to be an algebra. If we denote

$$\mathcal{A}(x) = \cap\{X \in \mathcal{A} : x \in X\}, \quad x \in U, \tag{57}$$

then $\mathcal{A}(x) \in \mathcal{A}$, it can be checked that $\{\mathcal{A}(x) : x \in U\} \subseteq \mathcal{A}$ forms a partition on U and $\mathcal{A}(x)$ is called the atom of $\mathcal{A}$ containing x.

Theorem 28. *Assume that $S = (\mathcal{P}(U), \cap, \cup, \sim, L, H)$ is a serial rough set algebra. Denote*

$$\mathcal{A} = \{X \in \mathcal{P}(U) : L(X) = X = H(X)\}. \tag{58}$$

Then $\mathcal{A}$ is a σ-algebra on U.

Proof. (1) Since S is a serial rough set algebra, we have $L(U) = U = H(U)$, thus $U \in \mathcal{A}$.

(2) If $X_n \in \mathcal{A}, n \in \mathbf{N}$, by the definition, we have $L(X_n) = X_n = H(X_n)$ for all $n \in \mathbf{N}$. Since S is serial, we have

$$L(\bigcup_{n \in \mathbf{N}} X_n) \subseteq H(\bigcup_{n \in \mathbf{N}} X_n) = \bigcup_{n \in \mathbf{N}} X_n. \tag{59}$$

On the other hand, by axiom (AL3), it follows that

$$\bigcup_{n \in \mathbf{N}} X_n = \bigcup_{n \in \mathbf{N}} L(X_n) \subseteq L(\bigcup_{n \in \mathbf{N}} X_n). \tag{60}$$

Thus, we conclude that $L(\bigcup_{n \in \mathbf{N}} X_n) = \bigcup_{n \in \mathbf{N}} X_n = H(\bigcup_{n \in \mathbf{N}} X_n)$, that is, $\bigcup_{n \in \mathbf{N}} X_n \in \mathcal{A}$.

(3) If $X \in \mathcal{A}$, that is, $L(X) = X = H(X)$, then, by the duality of L and H, we have

$$L(\sim X) = \sim H(X) = \sim X = \sim L(X) = H(\sim X). \tag{61}$$

Hence $\sim X \in \mathcal{A}$.

Therefore, we have proved that $\mathcal{A}$ is a σ-algebra on U.

Remark 11. In rough set theory, if $L(X) = X = H(X)$, then X is called a definable set [22]. Theorem 28 shows that the family of all definable sets forms a σ-algebra and the pair $(U, \mathcal{A})$ is a measurable space. It should be pointed out the condition that $S = (\mathcal{P}(U), \cap, \cup, \sim, L, H)$ is serial cannot be omitted.

Theorem 29. *Assume that $S = (\mathcal{P}(U), \cap, \cup, \sim, L, H)$ is a Pawlak rough set algebra. Denote*
$$\mathcal{A}_l = \{X \in \mathcal{P}(U) : L(X) = X\},$$
$$\mathcal{A}_L = \{L(X) : X \in \mathcal{P}(U)\},$$
$$\mathcal{A}_h = \{X \in \mathcal{P}(U) : H(X) = X\},$$
$$\mathcal{A}_H = \{H(X) : X \in \mathcal{P}(U)\}.$$
Then $\mathcal{A}_l = \mathcal{A}_L = \mathcal{A}_h = \mathcal{A}_H = \mathcal{A}$.

Proof. Since for a Pawlak rough set algebra it is well-known that

$$X \text{ is definable} \iff L(X) = X \iff H(X) = X. \tag{62}$$

Then we have $\mathcal{A}_l = \mathcal{A}_h$. Clearly, $\mathcal{A}_l \subseteq \mathcal{A}_L$. On the other hand, since L and H are Pawlak approximation operators, we have $L(L(X)) = L(X)$ for all $X \in \mathcal{P}(U)$, hence $\mathcal{A}_L \subseteq \mathcal{A}_l$. By the duality of L and H, we can conclude that $\mathcal{A}_h = \mathcal{A}_H$. Thus the proof is completed.

Remark 12. Theorem 29 shows that the Pawlak rough approximation operators map any set of the universe of discourse into a definable set. In general, if L and H are not Pawlak rough approximation operators, then the approximated sets may not be definable.

The following theorem shows that for a given $(\sigma\text{-})$algebra in a finite universe we can find a rough set algebra such that the family of all definable sets is no other than the given $(\sigma\text{-})$algebra.

Theorem 30. *Let U be a finite universe of discourse and $(U, \mathcal{A})$ a measurable space, then there exists a rough set algebra $(\mathcal{P}(U), \cap, \cup, \sim, L, H)$ such that*

$$\mathcal{A} = \{X \in \mathcal{P}(U) : L(X) = X = H(X)\}. \tag{63}$$

Proof. Define

$$L(X) = \cup\{Y \in \mathcal{A} : Y \subseteq X\}, \quad X \in \mathcal{P}(U), \tag{64}$$

$$H(X) = \cap\{Y \in \mathcal{A} : X \subseteq Y\}, \quad X \in \mathcal{P}(U). \tag{65}$$

Since U is finite, we have $L(X) \in \mathcal{A}$ and $H(X) \in \mathcal{A}$ for all $X \in \mathcal{P}(U)$. By the definition, we can see that $L(X) \subseteq X$. Since $U \in \mathcal{A}$, we conclude $L(U) = U$.

For any $X_1, X_2 \in \mathcal{P}(U)$, if $x \in L(X_1 \cap X_2)$, then, by the definition of L, we can find $Y \in \mathcal{A}$ such that $x \in Y \subseteq X_1 \cap X_2$. From $x \in Y \subseteq X_i$ we have $x \in L(X_i), i = 1, 2$. Thus $x \in L(X_1) \cap L(X_2)$. Consequently,

$$L(X_1 \cap X_2) \subseteq L(X_1) \cap L(X_2). \tag{66}$$

Conversely, if $x \in L(X_1) \cap L(X_2)$, then there exists $Y_1, Y_2 \in \mathcal{A}$ such that $x \in Y_i \subseteq X_i, i = 1, 2$. Denote $Y = Y_1 \cap Y_2$, of course, $x \in Y \subseteq X_1 \cap X_2$. Since $\mathcal{A}$ is an algebra, we have $Y \in \mathcal{A}$. Hence, $x \in L(X_1 \cap X_2)$. Consequently,

$$L(X_1) \cap L(X_2) \subseteq L(X_1 \cap X_2). \tag{67}$$

Combining Eqs. (66) and (67), we conclude

$$L(X_1 \cap X_2) = L(X_1) \cap L(X_2), \forall X_1, X_2 \in \mathcal{P}(U). \tag{68}$$

On the other hand, for any $X \in \mathcal{P}(U)$, we have

$$\begin{aligned}
\sim L(\sim X) &= \sim \cup\{Y \in \mathcal{A} : Y \subseteq \sim X\} \\
&= \sim \cup\{Y \in \mathcal{A} : Y \cap X = \emptyset\} \\
&= \cap\{\sim Y : Y \in \mathcal{A}, Y \cap X = \emptyset\} \\
&= \cap\{Z : \sim Z \in \mathcal{A}, (\sim Z) \cap X = \emptyset\} \\
&= \cap\{Z \in \mathcal{A} : X \subseteq Z\} = H(X).
\end{aligned}$$

By the duality of L and H, we can also conclude that $H(\emptyset) = \emptyset$ and $H(X_1 \cup X_2) = H(X_1) \cup H(X_2)$ for all $X_1, X_2 \in \mathcal{P}(U)$.

Thus we have proved that $(\mathcal{P}(U), \cap, \cup, \sim, L, H)$ is a rough set algebra.

Furthermore, by the definitions of L and H, we observe that

$$\{X \in \mathcal{P}(U) : L(X) = X = H(X)\} \subseteq \mathcal{A}. \tag{69}$$

Also, for $Y \in \mathcal{A}$, by the definitions of L and H, we conclude that $L(Y) = Y = H(Y)$. Consequently,

$$\mathcal{A} \subseteq \{X \in \mathcal{P}(U) : L(X) = X = H(X)\}. \tag{70}$$

Therefore, Eq. (63) holds.

8   Belief Structures of Rough Sets

The connections between rough set theory and the Dempster-Shafer theory of evidence have been studied by a number of authors [34,35,36,44,45,46,48,57]. In this section, we investigate belief structures in infinite universes of discourse and discuss the relationship between belief structures and approximation spaces.

8.1   Belief Structures and Belief Functions on Infinite Universes of Discourse

The basic representational structure in the Dempster-Shafer theory of evidence is a belief structure.

Definition 12. *Let W be a nonempty universe of discourse which may be infinite and $\mathcal{M} \subset \mathcal{P}(W)$ a countable subset of $\mathcal{P}(W)$. A set function $m : \mathcal{P}(W) \to [0,1]$ is referred to as a basic probability assignment or mass distribution if $m(A) > 0$ for all $A \in \mathcal{M}$ and 0 otherwise with additional properties* (M1) *and* (M2):

(M1) $m(\emptyset) = 0$, (M2) $\sum\limits_{X \in \mathcal{M}} m(X) = 1$,

A set $X \in \mathcal{M}$ is referred to as a focal element of m. The pair $(\mathcal{M}, m)$ is called a belief structure.

The value $m(A)$ represents the degree of belief that a specific element of W belongs to set A, but not to any particular subset of A. Associated with the belief structure $(\mathcal{M}, m)$, a pair of belief and plausibility functions can be defined.

Definition 13. *Let $(\mathcal{M}, m)$ be a belief structure on W. A set function* Bel $: \mathcal{P}(W) \to [0,1]$ *is referred to as a belief function on W if*

$$\text{Bel}(X) = \sum_{\{M \in \mathcal{M} : M \subseteq X\}} m(M), \quad \forall X \in \mathcal{P}(W). \tag{71}$$

A set function Pl $: \mathcal{P}(W) \to [0,1]$ *is referred to as a plausibility function on W if*

$$\text{Pl}(X) = \sum_{\{M \in \mathcal{M} : M \cap X \neq \emptyset\}} m(M), \quad \forall X \in \mathcal{P}(W). \tag{72}$$

Remark 13. Since $\mathcal{M}$ is a countable set, the change of convergence may not change the values of the infinite (countable) sums in Eqs. (71) and (72). Therefore, Definition 13 is reasonable.

8.2 Relationship between Belief Functions and Rough Sets on Infinite Universes of Discourse

The following theorem shows that any belief structure can associate with a probability approximation space such that the probabilities of lower and upper approximations induced from the approximation space yield respectively the corresponding belief and plausibility functions derived from the given belief structure.

Theorem 31. *Let $(\mathcal{M}, m)$ be a belief structure on W which may be infinite. If* Bel $: \mathcal{P}(W) \to [0, 1]$ *and* Pl $: \mathcal{P}(W) \to [0, 1]$ *are, respectively, the belief and plausibility functions defined in Definition 13, then there exists a countable set U, a serial relation R from U to W, and a normalized probability measure P on U (i.e., $P(\{u\}) > 0$ for all $u \in U$) such that*

$$\text{Bel}(X) = P(\underline{R}(X)), \quad \text{Pl}(X) = P(\overline{R}(X)), \quad \forall X \in \mathcal{P}(W). \tag{73}$$

Proof. With no loss of generality, we assume that $\mathcal{M}$ has infinite countable elements and we denote

$$\mathcal{M} = \{A_i \in \mathcal{P}(W) : i \in \mathbf{N}\}, \tag{74}$$

where $\sum_{i \in \mathbf{N}} m(A_i) = 1$. Let $U = \{u_i : i \in \mathbf{N}\}$ be a set having infinite countable elements, we define a set function $P : \mathcal{P}(U) \to [0, 1]$ as follows:

$$P(\{u_i\}) = m(A_i), \quad i \in \mathbf{N}, \tag{75}$$

$$P(X) = \sum_{u \in X} P(\{u\}), \quad X \in \mathcal{P}(U). \tag{76}$$

Obviously, P is a normalized probability measure on U.

We further define a binary relation R from U to W as follows:

$$(u_i, w) \in R \iff w \in A_i, \quad i \in \mathbf{N}, w \in W. \tag{77}$$

From R we can obtain the counter-image mapping $j : \mathcal{P}(W) \to \mathcal{P}(U)$ of R_s as follows:

$$j(A) = \{u \in U : R_s(u) = A\}, \quad A \in \mathcal{P}(W). \tag{78}$$

It is easy to see that $j(A) = \{u_i\}$ for $A = A_i$ and $\emptyset$ otherwise. Consequently, $m(A) = P(j(A)) > 0$ for $A \in \mathcal{M}$ and 0 otherwise. Note that $j(A) \cap j(B) = \emptyset$

for $A \neq B$ and $\bigcup_{A \in \mathcal{P}(W)} j(A) = U$. Then, by property (JL) in Section 3, we can conclude that for any $X \in \mathcal{P}(W)$,

$$
\begin{aligned}
P(\underline{R}(X)) &= P\Big(\bigcup_{\{A \in \mathcal{P}(W): A \subseteq X\}} j(A) \Big) = P\Big(\bigcup_{\{A \in \mathcal{P}(W): A \subseteq X, j(A) \neq \emptyset\}} j(A) \Big) \\
&= P\Big(\bigcup_{\{A \in \mathcal{M}: A \subseteq X\}} j(A) \Big) = \sum_{\{A \in \mathcal{M}: A \subseteq X\}} P(j(A)) \\
&= \sum_{\{A \in \mathcal{M}: A \subseteq X\}} m(A) = \mathrm{Bel}(X).
\end{aligned}
\tag{79}
$$

On the other hand, by property (JU) in Section 3, we have

$$
\begin{aligned}
P(\overline{R}(X)) &= P\Big(\bigcup_{\{A \in \mathcal{P}(W): A \cap X \neq \emptyset\}} j(A) \Big) = P\Big(\bigcup_{\{A \in \mathcal{P}(W): A \cap X \neq \emptyset, j(A) \neq \emptyset\}} j(A) \Big) \\
&= P\Big(\bigcup_{\{A \in \mathcal{M}: A \cap X \neq \emptyset\}} j(A) \Big) = \sum_{\{A \in \mathcal{M}: A \cap X \neq \emptyset\}} P(j(A)) \\
&= \sum_{\{A \in \mathcal{M}: A \cap X \neq \emptyset\}} m(A) = \mathrm{Pl}(X).
\end{aligned}
\tag{80}
$$

Theorem 32. *Assume that (U, W, R) is a serial approximation space, U is a countable set, and $(U, \mathcal{P}(U), P)$ is a normalized probability space. For $X \in \mathcal{P}(W)$, define*

$$
m(X) = P(j(X)), \quad \mathrm{Bel}(X) = P(\underline{R}(X)), \quad \mathrm{Pl}(X) = P(\overline{R}(X)).
\tag{81}
$$

Then $m : \mathcal{P}(W) \to [0,1]$ is a basic probability assignment on W and $\mathrm{Bel} : \mathcal{P}(W) \to [0,1]$ and $\mathrm{Pl} : \mathcal{P}(W) \to [0,1]$ are, respectively, the belief and plausibility functions on W.

Proof. Let

$$
j(A) = \{x \in U : R_s(x) = A\}, \quad A \in \mathcal{P}(W).
\tag{82}
$$

It is well-known that j satisfies properties (J1) and (J2), i.e.,

(J1) $A \neq B \implies j(A) \cap j(B) = \emptyset$,   (J2) $\bigcup_{A \in \mathcal{P}(W)} j(A) = U$.

Since R is serial, we can observe that $j(\emptyset) = \emptyset$, consequently,

$$
m(\emptyset) = P(j(\emptyset)) = P(\emptyset) = 0.
\tag{83}
$$

Moreover, let $\mathcal{M} = \{X \in \mathcal{P}(W) : m(X) > 0\}$, notice that P is a normalized probability measure on U, then it is easy to observe that $m(X) > 0$ if and only if $j(X) \neq \emptyset$. Hence, by (J1) and (J2), it can be seen that $\{j(A) : A \in \mathcal{M}\}$ forms a partition of U. Since U is countable, $\mathcal{M}$ is a countable subset of $\mathcal{P}(W)$. Therefore

$$
\sum_{A \in \mathcal{M}} m(A) = \sum_{A \in \mathcal{M}} P(j(A)) = P\Big(\bigcup_{A \in \mathcal{M}} j(A) \Big) = P(U) = 1.
\tag{84}
$$

Hence m is a basic probability assignment on W. And for any $X \in \mathcal{P}(W)$, according to properties (JL) and (J1) we have

$$
\begin{aligned}
\mathrm{Bel}(X) = P(\underline{R}(X)) &= P\Big(\bigcup_{\{A \in \mathcal{P}(W): A \subseteq X\}} j(A) \Big) \\
&= P\Big(\bigcup_{\{A \in \mathcal{P}(W): A \subseteq X, j(A) \neq \emptyset\}} j(A) \Big) = P\Big(\bigcup_{\{A \in \mathcal{M}: A \subseteq X\}} j(A) \Big) \\
&= \sum_{\{A \in \mathcal{M}: A \subseteq X\}} P(j(A)) = \sum_{\{A \in \mathcal{M}: A \subseteq X\}} m(A).
\end{aligned}
\tag{85}
$$

Therefore, we have proved that Bel is a belief function. Similarly, by properties (JU) and (J1), we can conclude that

$$
\begin{aligned}
\mathrm{Pl}(X) = P(\overline{R}(X)) &= P\Big(\bigcup_{\{A \in \mathcal{P}(W): A \cap X \neq \emptyset\}} j(A) \Big) \\
&= P\Big(\bigcup_{\{A \in \mathcal{P}(W): A \cap X \neq \emptyset, j(A) \neq \emptyset\}} j(A) \Big) = P\Big(\bigcup_{\{A \in \mathcal{M}: A \cap X \neq \emptyset\}} j(A) \Big) \\
&= \sum_{\{A \in \mathcal{M}: A \cap X \neq \emptyset\}} P(j(A)) = \sum_{\{A \in \mathcal{M}: A \cap X \neq \emptyset\}} m(A).
\end{aligned}
\tag{86}
$$

Therefore, Pl is a plausibility function.

Remark 14. According to Theorem 31, rough set theory may be treated as a basis for the evidence theory in some sense. Theorem 32 provides a potential application of the evidence theory to data analysis. For example, the Dempster-Shafer theory of evidence may be used to analyze uncertainty knowledge in information systems and decision tables (see e.g. [44,48]).

8.3 Properties of Belief and Plausibility Functions on Infinite Universes of Discourse

The following theorem presents the properties of belief and plausibility functions.

Theorem 33. *Let W be a nonempty set which may be infinite and $(\mathcal{M}, m)$ a belief structure on W. If $\mathrm{Bel}, \mathrm{Pl} : \mathcal{P}(W) \to [0, 1]$ are respectively the belief and plausibility functions induced from the belief structure $(\mathcal{M}, m)$. Then*
(1) $\mathrm{Pl}(X) = 1 - \mathrm{Bel}(\sim X), \quad X \in \mathcal{P}(W)$,
(2) $\mathrm{Bel}(X) \leq \mathrm{Pl}(X), \quad X \in \mathcal{P}(W)$,
(3) $\mathrm{Bel}(X) + \mathrm{Bel}(\sim X) \leq 1, \quad X \in \mathcal{P}(W)$,
(4) $\mathrm{Bel} : \mathcal{P}(W) \to [0, 1]$ *is a monotone Choquet capacity* [2] *of infinite order on W, i.e., it satisfies the axioms* (MC1)–(MC3) *as follows:*
(MC1) $\mathrm{Bel}(\emptyset) = 0$,
(MC2) $\mathrm{Bel}(W) = 1$.
(MC3) *For any $n \in \mathbf{N}$ and $\forall X_i \in \mathcal{P}(W)$, $i = 1, 2, \ldots, n$,*

$$
\mathrm{Bel}\Big(\bigcup_{i=1}^{n} X_i \Big) \geq \sum_{\emptyset \neq J \subseteq \{1,2,\ldots,n\}} (-1)^{|J|+1} \mathrm{Bel}\Big(\bigcap_{j \in J} X_j \Big).
$$

(5) Pl : $\mathcal{P}(W) \to [0,1]$ *is an alternating Choquet capacity of infinite order on*
W, *i.e., it satisfies the axioms* (AC1)–(AC3) *as follows:*
(AC1) $\mathrm{Pl}(\emptyset) = 0$,
(AC2) $\mathrm{Pl}(W) = 1$,
(AC3) *For any* $n \in \mathbf{N}$ *and* $\forall X_i \in \mathcal{P}(W)$, $i = 1, 2, \ldots, n$,

$$\mathrm{Pl}(\bigcap_{i=1}^n X_i) \le \sum_{\emptyset \ne J \subseteq \{1,2,\ldots,n\}} (-1)^{|J|+1} \mathrm{Pl}(\bigcup_{j \in J} X_j).$$

Proof. By Theorem 31, there exists a countable set U, a serial relation R from
U to W, and a normalized probability measure P on U such that

$$\mathrm{Bel}(X) = P(\underline{R}(X)), \quad \mathrm{Pl}(X) = P(\overline{R}(X)), \quad \forall X \in \mathcal{P}(W). \tag{87}$$

Then for any $X \in \mathcal{P}(W)$, by the dual properties of lower and upper approximation operators in Theorem 1, we have

$$\mathrm{Pl}(X) = P(\overline{R}(X)) = P(\sim \underline{R}(\sim X)) = 1 - P(\underline{R}(\sim X)) = 1 - \mathrm{Bel}(\sim X). \tag{88}$$

Thus property (1) holds.
(2) Notice that R is serial, then, according to property (LU0) in Theorem 2,
we have
$$\mathrm{Bel}(X) = P(\underline{R}(X)) \le P(\overline{R}(X)) = \mathrm{Pl}(X). \tag{89}$$

(3) follows immediately from (1) and (2).
(4) By property (L0) in Theorem 2, we have

$$\mathrm{Bel}(\emptyset) = P(\underline{R}(\emptyset)) = P(\emptyset) = 0, \tag{90}$$

that is, (MC1) holds. On the other hand, by property (L1) in Theorem 1, we
have
$$\mathrm{Bel}(W) = P(\underline{R}(W)) = P(U) = 1, \tag{91}$$

thus (MC2) holds.
For any $n \in \mathbf{N}$ and $\forall X_i \in \mathcal{P}(W)$, $i = 1, 2, \ldots, n$, by properties (L4) and (L2)
in Theorem 1, we have

$$\begin{aligned}
\mathrm{Bel}(\bigcup_{i=1}^n X_i) = P(\underline{R}(\bigcup_{i=1}^n X_i)) &\ge P(\bigcup_{i=1}^n \underline{R}(X_i)) \\
&= \sum_{\emptyset \ne J \subseteq \{1,2,\ldots,n\}} (-1)^{|J|+1} P(\bigcap_{j \in J} \underline{R}(X_j)) \\
&= \sum_{\emptyset \ne J \subseteq \{1,2,\ldots,n\}} (-1)^{|J|+1} P(\underline{R}(\bigcap_{j \in J} X_j)) \\
&= \sum_{\emptyset \ne J \subseteq \{1,2,\ldots,n\}} (-1)^{|J|+1} \mathrm{Bel}(\bigcap_{j \in J} X_j).
\end{aligned}$$

Thus (MC3) holds. Therefore, we have proved that Bel is a monotone Choquet
capacity of infinite order on W.

(5) Similar to (4), by Theorems 1 and 2, we have

(AC1) $\mathrm{Pl}(\emptyset) = P(\overline{R}(\emptyset)) = P(\emptyset) = 0$.

(AC2) $\mathrm{Pl}(W) = P(\overline{R}(W)) = P(U) = 1$.

(AC3) For any $n \in \mathbf{N}$ and $\forall X_i \in \mathcal{P}(W)$, $i = 1, 2, \ldots, n$, by properties (U4) and (U2) in Theorem 1, we have

$$
\begin{aligned}
\mathrm{Pl}(\bigcap_{i=1}^{n} X_i) = P(\overline{R}(\bigcap_{i=1}^{n} X_i)) &\leq P(\bigcap_{i=1}^{n} \overline{R}(X_i)) \\
&= \sum_{\emptyset \neq J \subseteq \{1,2,\ldots,n\}} (-1)^{|J|+1} P(\bigcup_{j \in J} \overline{R}(X_j)) \\
&= \sum_{\emptyset \neq J \subseteq \{1,2,\ldots,n\}} (-1)^{|J|+1} P(\overline{R}(\bigcup_{j \in J} X_j)) \\
&= \sum_{\emptyset \neq J \subseteq \{1,2,\ldots,n\}} (-1)^{|J|+1} \mathrm{Pl}(\bigcup_{j \in J} X_j).
\end{aligned}
$$

Thus we have concluded that Pl is an alternating Choquet capacity of infinite order on W.

From Theorem 33 we can see that semantics of the original Dempster-Shafer theory of evidence is still maintained.

9 Conclusion

In this paper, we have reviewed and studied a general framework for the study of rough sets in infinite universes of discourse. We have reviewed the constructive concepts of lower and upper approximation operators and presented the essential properties of the approximation operators. We have also established different set of independent axiomatic sets to characterize various types of rough set approximation operators. Under the view point of axiomatic approach, rough set theory my be viewed as an extension of set theory with two additional unary operators. We have investigated some important mathematical structures of rough sets. We have analyzed and compared the topological structures, measurable structures, and belief structures of rough sets. This work will help us to gain much more insights into the mathematical structures of approximation operators.

Acknowledgement

This work was supported by grants from the National Natural Science Foundation of China (Nos. 61075120, 60673096 and 60773174), and the Natural Science Foundation of Zhejiang Province (No. Y107262).

References

1. Birkhoff, G.: Rings of sets. Duke Mathematical Journal 3, 443–454 (1937)
2. Choquet, G.: Theory of capacities. Annales de l'institut Fourier 5, 131–295 (1954)

3. Chuchro, M.: On rough sets in topological Boolean algebras. In: Ziarko, W. (ed.) Rough Sets, Fuzzy Sets and Knowledge Discovery, pp. 157–160. Springer, Heidelberg (1994)

4. Chuchro, M.: A certain conception of rough sets in topological Boolean algebras. Bulletin of the Section of Logic 22(1), 9–12 (1993)

5. Comer, S.: An algebraic approach to the approximation of information. Fundamenta Informaticae 14, 492–502 (1991)

6. Dempster, A.P.: Upper and lower probabilities induced by a multivalued mapping. Annals of Mathematical Statistics 38, 325–339 (1967)

7. Fagin, R., Halpern, J.Y.: Uncertainty, belief, and probability. Computational Intelligence 7, 160–173 (1991)

8. Halmos, P.R.: Measure Theory. Van Nostrand-Reinhold, New York (1950)

9. Inuiguchi, M., Hirano, S., Tsumoto, S. (eds.): Rough Set Theory and Granular Computing. Springer, Berlin (2003)

10. Kelley, J.L.: General Topology. Van Nostrand, New York (1955)

11. Kondo, M.: Algebraic approach to generalized rough sets. In: Ślezak, D., Wang, G., Szczuka, M.S., Düntsch, I., Yao, Y. (eds.) RSFDGrC 2005. LNCS (LNAI), vol. 3641, pp. 132–140. Springer, Heidelberg (2005)

12. Kondo, M.: On the structure of generalized rough sets. Information Sciences 176, 586–600 (2006)

13. Kortelainen, J.: On relationship between modified sets, topological space and rough sets. Fuzzy Sets and Systems 61, 91–95 (1994)

14. Lai, H., Zhang, D.: Fuzzy preorder and fuzzy topology. Fuzzy Sets and Systems 157, 1865–1885 (2006)

15. Lin, T.Y.: Neighborhood systems—application to qualitative fuzzy and rough sets. In: Wang, P.P. (ed.) Advances in Machine Intelligence and Soft-Computing. Department of Electrical Engineering, Duke University, Durham, NC, pp. 132–155 (1997)

16. Lin, T.Y.: Granular computing on binary relations II: Rough set representations and belief functions. In: Polkowski, L., Skowron, A. (eds.) Rough Sets in Knowledge Discovery: 1. Methodolodgy and Applications, pp. 122–140. Physica-Verlag, Heidelberg (1998)

17. Lin, T.Y., Liu, Q.: Rough approximate operators: axiomatic rough set theory. In: Ziarko, W. (ed.) Rough Sets, Fuzzy Sets and Knowledge Discovery, pp. 256–260. Springer, Heidelberg (1994)

18. Lin, T.Y., Yao, Y.Y., Zadeh, L.A. (eds.): Data Mining, Rough Sets and Granular Computing. Physica-Verlag, Heidelberg (2002)

19. Pawlak, Z.: Information systems, theoretical foundations. Information Systems 6, 205–218 (1981)

20. Pawlak, Z.: Rough sets. International Journal of Computer and Information Science 11, 341–356 (1982)

21. Pawlak, Z.: Rough probability. Bulletin of the Polish Academy of Sciences: Mathematics 32, 607–615 (1984)

22. Pawlak, Z.: Rough Sets: Theoretical Aspects of Reasoning about Data. Kluwer Academic Publishers, Boston (1991)

23. Pei, D.W., Xu, Z.-B.: Rough set models on two universes. International Journal of General Systems 33, 569–581 (2004)

24. Peters, J.F., Pawlak, Z., Skowron, A.: A rough set approach to measuring information granules. In: Proceedings of COMPSAC, pp. 1135–1139 (2002)

25. Peters, J.F., Skowron, A., Synak, P., Ramanna, S.: Rough sets and information granulation. In: De Baets, B., Kaynak, O., Bilgiç, T. (eds.) IFSA 2003. LNCS, vol. 2715, pp. 370–377. Springer, Heidelberg (2003)
26. Polkowski, L.: Mathematical morphology of rough sets. Bulletin of the Polish Academy of Sciences: Mathematics 41(3), 241–273 (1993)
27. Polkowski, L.: Metric spaces of topological rough sets from countable knowledge bases. In: FCDS 1993: Foundations of Computing and Decision Sciences, pp. 293–306 (1993)
28. Polkowski, L.: Approximate mathematical morphology: Rough set approach. In: Rough Fuzzy Hybridization: A New Trend in Decision-Making, pp. 471–487. Springer, Heidelberg (1999)
29. Polkowski, L.: Rough Sets: Mathematical Foundations. Springer, Heidelberg (2002)
30. Pomykala, J.A.: Approximation operations in approximation space. Bulletin of the Polish Academy of Sciences: Mathematics 35, 653–662 (1987)
31. Qin, K.Y., Yang, J., Pei, Z.: Generalized rough sets based on reflexive and transitive relations. Information Sciences 178, 4138–4141 (2008)
32. Ruspini, E.H.: The Logical Foundations of Evidential Reasoning. Technical Note 408, SIR International, Menlo Park, CA (1986)
33. Shafer, G.: A Mathematical Theory of Evidence. Princeton University Press, Princeton (1976)
34. Skowron, A.: The relationship between the rough set theory and evidence theory. Bulletin of Polish Academy of Science: Mathematics 37, 87–90 (1989)
35. Skowron, A.: The rough sets theory and evidence theory. Fundamenta Informatica 13, 245–262 (1990)
36. Skowron, A., Grzymala-Busse, J.: From rough set theory to evidence theory. In: Yager, R., et al. (eds.) Advances in the Dempster-Shafer Theory of Evidence, pp. 193–236. Wiley, New York (1994)
37. Skowron, A., Stepaniuk, J.: Information granules: towards foundations of granular computing. International Journal of Intelligent Systems 16, 57–85 (2001)
38. Skowron, A., Świniarski, R.W., Synak, P.: Approximation spaces and information granulation. In: Peters, J.F., Skowron, A. (eds.) Transactions on Rough Sets III. LNCS, vol. 3400, pp. 175–189. Springer, Heidelberg (2005)
39. Slowinski, R., Vanderpooten, D.: A Generalized definition of rough approximations based on similarity. IEEE Transactions on Knowledge and Data Engineering 12(2), 331–336 (2000)
40. Thiele, H.: On axiomatic characterisations of crisp approximation operators. Information Sciences 129, 221–226 (2000)
41. Vakarelov, D.: A modal logic for similarity relations in Pawlak knowledge representation systems. Fundamenta Informaticae 15, 61–79 (1991)
42. Wasilewska, A.: Conditional knowledge representation systems—model for an implementation. Bulletin of the Polish Academy of Sciences: Mathematics 37, 63–69 (1990)
43. Wiweger, R.: On topological rough sets. Bulletin of Polish Academy of Sciences: Mathematics 37, 89–93 (1989)
44. Wu, W.-Z.: Attribute reduction based on evidence theory in incomplete decision systems. Information Sciences 178, 1355–1371 (2008)
45. Wu, W.-Z., Leung, Y., Zhang, W.-X.: Connections between rough set theory and Dempster-Shafer theory of evidence. International Journal of General Systems 31(4), 405–430 (2002)

46. Wu, W.-Z., Mi, J.-S.: An interpretation of belief functions on infinite universes in the theory of rough sets. In: Chan, C.-C., Grzymala-Busse, J.W., Ziarko, W.P. (eds.) RSCTC 2008. LNCS (LNAI), vol. 5306, pp. 71–80. Springer, Heidelberg (2008)
47. Wu, W.-Z., Zhang, W.-X.: Neighborhood operator systems and approximations. Information Sciences 144, 201–217 (2002)
48. Wu, W.-Z., Zhang, M., Li, H.-Z., Mi, J.-S.: Knowledge reduction in random information systems via Dempster-Shafer theory of evidence. Information Sciences 174, 143–164 (2005)
49. Wybraniec-Skardowska, U.: On a generalization of approximation space. Bulletin of the Polish Academy of Sciences: Mathematics 37, 51–61 (1989)
50. Yang, X.-P., Li, T.-J.: The minimization of axiom sets characterizing generalized approximation operators. Information Sciences 176, 887–899 (2006)
51. Yao, Y.Y.: Two views of the theory of rough sets in finite universes. International Journal of Approximate Reasoning 15, 291–317 (1996)
52. Yao, Y.Y.: Constructive and algebraic methods of the theory of rough sets. Journal of Information Sciences 109, 21–47 (1998)
53. Yao, Y.Y.: Generalized rough set model. In: Polkowski, L., Skowron, A. (eds.) Rough Sets in Knowledge Discovery 1. Methodology and Applications, pp. 286–318. Physica-Verlag, Heidelberg (1998)
54. Yao, Y.Y.: Relational interpretations of neighborhood operators and rough set approximation operators. Information Sciences 111, 239–259 (1998)
55. Yao, Y.Y.: Information granulation and rough set approximation. International Journal of Intelligent Systems 16, 87–104 (2001)
56. Yao, Y.Y., Lin, T.Y.: Generalization of rough sets using modal logic. Intelligent Automation and Soft Computing 2, 103–120 (1996)
57. Yao, Y.Y., Lingras, P.J.: Interpretations of belief functions in the theory of rough sets. Information Sciences 104, 81–106 (1998)
58. Zadeh, L.A.: Fuzzy sets and information granularity. In: Gupta, M.M., Ragade, R.K., Yager, R.R. (eds.) Advances in Fuzzy Set Theory and Applications, pp. 3–18. North-Holland, Amsterdam (1979)
59. Zadeh, L.A.: Towards a theory of fuzzy information granulation and its centrality in human reasoning and fuzzy logic. Fuzzy Sets and Systems 19, 111–127 (1997)
60. Zakowski, W.: On a concept of rough sets. Demonstratio Mathematica 15, 1129–1133 (1982)
61. Zhang, H.-P., Ouyang, Y., Wang, Z.D.: Note on Generalized rough sets based on reflexive and transitive relations. Information Sciences 179, 471–473 (2009)
62. Zhou, L., Wu, W.-Z., Zhang, W.-X.: On intuitionistic fuzzy rough sets and their topological structures. International Journal of General Systems 38(6), 589–616 (2009)
63. Zhu, W.: Topological approaches to covering rough sets. Information Sciences 177, 1499–1508 (2007)

Quadtree Representation and Compression of Spatial Data

Xiang Yin[1], Ivo Düntsch[1,*], and Günther Gediga[2,**]

[1] Brock University, St. Catharines, Ontario, Canada
[2] Universität Münster, Germany

Abstract. Granular computing is closely related to the depth of the detail of information with which we are presented, or choose to process. In spatial cognition and image processing such detail is given by the resolution of a picture. The quadtree representation of an image offers a quick look at the image at various stages of granularity, and successive quadtree representations can be used to represent change. For a given image, the choice of quadtree root node plays an important role in its quadtree representation and final data compression. The goal of this paper is to present a heuristic algorithm for finding a root node of a region quadtree, which is able to reduce the number of leaf nodes when compared with the standard quadtree decomposition. The empirical results indicate that the proposed algorithm improves the quadtree representation and data compression when compared to the traditional method.

1 Introduction

Owing to an ever–increasing requirement of information storage, spatial data representation and compression have become one of the popular and important issues in computer graphics and image processing applications. How to represent the whole information contained in a given image by using less time and less space is a concern of many researchers. Nowadays, in order to reduce the storage requirements of (a sequence of) images, the design of efficient data representations has been studied extensively.

Hierarchical data structures [1] are becoming increasingly important techniques in the representation of spatial data, which mainly focus on the data at various levels-of-detail. Such focusing provides an efficient representation and an improved execution time [2], and it is especially useful for performing set operations.

Quadtrees, introduced in the early 1970s [3], are one of such data structures based on the principle of recursive decomposition of space, and have since become a major representation method for spatial data.

Quadtrees are of great interest, in general, because they enable us to solve problems in a manner that focuses the work on the areas where the spatial data is of the largest density. For many situations, the amount of work is proportional to the number of aggregated blocks (e.g. nodes in the tree-structure) rather than to the actual number of pixels

---

* The author gratefully acknowledges support by the Natural Science and Engineering Council of Canada.
** Günther Gediga is also adjunct professor at Brock University.

J.F. Peters et al. (Eds.): Transactions on Rough Sets XIII, LNCS 6499, pp. 207–239, 2011.

in the original image. In such a way, they have the potential of reducing the requirement of storage space and improving the efficiency of execution time.

In many applications, by using quadtrees to implement the representation and compression of spatial data, most researchers have applied their central work to the three aspects: Build the quadtree data structure, encode all the tree nodes, and compress the coding of nodes.

In the present paper we shall investigate how to provide a better compression result for given images. Previous research has told us that the compression effect has been dominated by encoding, approximation and compression methods selected [4]. Undeniably, these are all important determining factors. Nevertheless, these factors are based on a common precondition, that is, we have built up the quadtree structure already. In essence, previous work has not accounted for two essential questions:

1. According to the given image, how do we build up the corresponding quadtree structure?
2. Is this quadtree structure an optimal representation for the given image?

Indeed, the problem of choosing a quadtree root node for a given image is still an open problem with no general solution. One goal of this paper will aim to extend traditional considerations into this new ground.

The remainder of this paper is organized as follows:

Sections 2 deals with the concepts of quadtrees and the relationship between quadtrees and rough sets. Section 3 constitutes the heart of the paper. A more extensive analysis of the significance of the choice of the quadtree root node will be given. Then, special attention is given to the explanation of the design idea of our *CORN* (Choosing an Optimal Root Node) algorithm. We will use a running example to describe each step of the algorithm in detail. Section 4 presents experimental results of the use of our CORN algorithm, and an interpretation of their significance. Section 5 concludes our paper with a set of open questions for future research.

2   Quadtrees and Rough Sets

2.1   Overview of Quadtrees

A quadtree is a class of hierarchical data structures which contains two types of nodes: non-leaf nodes and leaf nodes, or internal nodes and external nodes. Owing to its "Divide and Conquer" strategy [5], it is most often used to partition a two dimensional space by recursively subdividing it into four quadrants or regions.

Nowadays, in the fields of computer graphics and image processing, a quadtree is always regarded as a way of encoding data that enables one to reduce storage requirements by avoiding to subdivide the same areas rather than storing values for each pixel. At this point, a quadtree is considered as a spatial index which recursively decomposes an image into square cells of different sizes until each cell has the same value.

According to Samet, types of quadtrees can be classified by following three principles:

"(1) the type of data that they are used to represent, (2) the principle guiding the decomposition process, and (3) the resolution (variable or not)." [6]

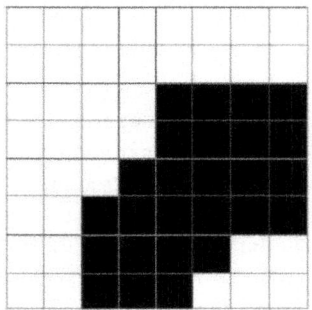 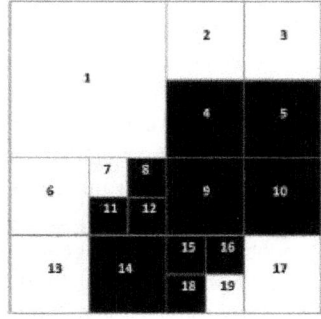

Fig. 1. A region I of cells and its decomposition

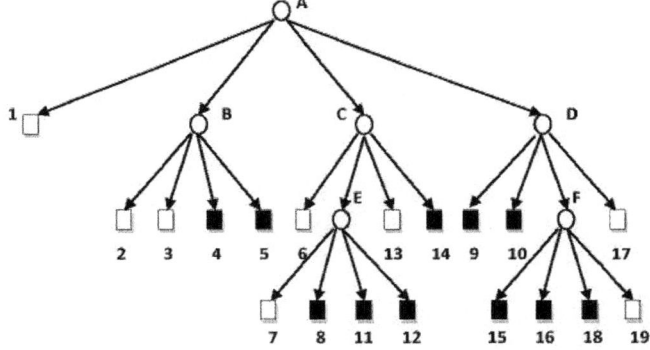

Fig. 2. The complete quadtree representation of I

In our paper, we mainly use region quadtrees. Therefore, in the remainder of this section, we will outline this type of quadtree.

The scenario for a region quadtree is as follows: We are given a window consisting of black cells situated in a fixed image area A of dimension $2^m \times 2^m$. A is recursively partitioned into equal sized quadrants until each quadrant consists entirely of unicolored cells. The process can be represented by a tree each non-leaf node of which has four children, corresponding to the four quadrants NW, NE, SW, and SE. Each descendant of a node g represents a quadrant in the plane whose origin is g and which is bounded by the quadrant boundaries of the previous step.

Consider, in a simple example, the region I placed into an area of dimension $2^3 \times 2^3$, shown in Figure 1 [7]. The NW quadrant is entirely made up of white cells, thus leading to a white leaf node which terminates this branch. The NE quadrant contains cells of different color; therefore an intermediate (grey) node is placed into the center of the NE quadrant, which is divided into four quadrants. Each quadrant is homogeneous in color, and thus, four leaf nodes are produced, two of which are white and two of which are black. The process continues until the whole image I is decomposed into square regions. The complete quadtree for the image of Figure 1 is shown in Figure 2.

It is not hard to see that the quadtree has a number of useful features:

- By virtue of its hierarchical nature, a quadtree especially facilitates the performance of set operations such as the union and intersection of several images. For an overview with more detail we invite the reader to consult [8,9].
- A quadtree provides an efficient data structure for computer graphics. In particular, it is easy to transform a detailed shape description into its original image form.
- The prime motivation for the development of the quadtree structure is the desire to reduce the amount of space necessary to store data through the use of aggregation of homogeneous blocks. More generally, one of the striking results of this aggregation is to decrease the executing time of a number of operations, such as connected component labeling [10] and component counting [11].

Undoubtedly, the quadtree representation also has some hardly avoidable problems: The largest drawback of the quadtree representation is the sensitivity of its storage requirements to its position. Owing to its changeable root node position, they may lead to many different kinds of tree structure. Thus, it is difficult to find a uniform algorithm for building a quadtree structure with minimal space requirements Indeed, this problem is especially obvious when the number and size of images are large.

Therefore, our aim is to design a new algorithm which enables us to make an optimal choice for the root node of a quadtree. More importantly, by this algorithm we will obtain a fixed quadtree structure and minimize the number of quadtree leaf nodes, so that it will finally result in a better quadtree representation and (lossless) data compression of the original image.

2.2 Alternative Types of Quadtree Representation

In order to represent quadtrees, there are two major approaches: pointer-based quadtree representation and pointerless quadtree representation.

In general, the pointer-based quadtree representation [12] is one of the most natural ways to represent a quadtree structure. In this method, every node of the quadtree will be represented as a record with pointers to its four sons. Sometimes, in order to achieve special operations, an extra pointer from the node to its father will also be included.

In the remainder of this part, we will briefly describe two common node description methods of the pointer-based quadtree representation:

- Each node is stored as a record with six fields: The first four fields contain its four sons, labeled NW, NE, SW, and SE; the fifth field is the pointer to its father; the sixth field would be the node type. For example, it may depict the block contents of the image which the node represents, that is, black (the point contains data), white (null), and grey (non-leaf point).
- Each node is stored as a record with eight fields: The first four fields contain pointers to the node's four sons; the fifth field is for the node type; the sixth field is the description of the node. For example, in the city map, it could be the name of a road. The last two fields are X coordinate and Y coordinate of the node, which can be used to obtain the relationship between each two nodes in the quadtree.

However, for further analysis, we can observe that the pointer-based quadtree representation exhibits some inevitable problems when considering space requirements for recording the pointers and internal nodes. For example, suppose we get a corresponding quadtree representation of a given image, and B and W indicate the number of black and white leaf nodes in the quadtree, respectively. In this way, $(B+W-1)/3$ additional nodes are necessary for the internal (grey) nodes, and each node also requires additional space for the pointers to its sons, and possibly another pointer to its father. As a result, this additional expense leads to an intolerable problem when dealing with images that are very complex and have large size. In other words, in some applications, the images may be so large that the space requirements of their quadtree representations will exceed the amount of memory that is available. Consequently, considerable attention is concentrated on another quadtree representation method, that is, pointerless quadtree representation.

In contrast with the pointer-based quadtree representation, the pointerless version of a quadtree and its variants have been known for many years [2,13]. The benefit of this kind of quadtree representation is to define each node of the tree as a unique index. By virtue of the regularity of the subdivision, it is possible to compute the location code of each node in the tree entirely in local memory rather than accessing the data structure in global memory. In other words, once the location code is known, the actual node containing the point can be accessed through a small number of accesses to global memory (e.g., by hashing). Much work has been done on employing this idea for representing the spatial data, and we refer the reader to [14] for further details.

The pointless quadtree representations can be grouped into two categories: The first regards each image as a collection of leaf nodes; the second represents the image in the form of a traversal of the nodes of its quadtree. Some general nodes encoding methods based on these two pointerless quadtree representation categories will be briefly discussed below:

SONTYPE4. In the spirit of the first type of pointerless quadtree representation, each leaf node is represented by a sequence of directional codes which correspond to an encoding of path from the root node of the tree to the leaf node itself. In essence, in order to reduce space requirement, we only need to be concerned with the directional codes of all the black nodes, since a binary image is represented and only grey nodes are internal nodes, which means, all white nodes can be dominated by the black nodes.

Even though this method was not published until 1982 [13], the core idea of SONTYPE4 had already been mentioned by Klinger and Dyer [15] as early as 1976.

The coding sequence in SONTYPE4 is defined as follows: Each node in the quadtree is represented by an n-element sequence $\{q_i\} = \{q_{n-1}, \ldots, q_1, q_0\}$ (n is the number of levels from the objective node to the root node of the quadtree) constructed from the digits $\{0, 1, 2, 3, 4\}$. Let $\{x_i\}$ represent the path of nodes from the root node of the quadtree to x_m, that is, the desired node, such that x_n = root node of the quadtree and $x_i = Father(x_{i-1})$. Codes 0, 1, 2, and 3 correspond to quadrants NW, NE, SW, and SE, respectively, and 4 denotes a don't care node. In this way, the encoding of the locational code for node x_m is given by q_n where q_i is defined by

$$q_i = \begin{cases} 0 & : \quad i = m \\ 5 \times q_{i-1} + \text{SONTYPE4}(x_i) & : \quad m < i \leq n. \end{cases}$$

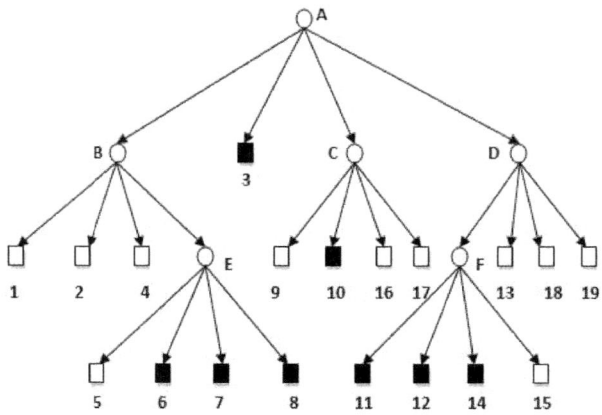

Fig. 3. Quadtree representation for SONTYPE4 method

As an illustration of this encoding method, node 7 in Figure 3 would be encoded into the sequence of directional codes $\{q_i\} = \{0, 3, 2\}$, that is, $q = 0 \times 5^2 + 3 \times 5^1 + 2 \times 5^0 = 17$.

DF-expression. The DF-expression was proposed by Kawaguchi and Endo [16]. Briefly, it is a compacted array that represents an image in the form of a pre-order tree traversal (i.e., depth first) of the nodes in the quadtree. The result of the DF-expression method is a sequence consisting of three symbols 'B', 'W', and '(' corresponding to black, white and grey nodes, respectively. As an example, the Figure 4 has ((BW(WBWWBWB((WBBBBBW as its DF-expression. To get more insight into the DF-expression method, we suggest that the reader consult the papers [17,18].

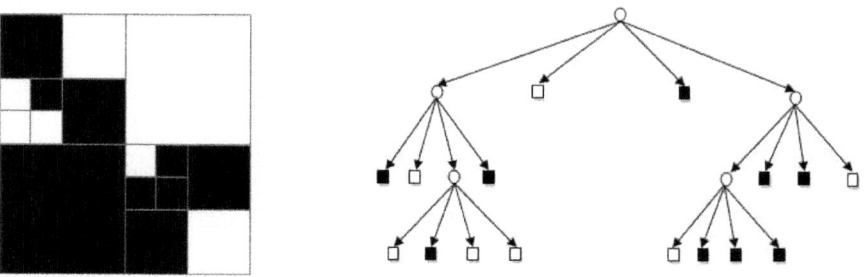

Fig. 4. A binary image and its quadtree representation for DF-expression method

In addition, the size of the DF-expression is always dependent on the number of nodes in the quadtree representation. In principle, given a $2^n \times 2^n$ binary image, each alphabet of the DF-expression can be represented by using one or two bits for each internal and external node, for example, the symbols '(', 'B' and 'W' are encoded by binary codes 10, 11 and 0, respectively. In this way, the number of bits required for representing Figure 4 is equal to 35.

2.3   Quadtree Approximation Methods

Concepts of approximation.   An approximation (represented by the symbol $\approx$) is an inexact representation of something that is still close enough to be useful.

In essence, approximation methods may be applied when incomplete or noisy information prevents us from getting exact representations. Many problems in real world applications are either too complex to solve, or too difficult to solve accurately. Thus, even when the exact representation is available, an approximation may yield a sufficiently good solution by reducing the complexity of the problem significantly (usually in time and space). For instance, suppose we have a picture of a landscape; from the standpoint of our eyes, we are only concerned with the whole contents inside this picture rather than the details of each object. For example, we might be attracted to the whole structure of a beautiful tree, while we will never take account of how many branches and leaves it contains.

The type of approximation used is mainly dependent on the available information, the degree of accuracy required, the sensitivity of the problem to the data, and the savings (usually in time and effort) that can be achieved. Although approximation is most often applied to numbers, it is also frequently applied to such domains as mathematical functions, visual perceptions, information systems and physical laws. However, in the paper, we will only aim to provide approximation methods for optimizing the quadtree representation.

Approximation spaces.   As an illustration of approximation spaces in information system [19], the granularity of information can be indicated by equivalence relations on a set U of objects, depending on the classes of which objects are discernible. With each equivalence relation θ, we associate a partition P_θ of U by defining that $a, b \in U$ are in the same class of P_θ, if and only if $a\theta b$. The classes of P_θ satisfy the following requirement:

$$\theta a = \{b \in U : a\theta b\}$$

Let U be a set, and θ is an equivalence relation on U. Then, the pair $< U, \theta >$ will be called *an approximation space*, and the relation θ is called *an indiscernibility relation*, which means the knowledge of the objects in U extends only up to membership in the classes of θ. The idea now aims to approximate our knowledge with respect to a subset X of U modulo the indiscernibility relation θ: For any $X \subseteq U$, we say that

$$\underline{X} = \bigcup\{\theta x : \theta x \subseteq X\}$$

is *the lower approximation or positive region of X*, in other words, the complete set of objects which can be positively (i.e., unambiguously) classified as belonging to target set X; and

$$\overline{X} = \bigcup\{\theta x : x \in X\}$$

is *the upper approximation or possible region of X*, in other words, the complete set of objects that are possibly members of the target set X.

If $X \subseteq U$ is given by a property P and $x \in U$, then

- $x \in \underline{X}$ means that x *certainly has property P*,
- $x \in \overline{X}$ means that x *possibly has property P*,
- $x \in U \setminus \overline{X}$ means that x *definitely does not have property P*.

In particular, *the region of uncertainty or boundary region* is defined as: $\overline{X} \setminus \underline{X}$; and $\underline{X} \cup -\overline{X}$ is called *the area of certainty*.

Based on these definitions, a pair of the form $< \overline{X}, \underline{X} >$ is called *a rough set*. At this point, the accuracy of rough set representation of the set X can be given [20] by the following:

$$0 \leq \alpha(X) = \frac{|\underline{X}|}{|\overline{X}|} \leq 1$$

$\leq \alpha(X)$ describes the ratio of the number of objects which can be positively placed in X to the number of objects that can possibly be placed in X. It also provides a measure of how closely the rough set is approximating the target set.

Approximation applied to quadtrees. By virtue of its hierarchical structure, in graphics and image processing fields the quadtree leads itself to serve as an image approximation machine. Similarly, approximation methods are usually regarded as a tool to optimize the quadtree representation. At this point, by truncating a quadtree (i.e., ignoring all nodes below a certain level or neglecting some nodes satisfying an optimization strategy), a crude approximation is able to be realized. For further clarification, in the remainder of this section, we will outline two different quadtree approximation methods:

Hierarchical approximation method

A sequence of inner and outer approximations to an image defined by Ranade, Rosenfeld and Samet [21], is used for the hierarchical approximation. According to their definitions, the inner approximation consists of treating grey nodes as white nodes, whereas the outer approximation treats them as black nodes. Based on these considerations, a more accurate definition is given by Samet [4]:

> "Given an image I, the inner approximation, $IB(k)$ is a binary image defined by the black nodes at levels $\geq k$; the outer approximation, $OB(k)$ is a binary image defined by black nodes at levels $\geq k$ and the grey nodes at level k."

As a simple example, the region I is placed into an area of dimension $2^3 \times 2^3$. The quadtree decomposition and representation of region I are demonstrated by Figure 5 and Figure 6. According to Samet's definitions, Figure 7 shows $IB(2)$ and $OB(2)$ for Figure 6; and Figure 8 shows $IB(1)$ and $OB(1)$ for Figure 6.

At this point, suppose we use $\subseteq$ and $\supseteq$ to indicate set inclusion in the sense that $A \subseteq B$ and $B \supseteq A$ imply that the space spanned by A is a subset of the space spanned by B. It is easy for us to obtain a conclusion that $IB(n) \subseteq IB(n-1) \subseteq \ldots \subseteq IB(0) = I$ and $OB(n) \supseteq OB(n-1) \supseteq \ldots \supseteq OB(0) = I$.

In addition, the hierarchical-based quadtree approximation method has also been used in transmission of binary and grey-scale images. In essence, a number of pyramid-based approaches [22] to solve this problem have already been proposed in 1979 by

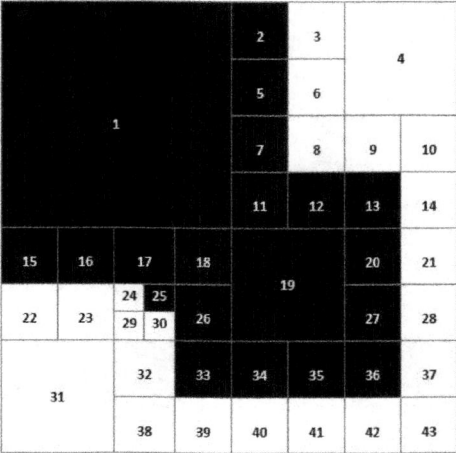

Fig. 5. Quadtree decomposition of image *I*

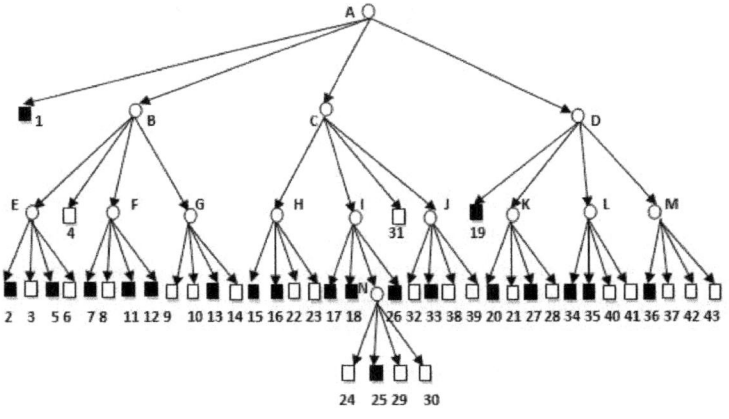

Fig. 6. Quadtree representation of image *I*

Sloan and Tanimoto [23]. They point out that the largest drawback of this kind of quadtree approximation is that the redundant information may take up to 1/3 more spatial space. Thus, it may lead to no compression. In order to alleviate this problem, they indicate two optimization restrictions as follows:

– For each node in the quadtree representation, a level number and its coordinate are suggested to be included. However, this information is transmitted only if it is distinct from the value of the node's predecessor.
– The receiver is encouraged to deduce the node's value by analyzing its predecessor and three sibling nodes, so that it can lead to decreased space requirement for implementing transmission operation.

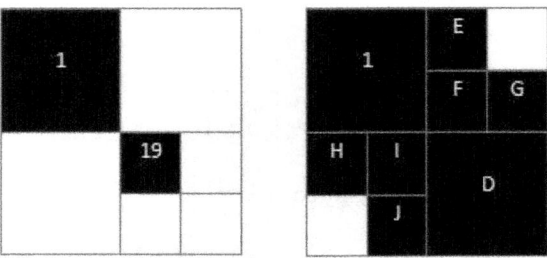

Fig. 7. IB(2) and OB(2) for image *I*

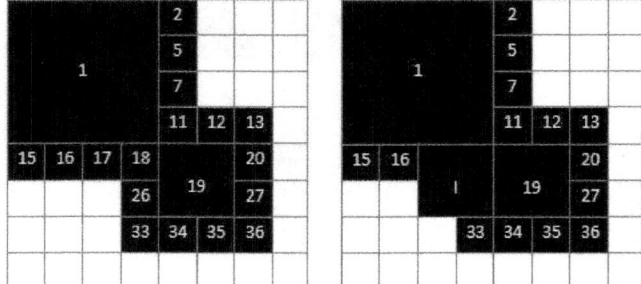

Fig. 8. IB(1) and OB(1) for image *I*

Knowlton addresses a similar problem (transmission of binary and grey-scale images) by using a binary tree version of a quadtree [24]. In his viewpoint, an image is repeatedly split into two halves alternating between horizontal and vertical splits. In particular, he puts forward that all information needed to be transmitted is the composite value for the root node of the quadtree and the successive sets of differentiators. Therefore, the sequence of transmission is a breadth-first traversal of the binary tree with differentiator values. For a comprehensive review of these two methods we invite the reader to consult [23,24].

Another quadtree approximation method which we will discuss in the following part makes no use of above techniques.

Forest-based approximation method

A forest of quadtrees [25] is a decomposition of a quadtree into a collection of subquadtrees. Each of subquadtree corresponds to a maximal square, which is identified by refining an internal node to indicate some information about its subtrees. At this point, the forest is a refinement of a quadtree data structure used to develop a sequence of approximations to a binary image, and provides space savings over regular quadtrees by concentrating on vital information [26].

Before expanding the discussion of the forest-based approximation method, we need to be aware of some concepts: GB node, GW node, and black forest. A grey node (internal node) is said to be of type GB if at least two of its sons are black nodes or of type GB. Otherwise, the node is said to be of type GW.

For example, in Figure 6, the nodes E, F, H, I, K, and L are of type GB, and nodes A, B, C, D, G, J, M, and N are of type GW. Naturally, we can draw a conclusion that each black node or an internal node with a label GB can be regarded as a maximal square.

Based on these concepts, a black forest is defined as follows (we can define a white forest in the same way):

- It contains the minimal set of maximal squares.
- All the black or GB nodes in each maximal square are not included in any other square.
- The squares in this minimal set would cover the whole black area of the original image.

Therefore, the black forest of Figure 6 is {A}. By giving another example, the black forest corresponding to Figure 9 is {F, 10, 16, 25, 27, M, 38}.

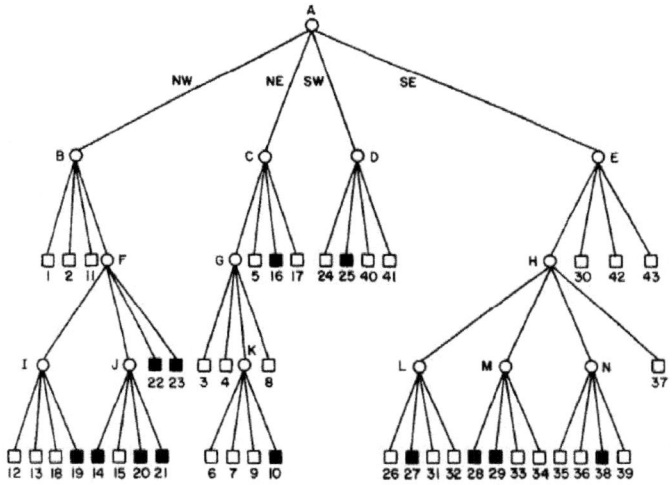

Fig. 9. An example of a black forest

In essence, a forest can also be used to achieve a better approximation for the quadtree representation. If we regard the elements of the forest as black and all remaining nodes as white, we can employ a forest to approximate images. It is valuable to sort the nodes of the forest based on their codes, e.g. we can use the locational codes given by q_i which we introduced in Section 3.1.2.1 to implement nodes encoding. For example, in Figure 9, the nodes will appear in the order 25, 16, F, M, 27, 38, and 10. This order is a partial ordering $(S, \geq)$ such that $S_i \geq S_{i+1}$ means the block subsumed by S_i is $\geq$ in size than the block subsumed by S_{i+1}. In this way, for a breadth-first traversal we only need to deal with the nodes in an order that satisfies the above subsumption relation.

Samet gives an explicit explanation for the FBB, FWW, and FBW approximation methods, which reasonably lead to compression in the sense. In other words, they reduce the amount of spatial data that is required by encoding the image, as well as the transmission. For an overview of definitions of these methods and the relationship between them, we advise the reader to refer to [4]. Indeed, aside from its superiority with respect to the quality of the resulting approximation, the forest-based approximation has a number of interesting properties:

- It is obvious that the forest-based approximation method is biased in favor of approximating objects with the shape of a "panhandle", whereas the hierarchical-based approximation method is insensitive to them.
- Owing to the fact that the elements of the forest are encoded by locational codes, such approximation can lead to savings of space requirement whenever the situation satisfies that three out of four sons of any grey node have the same type (i.e., black nodes and GB nodes, respectively, white nodes and GW nodes).
- The total number of nodes in the approximation sequence will never exceed the minimum number of black or white nodes in the original quadtree representation. In other words, we can guarantee that this approximation method is always at least as good or better than encoding the quadtree by listing its black nodes (or white nodes). In this way, this kind of approximation will result in a more efficient compression. Even as larger images are used, the compression effect will become more conspicuous.

2.4   Quadtrees and Rough Sets

In the rough set model [20], objects are described by a lower approximation and an upper approximation induced by an equivalence relation θ on the base set U; both approximations are unions of equivalence classes of θ.

In Figure 10, the cells represent the classes of θ. The set to be approximated is the area of the closed ellipse X; the union of all cells completely contained in X is the lower approximation $\underline{X}_\theta$, and the union of all cells that intersect X is the upper approximation $\overline{X}_\theta$. Therefore, the plane is partitioned into three disjoint regions, namely, the "Yes" region $\underline{X}_\theta$, the "Maybe" region $\overline{X}_\theta \setminus \underline{X}_\theta$, and the "No" region $U \setminus \overline{X}_\theta$.

In the next section we will give a solution about how data as in Figure 10 can be efficiently represented using as little data space as possible. Our motivation is the investigation of J. J. Gibson's approach to ecological perception by two of the present authors [27]. According to Gibson, the main task of perception is to recognize the invariant parts within a variant world:

> "We perceive that the environment changes in some respects and persists in others. We see that it is different from time to time, even moment to moment, and yet that it is the same environment over time. We perceive *both* the change and the underlying non–change. My explanation is that the perceptual systems work by detecting invariants in the flux of stimulation but are also sensitive to the flux itself". [28]

If we think of U as a plane region and the cells as pixels output by a sensor, then θ fixes the granularity at which the region is perceived; each granule may consist of a class of an information system which describes the nature of the pixel – color, hue, intensity, shade, or just on/off. Table 1 is an excerpt of such a situation. The "container" U represents a snapshot (freeze) of the visual field which is composed of a sequence of such containers. Here we interpret 1 as "true" or "present", and 0 as "false" or "not present"; a question mark in a cell indicates ignorance about the cell content.

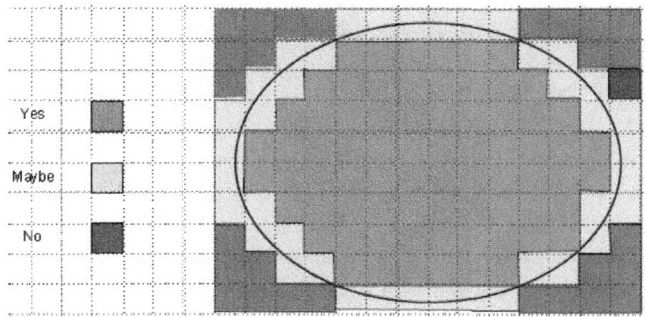

Fig. 10. Rough set approximation

Briefly, the logical background is as follows: We assume that an agent a (animal, human, robot) knows ("sees") some or all properties of some or all bits; in other words, we are setting up an epistemic system with one agent a and a knowledge operator K. The atomic sentences are of the form `attribute value = constant`, e.g. "bit is on" or "color is red". For each atomic proposition p we consider $K(p)$, $K(\neg p)$ and $\neg K(p) \wedge \neg K(\neg p)$; in other words, "$a$ knows that p", "a knows that $\neg p$", and "a has no knowledge of the truth of p". We suppose that K satisfies the axiom T, i.e. $K(p) \implies p$; in other words, we suppose that the agent's observations are correct. The natural evaluation e of the propositional variables is given by the cells under consideration.

The agent, therefore, presents a matrix consisting of r rows and s columns each of which represents $K(p)$ for an atomic proposition p; these, in turn, are grouped by the un–binarized attribute they belong to. There are some semantical considerations to take care of – e.g. if $K(p) = 1$, then the agent knows that $K(q) = 0$ for all other atomic sentences obtained from the attribute.

In the simplest case we suppose that the information given by a cell is just an on/off state, given by only one symmetric (binary) attribute which we may or may not be able to perceive. This situation is a strong simplification – in more realistic situations, each cell may contain a whole information system as in Table 1. For an overview of representation of imprecision in finite resolution data and its connections to rough set theory we invite the reader to consult [29].

More importantly, by applying the idea in rough sets theory to express the quadtree representation (e.g., Fig 2), we can see the final complete quadtree representation

Table 1. Bitstrings of a perceived field

Cell	Status On	Color red	green	...	Intensity high	medium	low
$\langle 0,0 \rangle$	1	1	0		?		
$\langle 0,1 \rangle$	1	?			0	1	0
$\langle 0,2 \rangle$	0	0			0		
$\langle 0,3 \rangle$	?	?			?		
...	...	...			...		

consists of eight white leaf nodes and eleven black leaf nodes; there are six interme-
diary ("grey") nodes. As a hierarchical decomposition method, the quadtree structure
lends itself well to varying granularity, since each level of the tree is a refinement of
the previous one. This is well within the rough set paradigm; indeed, as early as 1982
Ranade et al [21] introduced an approximation of images which can be regarded as a
special case of Pawlak's approach: each level j of the quadtree corresponds to an equivalence relation θ_j; θ_0 is the universal relation, and θ_m is the identity (the pixel level).
The inner approximation of I at level j consists of all black leaf nodes obtained up to
level m, and the outer approximation additionally contains the grey nodes for this level.
As an example, Table 2 shows the upper and lower approximations of the image of
Figure 2 with respect to the resolution corresponding to the levels of the quadtree.

Table 2. Rough approximation and quadtrees

Level	Lower approximation	Boundary
0	∅	A
1	∅	B,C,D
2	$4,5,14,9,10$	E,F
3	$4,5,14,9,10,8,11,12,15,16,18$	∅

3   Algorithm Design

In this section, we will aim to provide some solutions for two essential questions mentioned in Section 1. Furthermore, a brand-new heuristic algorithm will be designed for
deciding the root node of a region quadtree, which effectively reduces the amount of
leaf nodes in the quadtree representation when compared with the standard quadtree
decomposition.

3.1   Problem Analysis

In Section 1 we have pointed out that the compression rate for a given image is not only
influenced by the encoding, approximation, and compression methods we choose, but
also has a compact relation with its quadtree representation according to the original
image. Even when viewing all these factors, in our opinion the former are only the
external factors, and the latter are an internal factor which has a greater impact on the
compression result.

In the remainder of this section, we will aim to answer three main questions:

  – Why does the root node choice of a quadtree have an extremely important meaning
    for image compression?
  – What criteria do we use to evaluate the selection of a quadtree root node?
  – How do we find a good root node choice for a quadtree?

3.2   Influential Factors of Quadtree Cost

Past experience has taught us that the space cost of quadtrees is measured by the number
of their leaf nodes, since the number of nodes in a quadtree is directly proportional to

the number of its leaves. This proof procedure has been demonstrated by Horowitz and Sahni [30]. Therefore, it may be noted that minimizing the total number of leaf nodes of a quadtree is equivalent to minimizing its total number of nodes (leaf nodes and non-leaf nodes).

In essence, two factors often greatly influence the total number of leaf nodes of a quadtree: The size of the coordinate grid and the position of the image on the grid. Figure 11 clearly demonstrates that reducing the grid size may result in less cost of a quadtree; and Figure 12 shows that we may obtain a cheaper quadtree cost by moving the image position.

Furthermore, these two crucial factors are just decided by the root node choice of the quadtree, which is the main reason why the root node choice is of great significance in our consideration.

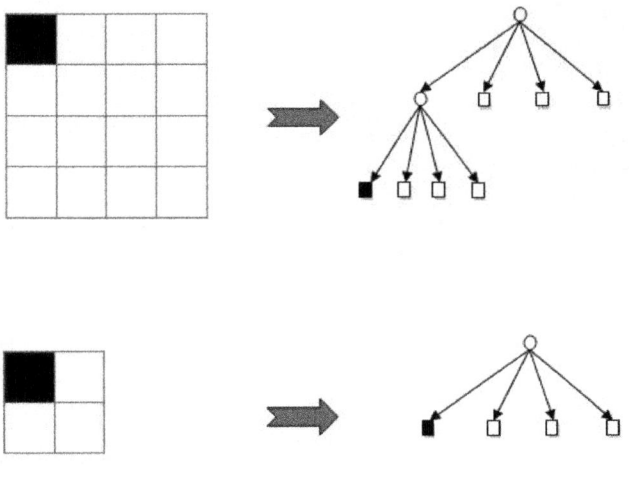

Fig. 11. The cost of quadtree influenced by grid size

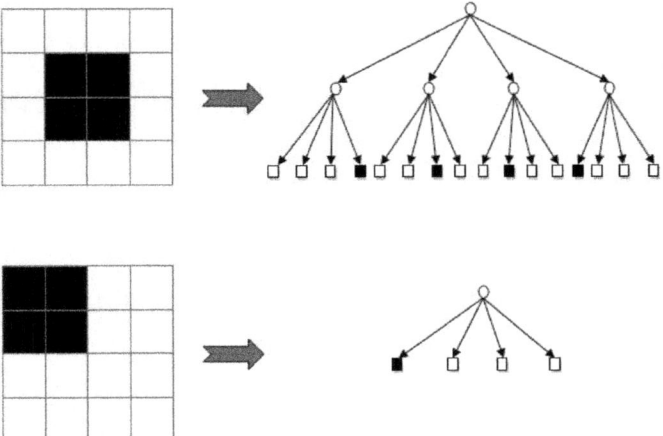

Fig. 12. The cost of quadtree influenced by image position

3.3   Significance of Quadtree Root Node Choice

As mentioned in last section, it may be argued that in the visual field, the choice of the base frame is somewhat arbitrary and, as a consequence, the root node of the quadtree representation of an image I may be to some extent variable. Indeed, if we aim to minimize the number of black nodes of a quadtree representation of I, then the choice of the root node will influence the size of the quadtree. A striking example is the case, when an optimal position with respect to a central root node is shifted by one pixel, see, for example, the discussion in [31]. We shall illustrate this with another example [6]. An image and its bounding box are shown in Figure 13.

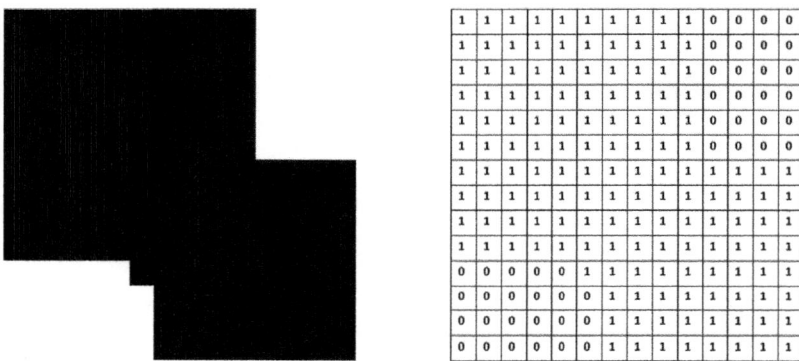

Fig. 13. A region I and its bounding box

Following [6], we place the image into an 8×8 grid as shown in Figure 14; the corresponding quadtree can be found in Figure 15. It contains 43 leaf nodes, 20 of which are black.

If we have the freedom to choose where the image I is placed into an area A, then a smaller number of leaf nodes can be obtained. Figure 16 shows a different placement, and Figure 17 depicts the corresponding quadtree. This second quadtree contains only 34 leaf nodes, 14 of which are black.

The analysis above clearly demonstrates the significance of the quadtree root node choice. These results indicate that a good choice of the quadtree root node will result in great improvement for the quadtree representation, as well as the final image compression. In particular, when dealing with an image of large size, this benefit would be much more conspicuous. However, in the meantime, a new and more severe problem is generated, which involves the difficulty of evaluating among the different quadtree root node choices for the same image. Thus, these observations now lead to the following question:

- Suppose we are given an image of black cells with bounding box dimension $n \times m$. Which position of the root node will minimize the number of black leaf nodes in the quadtree decomposition of A? In other words, how do we make a choice among a number of candidate quadtree root nodes?

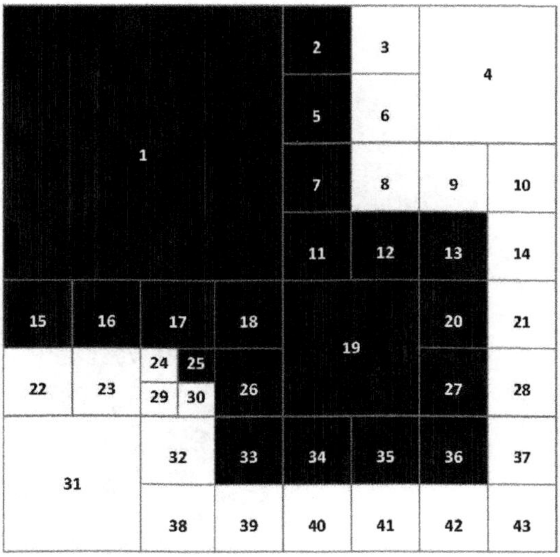

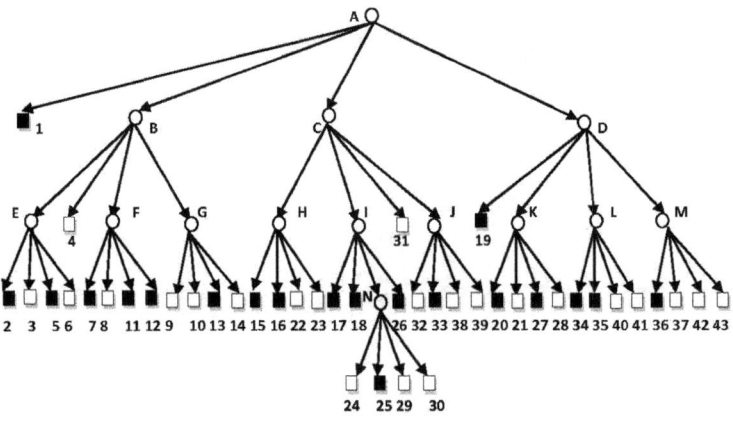

Fig. 14. 1st decomposition of *I*

Fig. 15. 1st quadtree representation of *I*

For further discussion, we would like to emphasize that the goal of this paper is to investigate how the spatial data contained in a given image could be efficiently represented in the quadtree structure by using as little data space as possible. Since the difference in the storage requirements is so significant, at times it may be deemed worthwhile to minimize the space requirement of the image stored in the quadtree, especially when the size of an image is large.

Given an image, the quadtree that contains the least number of leaf nodes among all the possible quadtrees that represent the same image is called the normalized quadtree of the image [32]. Therefore, our final motivation is changed to design an algorithm to choose a suitable root node under certain specific criteria, so that it will lead to a

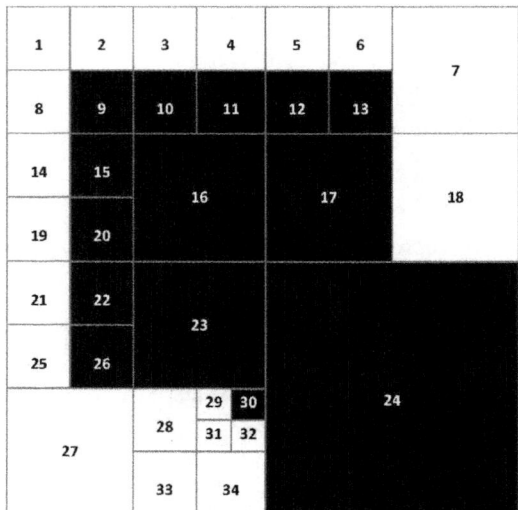

Fig. 16. 2nd decomposition of I

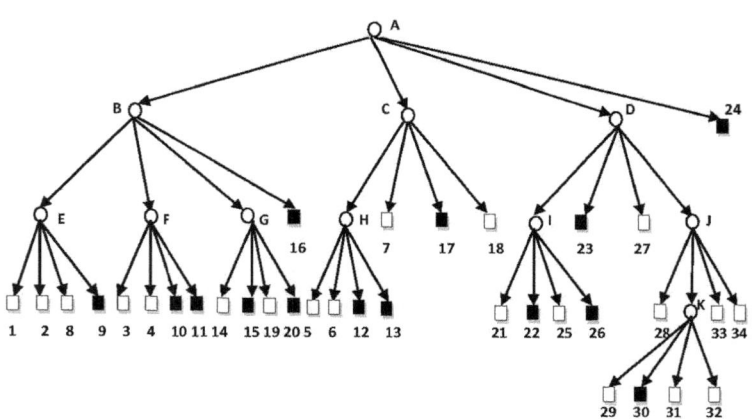

Fig. 17. 2nd quadtree representation of I

unique quadtree structure, and finally build a normalized quadtree representation. In what follows, we shall exhibit these criteria for the quadtree root node choice.

3.4   Criteria for the Root Node Selection

For brevity, we call a (closed) square of size $2^k \times 2^k$ containing only black cells a *block*. We say that two blocks s,t are *in contact* if $s \cap t \neq \emptyset$, i.e.,

– s and t contain a common cell (Figure 18(a)), or
– s and t contain a common edge of a cell (Figure 18(b)), or
– s and t contain a common corner point (Figure 18(c)).

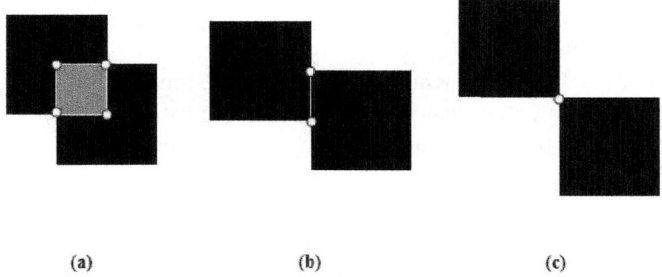

Fig. 18. Two Blocks in contact

A set S ($S = \{s_1, s_2, \ldots, s_n\}$) of blocks is called a *contact set*, if $s_1 \cap s_2 \cap \ldots \cap s_n \neq \emptyset$, i.e., the blocks in S have at least one common point of contact.

Furthermore, if A is a point (i.e., the intersection of two orthogonal edges), the *neighbor number n_A of A* is the sum of the size of all blocks of which A is a corner point.

In order to make an assessment for each candidate quadtree root node, in the sequel we present three main criteria in order of priority for selecting a root node A:

1. The choice of A minimizes the size of the image space $2^t \times 2^t$ containing I.
   Let $Image_x(Image_y)$ be the length of horizontal (vertical) side of its enclosing rectangle. Defining k such that

   $$2^k < max(Image_x, Image_y) \leq 2^{k+1},$$

   thus, the optimal grid size (minimum size) is either $2^{k+1} \times 2^{k+1}$ or $2^{k+2} \times 2^{k+2}$. It is very easy to prove this conclusion: It is clear that the image will not fit in a grid of size less than $2^{k+1} \times 2^{k+1}$; equally, a grid of size greater than $2^{k+2} \times 2^{k+2}$ is also non-optimal because an image of size less than or equal to $2^{k+1} \times 2^{k+1}$ can be completely covered in a quadtree of a $2^{k+2} \times 2^{k+2}$ grid. So, the optimal grid size can only be $2^{k+1} \times 2^{k+1}$ or $2^{k+2} \times 2^{k+2}$. (In some special cases, all the candidate root nodes cannot expand the original image into the size of $2^{k+1} \times 2^{k+1}$, so that we have to expand the original image to the size $2^{k+2} \times 2^{k+2}$.)

   In addition, it may be noted that the extra space spent on the larger image-expansion would far exceed the space saved from reducing some black leaf nodes of the quadtree representation under this situation, especially, with images of a large size.

   Thus, the primary standard to evaluate the root node choice is whether it can minimize the size of the extended image. For example, suppose we have a given binary image with a bounding box of dimension 13×14. In this case, we prefer choosing a root node by which the original image is only needed to expand to the size of 16×16, and excluding all the candidate choices which have to expand the image size to 32×32, even though they maybe contain less black leaf nodes corresponding to their quadtree representations.

2. A block of maximal size will be one leaf node in the final quadtree representation.
   In our viewpoint, the largest black square in the original image is the most important factor to influence the number of black leaf nodes in the final quadtree representation. As a simple example, if there is an image with the size 100×100 pixels

(bits), and the largest square of which is 32×32. In such situation, if after image-expansion, we only need to use one leaf node to represent this square, namely, 1023 cells space would directly be saved.

3. The size of neighbors of A which are blocks is maximized.
   This criterion guarantees that we could combine most black blocks in the original image when we build up its corresponding quadtree representation.

3.5 Choosing a Root Node

Minimizing the number of black leaf nodes will result in a better (lossless) compression of the original image, and we will use this number as a metric. Associating with the evaluation standards mentioned above, in this section we will outline a heuristic algorithm which decreases the number of black nodes required to represent an image.

The cells of the bounding box are scanned from left to right, row by row, starting at the upper left hand corner of the bounding box of I. The pseudocode of the CORN [1] algorithm proceeds as follows:

1. **repeat**
   Progress to the next unvisited black cell b_i and record its position.
   Find a block s_i of maximal size (i.e., $2^n \times 2^n, n \geq 0$) that
       (a) contains b_i, and
       (b) maximizes the number of unvisited cells.
   Change black color of the cells of s_i into, say, red.
   until all black cells have been visited.
   {At this stage, we have a set $B = \{b_1, \dots, b_n\}$ of points and a set of blocks $S = \{s_1, \dots, s_n\}$.}
2. Find the maximal blocks, say, $s_1, \dots, s_n$ corresponding to each cell b_i ($i = 0, 1, 2, \dots, n$).
3. Find all maximal contact sets $S_1, \dots, S_m$ from S containing at least one of $s_1, \dots, s_k$ and for each S_j the set C_i of points in the intersection.
4. Order the points in $\bigcup\{C_i : 1 \leq i \leq m\}$ by their neighbor numbers in non-increasing order, say $A_1, A_2, \dots, A_k$ such that $n_{A_1} \geq n_{A_2} \geq \dots \geq n_{A_k}$.
5. Let t be the smallest number such that $n, m \leq 2^t$; in other words, a block with side length t is a smallest block that can contain I.
6. Let $i = 0, j = t$.
   repeat
       repeat Increase i by 1.
       Decompose the image with A_i as a root node. If the resulting block has side length 2^j then choose A_i as root node and stop.
       until $i = k$.
       {At this stage, none of the A_i will allow using 2^j as a side length, so we try the next smallest block.}
   Increase j by 1.
   until FALSE

---

[1] Choosing an **O**ptimal **R**oot **N**ode - we are aware that the name is somewhat optimistic.

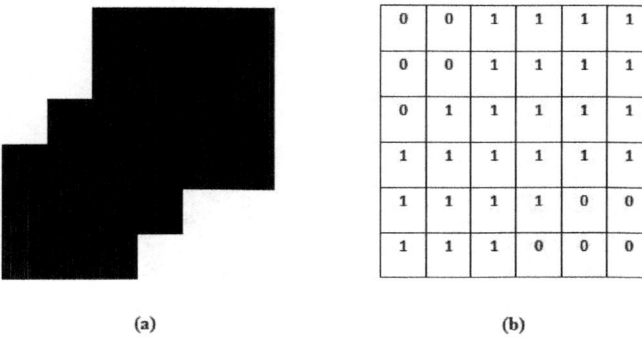

0	0	1	1	1	1
0	0	1	1	1	1
0	1	1	1	1	1
1	1	1	1	1	1
1	1	1	1	0	0
1	1	1	0	0	0

(a)   (b)

Fig. 19. A region and its binary array

1	2	3	4	5	6
7	8	9	10	11	12
13	14	15	16	17	18
19	20	21	22	23	24
25	26	27	28	29	30
31	32	33	34	35	36

Fig. 20. Bounding box of I

In order to explain each stage of the CORN algorithm in detail, we will use Figure 19(a) as a running example.

Suppose that the input consist of a binary array I with a bounding box of dimension $m \times n$; an entry "1" indicates a black cell and an entry "0" indicates a white cell, see Figure 19(b). According to the CORN algorithm, the procedure of searching its quadtree root node proceeds as follows:

1. Find the bounding box of I, see Figure 20.
2. Starting at the upper left hand corner of bounding box, scan row by row, from left to right, until finding the first unvisited black cell b_1 (In this example, b_1 is cell 3). Record the position of cell 3.
3. Find and record the block s_1, which consists of cells $\{3, \ldots, 6, 9, \ldots 12, 15, \ldots, 18, 21, \ldots, 24\}$. ($s_1$ is the green block shown in Figure 21(a).)
4. Change the cells of s_1 to red. (see Figure 21(b))
5. Scan the bounding box from cell 4, until finding cell b_2 (cell 14). Then, search and record the block s_2, which contains cells $\{14, 15, 20, 21\}$ (see Figure 22(a)). In same manner, change its color to red (Figure 22(b)).

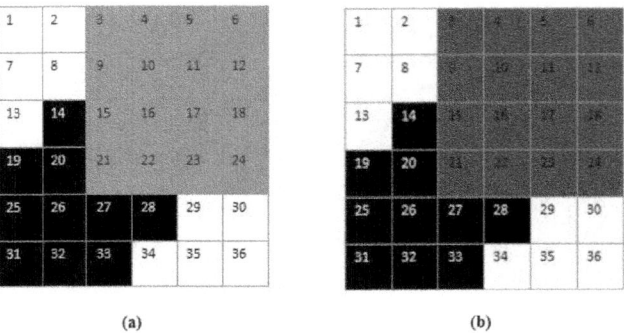

Fig. 21. Maximum square containing b_1

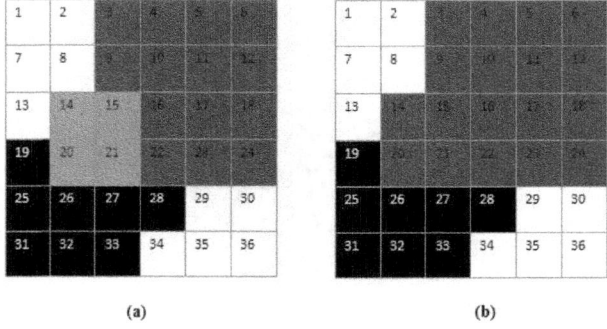

Fig. 22. Maximum square containing b_2

6. Since $s_1 \cap s_2 \neq \emptyset$, which means these two blocks are in contact. Therefore, put them into the contact set S_1, namely, now, $S_1 = \{s_1, s_2\}$.

7. Repeat above operations until all cells have been visited. (Figures 23 and 24 demonstrate the maximum squares for b_3, b_4, b_5, b_6. The resulting squares are shown in Table 3; the cell b_i is shown in bold face.)

8. Find the maximal square among $s_1, \ldots, s_6$ in comparison with their size. (At this stage, we get s_1 is the maximal square in the current image).

9. Find all the contact sets containing s_1. (In this example, there is only one contact set, namely, contact set $S_1 = \{s_1, s_2, s_3, s_4, s_5, s_6\}$).

10. Check the intersection part of S_1 and get the common point of contact squares is the NW corner of cell 27, see Figure 25.

11. Use the NST method to count the neighbors number of point A, that is, $n_A = 1 + 16 + 4 + 1 = 22$.

12. There is only one candidate root node choice, and point A satisfies three evaluation criteria defined in the section IV. Therefore, point A is the final quadtree root node choice.

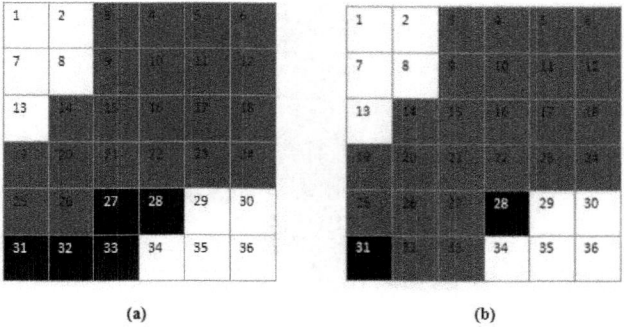

Fig. 23. Maximum squares containing b_3 and b_4

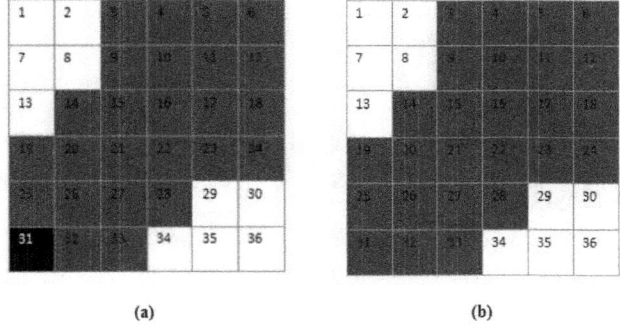

Fig. 24. Maximum squares containing b_5 and b_6

The final image decomposition and its corresponding quadtree representation are shown in Figure 26. We find that the total number of black nodes in the final quadtree representation is reduced from original 26 to 8.

Table 3. Maximum squares in bounding box of I

$s_1 : \{\mathbf{3}, \ldots, 6, 9, \ldots 12, 15, \ldots, 18, 21, \ldots, 24\}$
$s_2 : \{\mathbf{14}, 15, 20, 21\}$
$s_3 : \{\mathbf{19}, 20, 25, 26\}$
$s_4 : \{26, \mathbf{27}, 32, 33\}$
$s_5 : \{21, 22, 27, \mathbf{28}\}$
$s_6 : \{25, 26, \mathbf{31}, 32\}$

1	2	3	4	5	6
7	8	9	10	11	12
13	14	15	16	17	18
19	20	21	22	23	24
25	26	27	28	29	30
31	32	33	34	35	36

Fig. 25. Intersection point of S_1

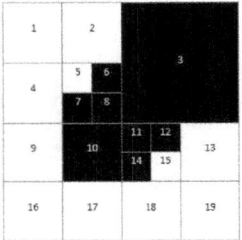 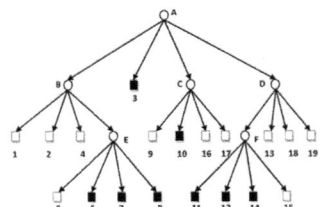

Fig. 26. Final image decomposition and quadtree representation of I

4   Empirical Results

In this section, some experiments are given to exhibit the efficiency of the CORN algorithm. These empirical results will be in comparison with the traditional method for demonstrating the power of our algorithm.

Traditionally, the root node is always placed in the center of the chosen image area or the image to be considered. Adhering to the traditional method, Example 1 puts the root node in the center of the black region of Figure 19(a). The image decomposition and its corresponding quadtree representation are shown in Figure 27 and Figure 28. At this point, the CORN algorithm reduces the number of relevant leaf nodes by more than half (see Figure 26). Similarly, in comparison with Figure 14 and Figure 16, the CORN algorithm also fully demonstrates it is an essential tool to offer an optimal root node choice and provide a good quadtree representation of the given image data.

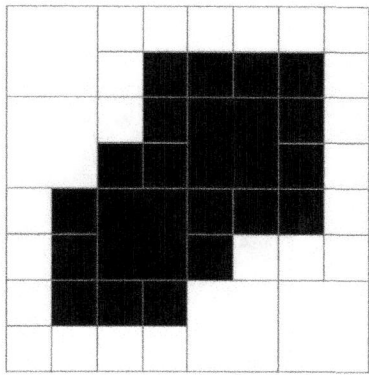

Fig. 27. Image decomposition of Example 1

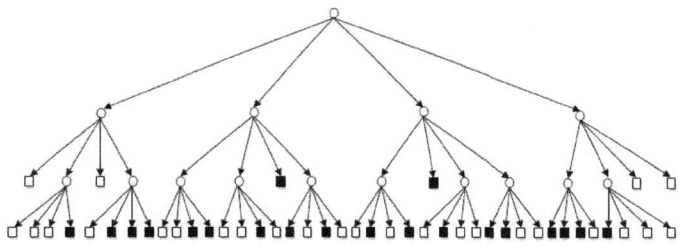

Fig. 28. Quadtree representation of Example 1

Furthermore, Figure 10 is a classical example in the rough set model. Figure 29 shows the decomposition of the inner approximation of the ellipse (i.e., the green region) in Figure 10 by using the traditional method, and Figure 30 performs its quadtree representation. However, a better result is obtained with the use of the CORN algorithm instead of the traditional method, the image decomposition and quadtree representation are shown in Figure 31 and Figure 32.

In the previous examples, there is only one connected black region in each example. Indeed, we are also interested in the performance of CORN algorithm when dealing with an image containing more than one connected regions. Figure 33 shows an image including a disconnected region. Figure 34 and Figure 35 respectively show the image decomposition and quadtree representation according to the traditional method. The result based on the CORN algorithm is displayed in Figure 36 and Figure 37, which reduces by nearly three quarters the number of relevant leaf nodes in comparison with the traditional method.

The worst case for a quadtree of a given depth in terms of storage requirement occurs when the region corresponds to a "checkerboard" pattern. In other words, each black cell in the original image will be expressed by one leaf node in the final quadtree

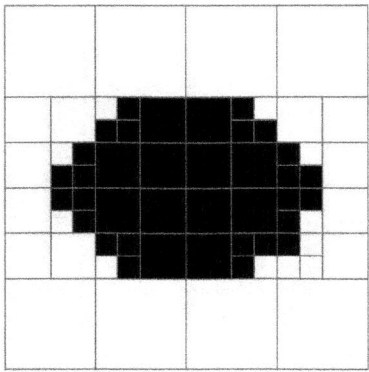

Fig. 29. Image decomposition of Example 3 by traditional method

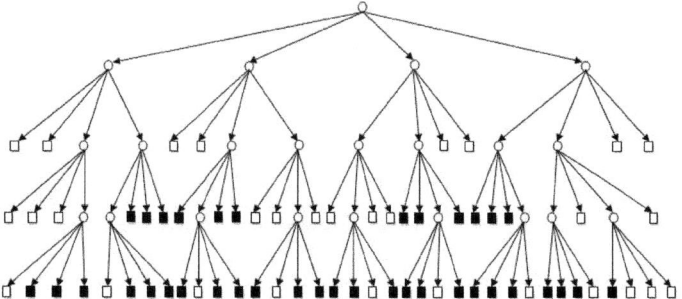

Fig. 30. Quadtree representation of Example 3 by traditional method

representation. The yellow region in Figure 10 is such a obvious example. In this case, the CORN algorithm and tradition method will handle the image in the same manner, that is, choosing its center point as the quadtree root node. The image composition and quadtree representation of this worst case are shown in Figure 38 and Figure 39, respectively.

According to the experimental results above, a brief indication about the reduction performance of the choice of the root node is shown in Table 4.

In essence, all the experimental results in Table 4 have indicated that the CORN algorithm is always at least as good or better than the traditional method with respect to the image decomposition and quadtree representation, as well as including the spatial data compression. However, it may be worthy of note that, until now, all experimental images we have chosen, are confined to a small size. At this point, this limitation may lead to the following doubt:

– Suppose we are given some images of much larger size. Does the CORN algorithm still outperform the traditional method?

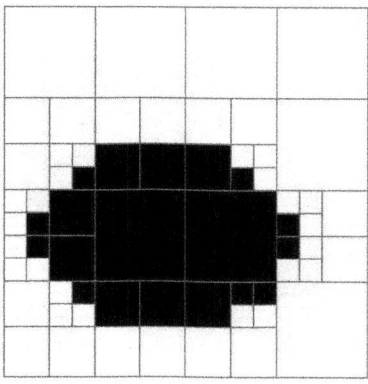

Fig. 31. Image decomposition of Example 3 by CORN algorithm

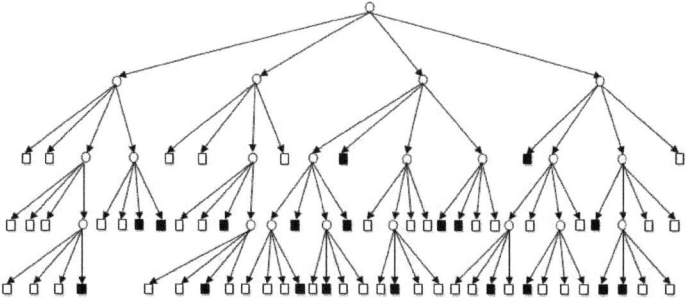

Fig. 32. Quadtree representation of Example 3 by CORN algorithm

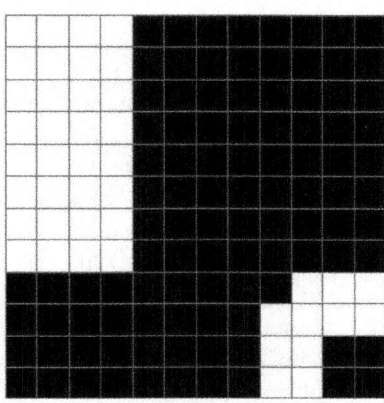

Fig. 33. An image containing a disconnected region

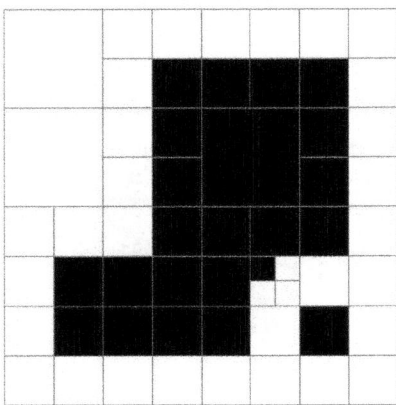

Fig. 34. Image decomposition of Example 4 by traditional method

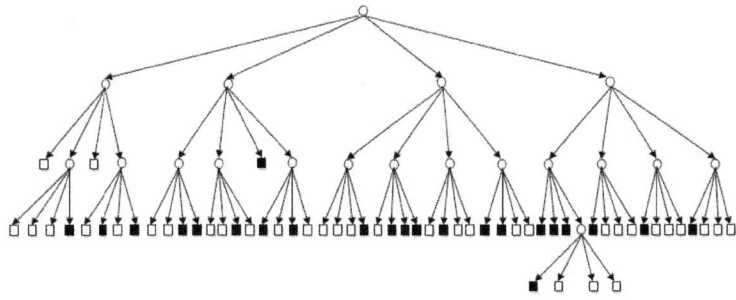

Fig. 35. Quadtree representation of Example 4 by traditional method

In order to indicate an answer this question, the CORN algorithm will be applied to two binary images from [18], p.230. Each image with size 256×256, is shown in Figure 40, where each one requires 65536 bits.

By applying the CORN algorithm to the two test images, in their final quadtree representations there are 651 black leaf nodes, 757 white leaf nodes and 469 grey nodes concerned with the image "Taiwan"; and for another image "Cloud", the number of black, white and grey nodes is 4093, 4116 and 2736, respectively. Then, choosing the DF-expression (outlined in Section 2.2) as the encoding method, the total number of bits required for compression performance is 2997 ($651 \times 2 + 757 \times 1 + 469 \times 2 = 2997$) and 17774 ($4093 \times 2 + 4116 \times 1 + 2736 \times 2 = 17774$), respectively.

In comparison with the results obtained by using the traditional method, which are shown in p.231 of [18], Table 5 lists the memory space needed to store the two binary

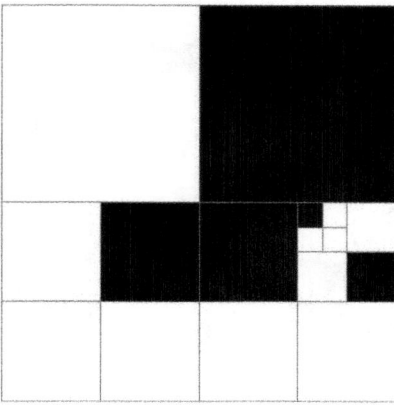

Fig. 36. Image decomposition of Example 4 by CORN algorithm

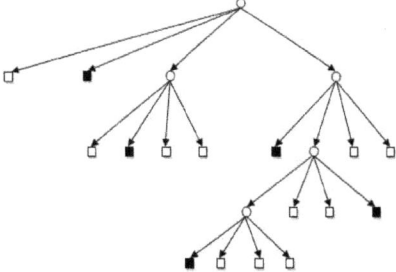

Fig. 37. Quadtree representation of Example 4 by CORN algorithm

Table 4. Leaf Nodes Reduction

Experiment	Method	Black	White	Total
Example 1	Center (Figure 27, 28)	20	26	46
	CORN (Figure 5.17)	8	11	19
Example 2	Center (Figure 4.4, 4.5)	20	23	43
	CORN (Figure 4.6, 4.7)	14	20	34
Example 3	Center (Figure 29, 30)	37	30	67
	CORN (Figure 31, 32)	19	45	64
Example 4	Center (Figure 34, 35)	23	35	58
	CORN (Figure 36, 37)	5	11	16
Example 5	Center (Figure 38, 39)	39	85	124
	CORN (Figure 38, 39)	39	85	124

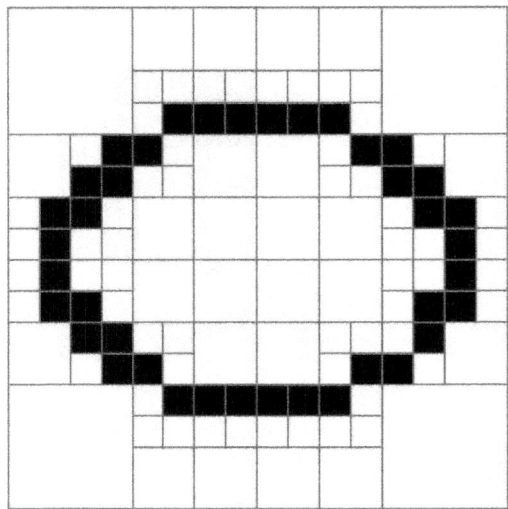

Fig. 38. Image decomposition of Example 5

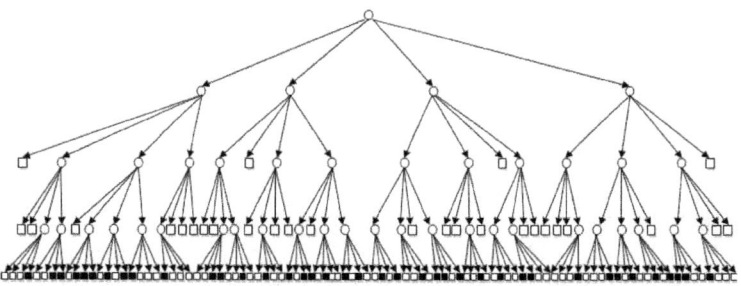

Fig. 39. Quadtree representation of Example 5

test images before and after processing by using the traditional method and CORN algorithm, followed by data compression rate C_r, which is defined as follows:

$$C_r = \frac{\text{number of bits needed to store original binary image}}{\text{number of bits needed to store compressed binary image}}$$

When viewing Table 5, we observe that the proposed CORN algorithm improves somewhat the quadtree representation and data compression when compared to the traditional method.

As a consequence, experimental results shown in this section reveal that, for all the test images, whether their contents are more complicated or simpler, the quadtree representations obtained from the CORN algorithm compare favorably with the traditional method.

(a) Taiwan.   (c) Cloud.

Fig. 40. Two binary images [18]

Table 5. Data compression rate comparison

Binary image	Method	No. of bits needed for original image	No. of bits needed for compressed image	C_r
Taiwan	Center	65536	3058	21.43
	CORN	65536	2997	21.88
Cloud	Center	65536	19024	3.44
	CORN	65536	17774	3.69

5   Outlook

The quadtree representation offers a hierarchical decomposition of visual information which can be implemented in a system or a robot. More generally, the quadtree serves well as a representation of a hierarchy of approximation spaces in the sense of Pawlak [20]. Due to the fact that the node building process is triggered by the difference of information and the geometrical structure of the plane, we are able to use the quadtree not only as a coding scheme, but as a simple pyramid representation of image data. Finding the optimal root node is an essential tool to offer a good representation of the given image data.

Due the hierarchical geometrical structure of the quadtree in NW-NE-SW-SE-blocks, we are enabled to analyze the information of the data structure at any stage of the hierarchy. The information presented at any stage of the hierarchy are rough sets: If we code 1 for "black", 0 for "white" and ? for "branch" (i.e. "grey") the rough set hierarchy of the first three levels of Figure 17 is given by:

```
Level 0: ?
Level 1: ???1
Level 2: ???1 ?010 ?10? 1111
...
```

The quadtree-rough-set representation offers a quick look at the image at an early stage using only a little amount of information. Successive quadtree representations can be used to represent change.

As the information can be used at any level of the representation, the change information is organized in hierarchical rough sets as well, and the approximation quality can be derived by α(level) without using information outside the level representation. Evaluation of the example shows

Level 0: $\alpha(0) = 0/1 = 0$
Level 1: $\alpha(1) = 1/4 = 0.25$
Level 2: $\alpha(2) = 7/13 = 0.538$

Our next tasks will be to optimize successive quadtree representations (SQTR) – in particular, to investigate the properties of the CORN algorithm – and to develop a rough logic for change in SQTR. The tasks will be focused on the detection of global change (e.g. moving environment) and local change (e.g. moving objects) within successive quadtree representations using the hierarchy of rough set representation. Solving these problems will enable us to formalize machine vision algorithms using a suitable (rough) logic based on quadtrees or even cut quadtrees. As patterns of "approaching" or "frames to contact" are then (roughly) definable on each level, a first step towards an ecological perception using rough quadtree representations is feasible and may end up in time efficient algorithms, due to the fact that "approaching" of an object or the number of "frames to contact" can be derived using representations with a medium approximation quality.

References

1. Samet, H.: Spatial data structures. Addison-Wesley, Reading (1995)
2. Samet, H.: Applications of Spatial Data Structures: Computer Graphics, Image Processing, and GIS. Addison-Wesley, Reading (1990)
3. Finkel, R., Bentley, J.: Quad trees: A data structure for retrieval on composite keys. Acta Informatica 4, 1–9 (1974)
4. Samet, H.: Data structures for quadtree approximation and compression. Image Processing and Computer Vision 28 (1985)
5. Aho, A.V., Hopcroft, J.E., Ullman, J.D.: The design and analysis of computer algorithms, vol. 6, pp. 77–78. Addison-Wesley, Reading (1974)
6. Samet, H.: The quadtree and related hierarchical data structure. Computing Surveys 16 (1984)
7. Shaffer, C.A., Samet, H.: Optimal quadtree construction algorithms. Computer Vision, Graphics, and Image Processing 37, 402–419 (1987)
8. Hunter, G.M.: Efficient computation and data structures for graphics. Phd. thesis, Princeton University (1978)
9. Hunter, G.M., Steiglitz, K.: Operations on images using quad trees. IEEE Trans. Pattern Anal. Mach. Intell., 145–153 (1979)
10. Rosenfeld, A., Pfaltz, J.L.: Sequential operations in digital image processing. J. ACM, 471–494 (1966)
11. Dyer, C.R.: Computing the Euler number of an image from its quadtree. Comput. Gr. Image Process. 13, 270–276 (1980)

12. Dehne, F., Ferreira, A.G., Rau-Chaplin, A.: Parallel processing of pointer based quadtrees on hypercube multiprocessors (1991)
13. Gargantini, I.: An effective way to represent quadtrees. Commun. ACM 25, 905–910 (1982)
14. Atalay, F.B., Mount, D.M.: Pointerless implementation of hierarchical simplicial meshes and efficient neighbor finding in arbitrary dimensions. In: Proc. International Meshing Roundtable (IMR), pp. 15–26 (2004)
15. Klinger, A., Dyer, C.R.: Experiments on picture representation using regular decomposition. Comput. Gr. Image Process. 5, 68–105 (1976)
16. Kawaguchi, E., Endo, T.: On a method of binary picture representation and its application to data compression. IEEE Trans. Pattern Anal. Mach. Intell. 2, 27–35 (1980)
17. Huang, C.Y., Chung, K.L.: Transformations between bincodes and the df-expression. Comput. Gr. 19, 601–610 (1995)
18. Yang, Y.H., Chuang, K.L., Tsai, Y.H.: A compact improved quadtree representation with image manipulations. Image and Vision Computing, 223–231 (1999)
19. Düntsch, I., Gediga, G.: Uncertainty measures of rough set prediction. Artificial Intelligence, 109–137 (1997)
20. Padwlak, Z.: Rough sets. Internat. J. Comput. Inform. Sci. 11, 341–356 (1982)
21. Ranade, S., Rosenfeld, A., Samet, H.: Shape approximation using quadtrees. Pattern Recognition 15, 31–40 (1982)
22. Tanimoto, S., Pavlidis, T.: A hierarchical data structure for picture processing. Comput. Gr. Image Process. 4, 104–119 (1975)
23. Sloan, K.R., Tanimoto, S.L.: Progressive refinement of raster images. IEEE Trans. Comput. 28, 871–874 (1979)
24. Knowlton, K.: Progressive transmission of grey-scale and binary pictures by simple, efficient, and lossless encoding schemes. Proceeding of the IEEE 68, 885–896 (1980)
25. Jones, L., Iyengar, S.S.: Representation of regions as a forest of quadtrees. In: Proceedings of the IEEE Conference on Pattern Recognition and Image Processing, Dallas, Tex., pp. 57–59 (1981)
26. Gautier, N.K., Iyengar, S.S., Lakhani, N.B., Manohar, M.: Space and time efficiency of the forest-of-quadtrees representation. Image and Vision Computing 3, 63–70 (1985)
27. Düntsch, I., Gediga, G.: Towards a logical theory of perception (2009) (preprint)
28. Gibson, J.J.: On the new idea of persistence and change and the old idea that it drives out, pp. 393–396. Lawrence Erlbaum Associates, Hillsdale (1982)
29. Worboys, M.: Imprecision in finite resolution spatial data. Geoinformatica 2, 257–280 (1998)
30. Horowitz, E., Sahni, S.: Fundamentals of data structures. Computer Science Press, RockVillie (1976)
31. Dyer, C.R.: The space efficiency of quadtrees. Computer Graphics and Image Processing 19, 335–348 (1982)
32. Ang, C.H., Samet, H.: A fast quadtree normalization algorithm. Pattern Recognition Letters 15, 57–63 (1994)

Solving the Attribute Reduction Problem with Ant Colony Optimization

Hong Yu[1,2], Guoyin Wang[1], and Fakuan Lan[1]

[1] Institute of Computer Science and Technology, Chongqing University of Posts and Telecommunications, Chongqing, 400065, P.R. China
wanggy@cqupt.edu.cn, agatelan@yahoo.com.cn
[2] Department of Computer Science, University of Regina, Regina, Saskatchewan Canada S4S 0A2
yuhongcq@yahoo.com.cn

Abstract. Attribute reduction is an important process in rough set theory. More minimal attribute reductions are expected to help clients make decisions in some cases, though the minimal attribute reduction problem (MARP) is proved to be an NP-hard problem. In this paper, we propose a new heuristic approach for solving the MARP based on the ant colony optimization (ACO) metaheuristic. We first model the MARP as finding an assignment which minimizes the cost in a graph. Afterward, we introduce a preprocessing step that removes the redundant data in a discernibility matrix through the absorption operator and the cutting operator, the goal of which is to favor a smaller exploration of the search space at a lower cost. We then develop a new algorithm R-ACO for solving the MARP. Finally, the simulation results show that our approach can find more minimal attribute reductions more efficiently in most cases.

Keywords: Attribute reduction problem, ant colony optimization, rough sets.

1 Introduction

Rough set theory, proposed by Pawlak [9] in 1982, is a valid mathematical tool to deal with imprecise, uncertain, and vague information. It has been developed and applied to many fields such as decision analysis, machine learning, data mining, pattern recognition, and knowledge discovery successfully.

In these applications, it is typically assumed that the values of objects are represented by an information table, with rows representing objects and columns representing attributes. The notion of a reduction plays an essential role in analyzing an information table [9]. In many cases, the minimal (optimal) attribute reduction is expected. Unfortunately, it is proven to be an NP-hard problem [20] to compute the minimal attribute reduction problem(MARP) of an information table. Thus, many heuristic methods have been proposed and examined for finding the set of all reductions or a single reduction [12,15,16,17,18,19,21,22,23,24]. In addition, many research efforts have shifted to the research of metaheuristics,

J.F. Peters et al. (Eds.): Transactions on Rough Sets XIII, LNCS 6499, pp. 240–259, 2011.
© Springer-Verlag Berlin Heidelberg 2011

such as genetic algorithm (GA)[1], ant colony optimization (ACO)[4,5,6,7], and more recently particle swarm optimization (PSO)[19].

Ant colony optimization (ACO) [2], is a novel nature-inspired metaheuristic for the solution of hard combinatorial optimization (CO) problems [2,3]. The main idea of ACO is to model the problem as a search for a minimum cost path in a graph. Artificial ants walk through this graph, looking for good paths. Each ant has a rather simple behavior so that it will typically only find rather poor-quality paths on its own. Better paths are found as the emergent result of the global cooperation among ants in the colony. ACO approach has been demonstrated by successful applications in a variety of hard combination optimization problems such as traveling salesman problems, vehicle routing problem, constraint satisfaction problem, machine learning, etc. It is noteworthy that an ACO-based algorithm [5], which is called AntRSAR, has been proposed for attribute reduction. The experimental results have shown that AntRSAR is a promising approach.

Since the MARP is an NP-hard problem, inspired by the character of ant colony optimization, more researchers [4,5,6,7] have focused on solving the problem with ACO. The model they used is similar to AntRSAR[5], in which the search space is a complete graph whose nodes represent conditional attributes, with the edges between them denoting the choice of the next conditional attribute. The search for the minimal attribute reduction is then an ant traversal through the graph where a minimum number of nodes are visited that satisfies the traversal stopping criterion. The ant terminates its traversal and outputs the attribute subset as a candidate of attribute reductions. The difference is mainly in the choice of the suitable heuristic desirability of traversing between attributes.

Actually, the heuristic desirability of traversing between attributes could be any subset evaluation function - for example, the rough set dependency measure [4,5,7], or an mutual information entropy based measure [6]. However, the relevant operations cost too much time because the operations are all in the space $|U| \times |C|$, where $|U|$ and $|C|$ are the cardinality of the objects and the cardinality of the attributes in the information table, respectively. On the other hand, the approaches are also heuristic. With the model, $O(|C| \cdot ((|U|^2 + |U|)/2))$ evaluations are performed in the worst case [4,5,6,7].

To combat the efficiency and gain more minimal attribute reductions, we will propose a new heuristic approach for solving the MARP based on ACO in this paper. We transfer the MARP to a constraint satisfaction problem firstly. The goal is to find an assignment which satisfies the minimum cost in a graph, the R-Graph, whose nodes represent the distinct entries of the discernibility matrix.

As we know, the notion of the discernibility matrix introduced by Skowron and Rauszer [10] is important in computing cores and attribute reductions. Many researchers have studied reduction construction by using the discernibility information in the discernibility matrix [15,18,22,24]. [24] proposes an elegant algorithm for constructing a reduction by using the discernibility matrix and an ordering of attributes. [22] proposes a reduction construction method based on discernibility matrix; the method works in a similar way as the classical Gaussian elimination method for solving a system of linear equations; the method obtains one minimum discernibility matrix and the union of all the elements in the minimum matrix produces a reduction.

In fact, there usually exists redundant data in the discernibility matrix. We can remove the redundant data from the matrix in order to reduce the exploration of the search space. In other words, to simplify the discernibility matrix before reduction can make improvement in time and space. Therefore, a new algorithm AMRS is proposed in Section 5, which can acquire the minimal reduction space(MRS) through absorption operations. Besides, we show that the reduction based on the simplified matrix MRS is equal to the reduction on the original discernibility matrix.

The rest of this paper is structured as follows. First, we introduce some definitions and terminologies about the minimal attribute reduction problem. In Section 3, we introduce the exist model to solve the MARP with ACO. Then, a new model R-Graph to solve the MARP with ACO is proposed in Section 4. Section 5 develops a preprocessing step by removing the redundant data in a discernibility matrix through the absorption operator. A new algorithm R-ACO for solving the MARP is given in Section 6. The experiment results in Section 7 show that the approach to solve the MARP with ACO can find more minimal reductions more efficiently in most cases. Some conclusions will be given in Section 8.

2   The Attribute Reduction Problem

This section introduces some notions such as information tables, discernibility matrices, reductions and so on.

2.1   Information Tables

An information table can be seen as a data table where columns are labelled by attributes, rows are labelled by objects and each row represents some piece of information about the corresponding object. Formally, an information table can be defined as follows [9] .

Definition 1. Information Table. An *information table* is the following tuple: $I = (U, Atr = C \cup D, V, f)$ where $U = \{x_1, x_2, \ldots, x_n\}$ is a finite non-empty set of objects, $C = \{a_1, a_2, \ldots, a_m\}$ is a finite non-empty set of attributes and also called the conditional attribute set, $D = \{d\}$ is the decision attribute set, V is the set of possible feature values, f is the information function, given an object and a feature, f maps it to a value $f : U \times Atr \rightarrow V$.

Example 1. Table 1 is an instance of a very simple information table, where the universe is $U = \{x_1, x_2, x_3, x_4, x_5\}$, the conditional attributes are a, b, c, d and e, and the decision attribute is f.

Table 1. An Information Table

U	a	b	c	d	e	f
x_1	1	0	0	1	0	0
x_2	0	1	1	0	1	1
x_3	0	0	1	1	0	1
x_4	1	1	1	0	0	0
x_5	0	1	0	0	1	0

2.2 Discernibility Matrices

The concept of a discernibility matrix [10] is an important one when it comes to the computation of cores, reductions and such in applications.

Definition 2. Discernibility Matrix. Given a consistent information table $I = (U, C \cup D, V, f)$, its *discernibility matrix* $M = (M_{i,j})$ is a $|U| \times |U|$ matrix, in which the element $M_{i,j}$ for an object pair (x_i, x_j) is defined as below.

$$M_{i,j} = \begin{cases} \{a \mid a \in C \wedge a(x_i) \neq a(x_j)\} & \text{if } d(x_i) \neq d(x_j) \\ \emptyset & \text{else} \end{cases} \tag{1}$$

Thus entry $M_{i,j}$ consists of all the attributes which discern elements x_i and x_j. Note that $M = (M_{i,j})$ is symmetric. The core can be defined below as the set consisting of all single element entries in M:

$$Core(C) = \{a \in C | M_{i,j} = \{a\}\} \tag{2}$$

The subset $R \subseteq C$ is a reduction of C if R is a minimal subset of C such that $R \cap M_{i,j} \neq \emptyset$ for any nonempty entry $M_{i,j}$ in M.

Example 2. Table 2 is the discernibility matrix of Table 1. For object pair (x_2, x_4), the entry $\{a, e\}$ indicates that either attribute a or e discerns the two objects. Obviously, the core is $\{c\}$.

Table 2. The Discernibility Matrix of Table 1

$M_{i,j}$	x_1	x_2	x_3	x_4	x_5
x_1	$\emptyset$	$\{a, b, c, d, e\}$	$\{a, c\}$	$\emptyset$	$\emptyset$
x_2	$\{a, b, c, d, e\}$	$\emptyset$	$\emptyset$	$\{a, e\}$	$\{c\}$
x_3	$\{a, c\}$	$\emptyset$	$\emptyset$	$\{a, b, d\}$	$\{b, c, d, e\}$
x_4	$\emptyset$	$\{a, e\}$	$\{a, b, d\}$	$\emptyset$	$\emptyset$
x_5	$\emptyset$	$\{c\}$	$\{b, c, d, e\}$	$\emptyset$	$\emptyset$

2.3 Discernibility Function

From a discernibility matrix, the notion of a discernibility function can be defined as below [10].

Definition 3. Discernibility Function. The *discernibility function* of a discernibility matrix is defined by:

$$f(M) = \bigwedge_{\substack{1 \le i \le |U|-1 \\ i+1 \le j \le |U|}} \left(\bigvee_{\substack{a_k \in M_{i,j} \\ M_{i,j} \ne \emptyset}} a_k \right). \tag{3}$$

The expression $\bigvee a_k (a_k \in M_{i,j})$ is the disjunction of all attributes in $M_{i,j}$, indicating that the object pair (x_i, x_j) can be distinguished by any attribute in $M_{i,j}$. The expression $\bigwedge(\bigvee a_k)$ is the conjunction of all $\bigvee a_k$, indicating that the family of discernible object pairs can be distinguished by a set of attributes satisfying $\bigwedge(\bigvee a_k)$.

By using the absorption and distribution laws, the discernibility function can be transformed to a disjunctive form as $f(M) = \bigvee R_q$, where R_q is a conjunction of some attributes. Each conjunctor $R_p = a_1 \wedge a_2 \wedge \cdots \wedge a_q$ is a reduction, denoted by $R_p = \{a_1, a_2, \ldots, a_q\}$ [10].

Take Example 2 as an example, the discernibility function of the information table can be denoted by $f(M) = \{a \vee b \vee c \vee d \vee e\} \wedge \{a \vee c\} \wedge \{a \vee e\} \wedge \{c\} \wedge \{a \vee b \vee d\} \wedge \{b \vee c \vee d \vee e\}$. The discernibility function $f(M)$ can be transformed to a disjunction form as $f(M) = \{a, c\} \vee \{a, b, c\} \vee \{a, d, c\} \vee \{a, e, c\} \vee \ldots$. We can acquire the minimal attribute reductions based on this Boolean calculation. For example, $\{a, c\}$ is a minimal attribute reduction here.

The problem of finding more minimal reductions of an information table has been the subject of many researches. The most basic solution to locating such a reduction is to simply generate all possible reductions and choose any with minimal cardinality. Obviously, this is an expensive solution to the problem and is only practical for very simple datasets. The computation is very complex when considering the scale of the problem since it is an NP-hard problem. Therefore, many research efforts have shifted to metaheuristics, such as the ant colony optimization.

3   ACO for Attribute Reduction

The existing searching space model with ACO to solve the attribute reduction problem will be explained in this section.

In general, an ACO algorithm can be applied to any combinational problem as far as it is possible to define [5]:

- *Appropriate problem representation.* The problem must be described as a graph with a set of nodes and edges between nodes.
- *Heuristic desirability of edges.* A suitable heuristic measure of the "goodness" of paths from one node to every other connected node in the graph.
- *Construction of feasible solutions.* A mechanism must be in place whereby possible solutions are efficiently created.
- *Pheromone updating rule.* A suitable method of updating the pheromone levels on edges is required with a corresponding evaporation rule.

- *Probabilistic transition rule.* The rule that determines the probability of an ant traversing from one node to the next in the graph.

3.1   Attribute-Graph for Attribute Reduction

As mentioned above, [4,5,6,7] have used the ACO approach to solve the attribute reduction problem. The problem is represented as the same model, called Attribute-Graph in this paper, which is a complete graph whose nodes represent conditional attributes, with the edges between them denoting the choice of the next conditional attribute. The search for the optimal attribute reduction subset is then an ant traversal through the graph where a minimum number of nodes are visited that satisfies the traversal stopping criterion.

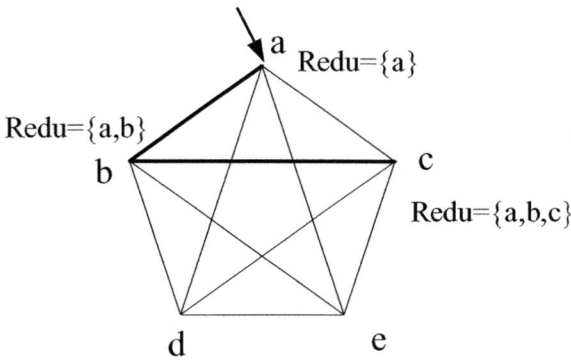

Fig. 1. Attribute-Graph of Table 1

Fig. 1 is the Attribute-Graph of Table 1, where C is a, b, c, d, e as used in Example 1. Each node is labeled with an attribute, and pheromone is distributed on the edges. The ant is currently at node a and has to make a choice among $\{b, c, d, e\}$. Here, b is the next one based on the transition rule, then c. The subset $Redu = \{a, b, c\}$ is determined to satisfy the traversal stopping criterion. The ant terminates its traversal and outputs this attribute subset as a candidate for attribute reduction. The subset $\{a, b, c\}$ is also called a path or a partial solution, which is described as thicker lines.

3.2   Pheromone Trails and Heuristic Information

The attribute reduction problem then is modeled as the problem of finding a path such that the set of selected attributes is a reduction and the cardinality of this set is minimized. Each edge is assigned a pheromone trail and heuristic.

After each ant has constructed a solution, pheromone trails are updated accordingly. Literatures [4,5,6,7] make some efforts on the pheromone updating rules.

A suitable heuristic desirability of traversing between attributes could be any subset evaluation function. For example, literatures [4,5,7] use the rough set dependency degree measure as the heuristic information. Formally, for $P, Q \subseteq C$, the dependency degree κ is defined as below.

$$\eta = \kappa = |POS_P(Q)|/|U| \tag{4}$$

where $POS_P(Q)$ is the positive region.

The difference between mutual information entropies is used to measure the heuristic information in [6] as following.

$$\eta = I(R \cup \{a\}; D) - I(R; D) \tag{5}$$

where $I(B; D) = -H(D|B)$ is the mutual information entropy.

However, the computation about the positive region or the entropy are all in the whole information table $U \times C$. In addition, the searching is in the complete graph Attribute-Graph, which is expensive in time.

3.3   Framework of a Basic ACO Algorithm

The overall process of ants finding reductions based on Attribute-Graph can be described in Algorithm 1, which is also a framework of a basic ACO algorithm.

The algorithm performs as follows: at each cycle, every ant constructs a solution and then pheromone trails are updated. The algorithm stops iterating when a termination condition is met. Generally, the termination condition may be a maximum number of cycles or a given time limit.

Algorithm 1. The Framework of a Basic ACO Algorithm

> **Input**   : the information table $I = (U, Atr = C \cup D)$
> **Output**: the set SP, which is a reduction
> **begin**
> > InitializePheromoneValues(τ);
> > InitializeHeuristicInformation(η);
> > // Construction
> > **while** *termination conditions not met* **do**
> > > **for** *each ant k* **do**
> > > > $SP_k = \emptyset$;//SP_k is the kth ant's partial solution
> > > > choose the v from the unvisited attribute nodes according to the probability transition rule(Equation 6);
> > > > Extend the SP_k until it is a reduction;
> > > **end**
> > > Update the pheromone trail.
> > **end**
> **end**

The heuristic desirability of traversal and edge pheromone levels are combined to form the so-called probabilistic transition rule, denoting the probability of an ant at attribute u choosing to travel to attribute v at time t. Then, we introduce the basic probabilistic transition rule is used in ACO algorithms as below[3].

$$P_{uv}^k(t) = \begin{cases} \dfrac{[\tau_{uv}(t)]^\alpha [\eta_{uk}(t)]^\beta}{\sum_{r \in allowed_k} [\tau_{ur}(t)]^\alpha [\eta_{ur}(t)]^\beta} & \text{if } v \in allowed_k \\ 0 & \text{otherwise} \end{cases} \tag{6}$$

where v denotes the attribute selected at the kth construction step, $allowed_k$ is the set of condition attributes that have not yet been selected, τ_{ur} and η_{ur} are the pheromone value and heuristic information of edge (u, r), respectively. α and β are two parameters which control the relative importance of pheromone trails and heuristic information.

4 R-Graph to Solve MARP with ACO

To combat the efficiency and gain more minimal attribute reductions, we will propose a novel approach for solving the MARP based on ACO in this section. We transfer the MARP to a constraint satisfaction problem firstly, so the goal is to find an assignment which satisfies the minimum cost in a graph.

4.1 Constraint Satisfaction Problem

Firstly, let us review the constraint satisfaction problem(CSP) [13].

A constraint satisfaction problem is defined by a triple (B, Dom, Con) such that $B = \{B_1, B_2, \ldots, B_k\}$ is a finite set of k variables, Dom is a function which maps every variable $B_p \in B$ to its domain $Dom(B_p)$, and Con is a set of constraints.

A solution of a CSP(B, Dom, Con) is an assignment $\mathcal{A} = \{< B_1, v_1 >, \cdots, < B_k, v_k >\}$, which is a set of variable-value pairs, where $v_i \in Dom(B_i)$. That is, an assignment is a set of variable - value pairs and corresponds to the simultaneous assignment of values to variables, respectively.

The cost of an assignment $\mathcal{A}$ is denoted by $cost(\mathcal{A})$. An optimal solution of a CSP(B, Dom, Con) is a complete assignment for all the variables in B, which satisfies all the constraints in Con with the minimal cost.

4.2 R-Graph for Attribute Reduction

Let us come back to the discernibility matrix M. If we take a uniform element $M_{i,j}$ as a variable B_p, then $Dom(B_p) = M_{i,j}$, and we denote $B = \cup B_p$. We can transfer an attribute reduction problem to a CSP and define the following model R-Graph to describe the attribute reduction problem.

A R-Graph associates a vertex with each value of the tuple $T = < c_1, \ldots, c_m >$ to be permuted, and T is a tuple of $|C| = m$ values, where C is the conditional attribute set of an information table $I = (U, C \cup D)$. There is an extra vertex corresponding to the nest, called *nest*, from which ants will start their paths. Hence, a R-Graph associated with a CSP can be formally defined as below.

Definition 4. R-Graph Associated with CSP. A *R-Graph associated with a CSP(B, Dom, Con)* is a complete oriented graph $G = (V, E)$ such that: $V = \{< B_i, v >| \ B_i \in B \ and \ v \in Dom(B_i)\}, E = \{(< B_i, v >, < B_j, u >) \in V^2 \mid B_i \neq B_j\}$.

That is, the B_i is the entry of the discernibility matrix, and the domain of B_i is the subset of the tuple $T = < c_1, \ldots, c_m >$. Obviously, we can use the 0-1 representation to encode the variables. For example, Fig.2 gives the R-Graph of Table 1. The R-Graph has six vertexes because there are six varying entries in the discernibility matrix (see Table 2). Vertex B_2 means entry $M_{1,3} = \{a, c\}$, and the domain of B_2 is $\{a, c\}$ which is the subset of $C = \{a, b, c, d, e\}$. Hence, we encode the B_2 as $\{10100\}$, where 1's means the B_2 has the value a and c, and 0's means there are no value b, d or e.

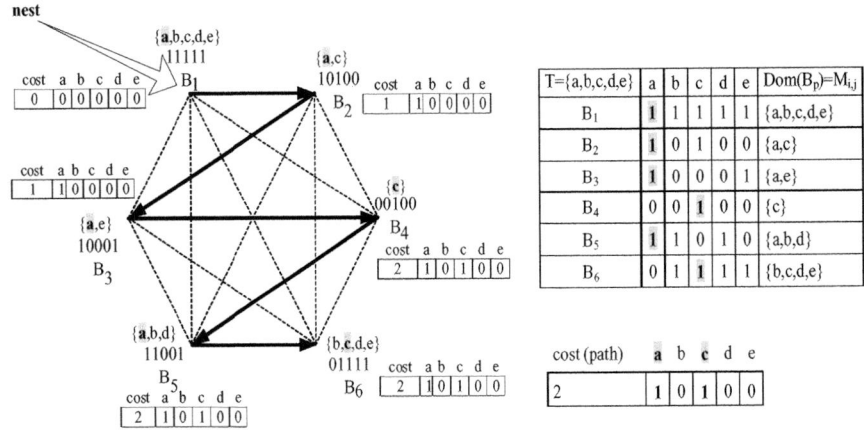

Fig. 2. R-Graph of Table 1

A path in a R-Graph $G = (V, E)$ is a sequence of vertices of V, which starts with the *nest*. Formally, the path is described as follows.

Definition 5. Path in a R-Graph. A *path in a R-Graph* associated with a CSP (B, Dom, Con) is $\pi = nest \rightarrow u_1 \rightarrow \ldots \rightarrow u_i \rightarrow \ldots \rightarrow u_P$, where $u_i \in Dom(B_i)$.

An ant will visit every vertex. Because every path starts with the *nest*, the path π is simply denoted as $\pi = u_1 \rightarrow \ldots \rightarrow u_i \rightarrow \ldots \rightarrow u_P$. We only consider elementary paths, which do not contain any cycles.

The goal of R-Graph is to find a complete assignment which minimizes the cost. Because the values in every B_i is in $T = < c_1, \ldots, c_m >$, ants going over the same values in the path will have a cost of 0. In other words, the more duplicate values selected in a path, the less cost a path is; the property can be used as in designing the heuristic information. Hence, we have the following definition.

Definition 6. Cost of a Path. The *cost of a path* π, denoted by $cost(\pi)$, is the number of the different values that appeared in the path π.

A solution of a R-Graph G is an assignment $\mathcal{A} = \{B_1 \leftarrow r_1, \cdots, B_i \leftarrow r_i, \cdots, B_k \leftarrow r_k\}$, where $r_i \in Dom(B_i)$. Ants deposit pheromone on edges of the R-Graph; the amount of pheromone laying on an edge (v_i, v_j) is denoted as $\tau_{v_i v_j}$. Ants start from the nest, and successively visit each of the other k vertices.

Here, all ants have a constraint condition such that they must visit every vertex.

According to [10], if we choose one attribute from every element of the discernibility matrix, then the set of the selected attributes can compose a super attribute reduction. Therefore, the path (assignment) of a R-Graph is also a super reduction of the information table. That is, we get a super reduction through the searching in the R-Graph now. However, our goal is to obtain more minimal reductions. Fortunately, ACO approach is good at finding the optimal solution through the colony intelligence. It is more likely to find the minimal reductions successfully as long as ants can find the optimal solutions.

4.3   Constructing a Solution

The model R-Graph is a graph whose nodes represent a uniform element of discernibility matrix. The node is a set of some attributes, not one attribute as used in Attribute-Graph model. The weight on the edges is 1 or 0 decided by the building searching path, not the significant which difficult to compute in the other model.

The path described as thicker lines (or shadow letters) in Fig.2 is a typical construction process, where 0-1 representation is used to encode the variables. For example, an ant starts from the *nest* with cost 0 to the variable (vertex) B_1, and the ant chooses the value a from the variable B_1 according to the initial information. Then the ant begins its travel. Firstly, the value a is chosen in the next variable B_2 with the cost 1. Then, the value a is chosen again in the B_3, and now cost is still 1 because a is appeared in the path. Then, the value c is chosen in the B_4, and now the cost will increase 1 since c is not in the path. At last, the solution(or called path, assignment) is $\mathcal{A} = \{B_1 \leftarrow a, B_2 \leftarrow a, B_3 \leftarrow a, B_4 \leftarrow c, B_5 \leftarrow a, B_6 \leftarrow c\}$.

In other words, we gain a reduction $f(M) = \{a\} \wedge \{a\} \wedge \{c\} \wedge \{a\} \wedge \{c\} = \{a\} \wedge \{c\} = \{a, c\}$. To represent with 0-1 representation simply, the path is denoted by 10100 with cost 2 corresponding to an ant from nest to food passes the values a and c. Recall that, $\{a, c\}$ is the minimal attribute reduction of Table 1.

In summary, if ants can find the minimal cost paths in a R-Graph, then ants can find the minimal attribute reductions of the information table. In fact, we prefer which values in the tuple T are selected in a path to how to build a path.

4.4   Model Comparison

As already mentioned in Section 3, the decision method, which is to decide whether an attribute can be added to the ant traversing path, used in the model

Attribute-Graph are all on the space $U \times C$. And we know, for each decision, we need to compute the entropy or the dependency function, and the compute time of each decision is $O(|U|^2 + |U|)/2$. The times of decisions is $|C|$. Therefore, the time complexity is $O(|C| \cdot (|U|^2 + |U|)/2)$ in [4,5,6,7].

By contraries, the searching space used in the R-Graph is no more than $|C|$ at each decision. There are no computation about the entropy or the dependency function, and the weight on the edges are only 1 or 0 depended on the searching path. Considering the worst case, the items of the discernibility matrix are different from each other, then the times of decisions will be $(|U|^2 + |U|)/2$. Therefore, the time complexity is $O((|U|^2 + |U|)/2) \cdot |C|)$.

However, there are usually many redundant items in the discernibility matrix, the compute times in the R-Graph is far less than $O((|U|^2 + |U|)/2) \cdot |C|)$ in most cases.

For example, the R-Graph in Fig.2 looks more complex than the Attribute-Graph one in Fig.1, because the R-Graph has 6 vertexes and the Attribute-Graph has 5 vertexes. However, we just need 2 vertexes to build the R-Graph of Table 1, which means great improvement in time and space. The approach will be described in the next section.

5   Discernibility Matrix Simplification

Paths in a R-Graph are in the reductions of the corresponding discernibility matrix. However, since there are many redundant data in the matrix, we may need to reduce the search space through removing the redundant elements of the matrix. A novel algorithm will proposed to acquire minimal reduction space in this section.

5.1   Minimal Reduction Space

Consider the reduction space S, which is the subset of discernibility matrix M with no uniform elements, denoted by $S = B = \cup B_p$, where $B_p = M_{i,j}$. Obviously, the reduction computation based on a discernibility matrix is the computation based on Boolean calculation [10].

Let $B_i \in S$, if $B_j \in S$, and $B_j \subseteq B_i$, B_i is called an absorbed discernibility attribute element by B_j. Namely, $B_j \cap B_i = B_j$. And we have the property:

Property 1. If $B_j \cap B_i = B_j$, then $(\vee(B_j)) \wedge (\vee(B_i)) = \vee(B_j), (\wedge(B_j)) \vee (\wedge(B_i)) = \wedge(B_j)$.

The property pertaining to Boolean logic can be used to reduce the reduction space, in other words, a minimal reduction space(MRS) can be acquired by removing all absorbed discernibility attribute items. Then, a novel algorithm AMRS(Acquire Minimal Reduction Space) is given in Algorithm 2 to simplify the discernibility matrix.

In addition, we know that the reduction based on the simplified matrix MRS is equal to the reduction on the original discernibility matrix M[22].

Algorithm 2. Acquire Minimal Reduction Space Algorithm

Input : the information table $I = (U, Atr = C \cup D)$, $|U| = n$, $|C| = m$.
Output: the minimal reduction space MRS.
begin
    $MRS \longleftarrow \emptyset$; $absorbed \longleftarrow 0$;
    for $i = 1$ *to* n **do**
        for $j = i + 1$ *to* n **do**
            if $d(x_i) \neq d(x_j)$ **then**
             | Compute $M_{i,j}$;
            end
            for *each* B_k *in* MRS *and* $B_k \neq \emptyset$ **do**
                if $B_k \subseteq M_{i,j}$ **then**
                 | //if $M_{i,j}$ be absorbed, i.e. $B_k \&\& M_{i,j} = B_k$
                 | $absorbed \longleftarrow 1$; //set the absorbed flag
                end
                else if $M_{i,j} \subseteq B_k$ **then**
                 | //if B_k be absorbed, i.e. $B_k \&\& M_{i,j} = M_{i,j}$
                 | $MRS \longleftarrow MRS - B_k$; //delete the B_k from the space MRS
                end
            end
            if $absorbed = 0$ **then**
                // if $M_{i,j}$ cannot be absorbed or B_k be absorbed
                $MRS \longleftarrow MRS + M_{i,j}$; // add $M_{i,j}$ to the space MRS
            end
            $absorbed \longleftarrow 0$; //reset
        end
    end
    Output MRS;
end

In order to make improvement in time, the absorption operator and cutting operator are used in Algorithm AMRS in this paper, which are defined as followed.

5.2   Absorption Operator

In order to judge whether a discernibility element is absorbed by another, we define the absorption operator $\&\&$, which is an and-operator one by one bit. Considering Fig. 2, $B_1 = \{a, b, c, d, e\} = \{11111\}$ and $M_{i,j} = M_{1,3} = \{a, c\} = \{10100\}$, then $B_k \&\& M_{i,j} = \{11111\} \&\& \{10100\} = \{10100\}$. Obviously, this leads to $B_k \&\& M_{i,j} = M_{i,j}$. We then have $M_{i,j} \subseteq B_k$, which means B_k should be removed.

The elements with $|B_i| = 1$ are the core attributes. We can remove the core from MRS and add it to the reduction later. The size of the $MRS[P][Q]$ will be extended to $(P + 1) \times (Q + 1)$, where each cell of the 0-th row stores the number of the 1's in the corresponding column. Each cell of the 0-th column stores the number of the 1's in the corresponding row. And we can move the row which has the most number of 1's to the first row. That is, $MRS[1][0]$ is the maximal one among $MRS[\ ][0]$. These values can be acquired easily through Algorithm AMRS, and they can be used as heuristic information in the next section.

Table 3 is the extended $MRS[P][Q]$ of Table 2, where the number in the parentheses means the number of 1's in the corresponding row or column.

Table 3. The $MRS[P+1][Q+1]$ of Table 2

$T = \{a,b,d,e\}$	$a(2)$	$b(1)$	$d(1)$	$e(1)$	$Dom(B_p)$
$B_1(3)$	1	1	1	0	$\{a,b,d\}$
$B_2(2)$	1	0	0	1	$\{a,e\}$

The R-Graph can be built from Table 3 instead of Table 2, the search space is greatly reduced.

5.3   Cutting Operator

The number of the 1's in the q-th column of the MRS is stored in the $MRS[0][q]$. In other words, if the a_q denotes the corresponding attribute of the q-th column, the $MRS[0][q]$ stores the number of the a_q in all the variables.

Let Q_{max} denotes the maximal one of the $MRS[0][q]$, that is, $Q_{max} = max(MRS[0][q])$ $(1 \le q \le Q)$. Let min_Redu denotes the cardinality of the minimal reduction, and C_0 is the core. Then we have the following property, called the cutting operator.

Property 2. $min\_Redu - |C_0| \le P - Q_{max} + 1$.

Observe the extended MRS, the number of attributes of the minimal reductions is $min\_Redu - |C_0|$ because the core is not included in the extended MRS. It is easy to prove the property by contradiction. That means, the cost of the minimal reduction is no more than the cost of the path which is formed by choosing the value, the corresponding attribute of the Q_{max}, in a variable as long as it has the value.

The property is used to decide whether the path could be a minimum cost path in Algorithm R-ACO in the next section. In other words, if the cardinality of the partial (local) solution is more than the $P - Q_{max} + 1$, it is unnecessary to continue the searching process. The cut operating can make improvement in time and space.

Considering Table 3, P is 2, Q is 4, Q_{max} is 2, and the corresponding attribute is a. Then, we have $min\_Redu - |C_0| \le 1$. That means if the cardinality of the local solution is more than 1, the search path should be stopped.

6   Description of Algorithm R-ACO

After the preprocessing step as described in the last section, a simplified discernibility matrix will be obtained. Then, we can construct the graph R-Graph based on the simplified discernibility matrix MRS. A novel Algorithm 3 (R-ACO: finding the set of attribute Reduction based on ACO) is proposed to find most of the minimal attribute reductions using the ant colony optimization based on the model R-Graph. Let us explain some ideas used in the algorithm.

6.1   Construction Graph and Initialization

As we have discussed in Section 5, we can construct the R-Graph based on the extended simplified discernibility matrix $MRS[P+1][Q+1]$. Artificial ants lay pheromone on the graph. Thus, we need to initialize the following problems.

- *Initialize the pheromone.* To initialize the pheromone, the constant e is used here.
- *Number of the ants.* As we have discussed in the subsection 5.2, the $MRS[1][0]$ means the variable B_1 has the maximal number of attributes. We initialize the number of ants is the $MRS[1][0]$, which ensure the ants can travel all possible paths. In other words, every node (value) in B_1 is assigned an ant.
- *Minimal cost of the visiting path.* We define *temp_mincost* to denote the cardinality of the current optimal solution, namely, the minimal cost of the visiting path. Therefore, it is initialized as $temp\_mincost = P - Q_{max} + 1$. Let SP_k denotes the local solution. If $cost(SP_k) > temp\_mincost$, we can stop the searching process according to Property 2.
- *Edge Covered.* In order to avoid convergence quickly and find more minimal solutions, we define $EdgeCovered_{uv}$ to denote the number of ants that selected E_{uv}. $EdgeCovered_{uv}$ is used in making choice of a value.

6.2   Selection of a Variable

When constructing an assignment, the order in which the variables are assigned is important. Some commonly used variable orderings are the following[11]: 1) most constraining first ordering, which selects an unassigned variable that is connected (by a constraint) to the largest number of unassigned variables; 2) most constrained first ordering, which selects an unassigned variable that is connected (by a constraint) to the largest number of assigned variables; 3) smallest domain ordering, which selects an unassigned variable that has the smallest number of consistent values with respect to the partial assignment already built; or 4) random ordering, which randomly selects an unassigned variable.

Since a solution path will visit every variable, we use the order of variables comes from MRS (the simplified discernibility matrix) directly with no additional computation in order to save time. To solve the MARP with R-Graph, we use the random ordering here. For example, the actually variable-path is expressed by the thicker line in Fig.2. That means there are no additional computation about how to select a variable, and there are no edges on the dotted lines actually. In other words, we get the variable path from the algorithm AMRS directly.

Algorithm 3. Finding the Set of Attribute Reduction Based on ACO

Input : $MRS[P+1][Q+1]$; // $2 \leq Q \leq |C|$
Output: The set SG whose element is an attribute reduction.
begin
    //Initialization:
    $temp\_mincost \longleftarrow P - Q_{max} + 1$; //come from Alg. AMRS and Pro. 2
    Artificial ants $ant \longleftarrow MRS[1][0]$; // the number of ants is ant
    $E_{uv}.pheromone \longleftarrow e$;// initialize the pheromone
    $EdgeCovered_{uv} \longleftarrow 0$; // the number of ants that selected E_{uv}
    $constringency \longleftarrow False$; // the termination conditions
    Initial q_0; max_cycles; //They are constants, $0 < q_0 < 10$.
            //Eg: $q_0 = 5$, $max\_cycles = 10$
    while ! $constringency$ **do**
        if $ant = MRS[1][0]$ **then**
          | Set the ant ants to each of the nodes in B_1;
        else
          | Set ant ants on nodes with larger pheromone values;
        end
        // construct solution:
        for $k = 1$ to ant **do**
          //local search
          $SP_k \longleftarrow B_{1k}$; //initialize the partial solution is the node B_{1k}
          for $i \in B_2$ to B_P **do**
            Produce a random value q; //$0 < q < 10$
            if $q > q_0$ **then**
              | Select a node v random from the set whose nodes have the
              | maximal heuristic information.
            else
              | $maxNode \longleftarrow max(P_{uv}^k)$; //Equation (7)
              | node $v \longleftarrow v \in maxNode \land v \in min(EdgeCovered)$.
            end
            $SP_k' \longleftarrow SP_k \cup \{v\}$; $EdgeCovered_{uv} + +$;
            if $(SP_k' \&\& SP_k)! = SP_k$ **then**
             | Compute new SP_k and the $cost(SP_k)$;
            end
            if $cost(SP_k) > temp\_mincost$ **then**
             | Delete SP_k; continue;
            end
            $\Delta\eta_v - -$; //The node v has been selected by the current ant.
          end
        end
        $SP_{min} \longleftarrow min\{SP_1, SP_2, \ldots, SP_k, \ldots, SP_{ant}\}$;
        //updating:
        if $|SP_{min}| < |SG|$ **then**
          | $SG \longleftarrow SP_{min}$;
          | $temp\_mincost \longleftarrow min(cost(SP_k[0]))$; //$1 \leq k \leq ant$
        else
          | $SG \longleftarrow SG \cup (SP_{min} - SG)$;
        end
        Update the pheromone of all nodes according to Equation (8);
        $max\_cycles - -$;
        if SG is stable or $max\_cycles = 0$ **then**
          | $constringency = True$;
        end
    end
    Output the set SG.
end

6.3   Choice of a Value

Once a variable has been selected, ants have to choose a value for it. The goal is to choose the "most promising" value for the variable, i.e., the one that will participate in the smallest number of attributes in the final complete assignment. However, if it is straightforward to evaluate the number of attributes with the already assigned variables, it is more difficult to anticipate the consequences of the choice of a value on the further assignments of the remaining unassigned variables. Hence, there are very few heuristics for guiding this choice.

The choice is decided by the transition probabilities, which is the basic ingredient of any ACO algorithm to construct solutions. The transition probabilities are defined as follows:

$$P_{uv}^k = \begin{cases} \frac{\tau_{uv}^\alpha (\eta_v + \Delta\eta_v)^\beta}{\sum_{r \in B_i} \tau_{ur}^\alpha (\eta_r + \Delta\eta_r)^\beta} & \text{if } v \in B_i \\ 0 & \text{otherwise} \end{cases} \qquad (7)$$

where B_i means the next selected variable, u is the current value, v is the next value in B_i, and η is the heuristic information.

Equation(7) is different from Equation(6). As we have discussed in Section 5, $MRS[0][j]$ is used as an initial η. In order to avoid convergence quickly, the heuristic information is adjusted by $\Delta\eta$ automatically.

The values of parameters α and β determine the relative importance value and heuristic information. There, $1 \leq \alpha \leq 5$ and $1 \leq \beta \leq 5$ are derived from experience.

To acquire more minimal reductions, the random probability is also used in the R-ACO algorithm. That is, we produce a random value q firstly. If $q > q_0$ (q_0 is a constant.), then select a node randomly from the set whose nodes have the maximal heuristic information. Otherwise, we need to compute Equation(7) to make the choice.

6.4   Pheromone Updating

After each ant has constructed a solution, pheromone trails are updated. Pheromone trails (τ) are updated by the following rule:

$$\begin{aligned} \tau &= (1 - \rho)\tau_{uv} + \Delta\tau_{uv}, & u \in B_{i-1}, v \in B_i \\ \Delta\tau_{uv} &= \sum_{k=1}^m \Delta\tau_{uv}^k \\ \Delta\tau_{uv}^k &= \begin{cases} \frac{Q}{cost(SP_k)} & u, v \in SP_k \\ 0 & \text{else} \end{cases} \end{aligned} \qquad (8)$$

where $\rho \in (0, 1]$ is a parameter for pheromone evaporation, Q is a constant, and SP is the partial solution.

6.5   Termination Conditions

The termination conditions used here as same as the basic ACO [2]. That is, when the solutions are stable or the max cycles are reached, the iterations are ended. In fact, the iterations perform no more than 5 times when the solutions are convergent in our experiments, even if the max cycles is initialized to 10.

Table 4. The EQ6100 Engine Fault Diagnosis Data Set

A	B	C	D	E	F	G	H	I	J	K	Symptom
2	4	1	4	2	1	4	2	2	1	1	piston knock
2	4	1	3	1	3	1	3	3	1	1	piston pin
2	2	2	3	4	1	1	4	4	1	1	valve
2	2	1	3	2	1	1	4	2	1	1	tappet
2	4	2	2	4	1	4	2	2	4	4	connecting rod bearing
4	3	2	2	3	1	4	2	1	3	4	crankshaft bearing
2	2	2	4	4	1	1	2	4	2	1	time gear
3	2	4	2	2	1	2	4	4	2	1	outbreak
4	2	4	2	4	1	2	4	3	1	1	premature firing
2	2	4	2	2	1	1	4	2	4	1	late firing
4	4	2	4	4	1	1	4	3	1	1	u1rdered firing

7   Experimental Results

In this section, we use the EQ6100 Engine fault diagnosis data set [8] and some standard UCI data sets [14] to highlight various features of the proposed algorithm R-ACO, which uses the R-Graph model for attribute reduction. The algorithm R-ACO is also compared with Algorithm RSACO [6], which uses the Attribute-Graph model for attribute reduction. In order to obtain the minimal reductions to reference, the algorithm in literature [15] is performed, which can compute all the reductions of an information table. The three algorithms are performed by C++ language on the computer whose CPU is Pentium4/2.40GHz and the memory is 512MB.

7.1   EQ6100 Engine Fault Diagnosis Data Set

The EQ6100 Engine fault diagnosis data set is used to to show that it has a wider significance to acquire more minimal reductions in some application fields.

The data set is used to decide which kind of fault arose according to the 11 categories of abnormal sounds. For example, the conditional attributes $A - E$ means the level of abnormal sounds when the engine in the following states, such as start-up in high temperature, start-up in low temperature, running in high speed, running in low speed and acceleration processing, respectively. Table 4 describes the data set.

According to the evaluations of experts, the "No", "mild", "clarity" and "significantly" are used to describe the level of the abnormal sounds and denoted by "1", "2", "3" and "4" by discreting, respectively. From Table 4, we can determine which attribute playing a major role and which one playing a secondary role when distinguishing the various abnormal sounds of the engine.

The algorithm R-ACO obtains six minimal reductions of the data set, they are $SG = \{\{D, G, I\}, \{D, H, I\}, \{B, D, I\}, \{C, D, I\}, \{D, E, I\}, \{D, I, K\}\}$. Actually, there are three other minimal reductions as $SG' = \{\{C, H, I\}, \{A, D, E\}, \{D, E, J\}\}$. Obviously, there are no core in this data set. However, according to

Table 5. Comparison of the CPU time and Results of the Algorithms

Database	$|U|$	$|C|$	Algorithm R-ACO			Algorithm RSACO			Algorithm 4
			$|SG|$	$|att\_R|$	CPU(s)	$|SG|$	$|att\_R|$	CPU(s)	$|att\_minR|$
ZOO	101	17	6	11	0.003	3	11	6.570	11
Car	1728	6	1	6	0.000	1	6	53.589	6
Soybean-large	307	35	4	11	0.857	1	11	207.809	9
LED24	1000	24	2	18	6.361	2	16	1144.455	16
Australian	690	14	2	3	0.076	2	3	38.017	3
Tic-tac-toe	958	9	6	8	0.016	2	8	39.804	8
statlog(germa)	1000	20	3	10	1.240	1	10	641.651	7
Mushroom	4062	22	1	1	0.045	1	1	829.314	1

analysing the set of the minimal reductions, we can find that the attributes D, E and I are the key attributes for the monitor to determine which fault arose. Whereas, the helpful results cannot be obtained if there is only one minimal reduction. And the results we get is consistent with the industrial reality [8].

7.2   Standard UCI Data Sets

In this section, we do some comparative experiments with the standard data sets from UCI repository [14]. More results of the experiments are shown in Table 5. $|U|$ and $|C|$ are the cardinality of the universe and the conditional attribute set, respectively. $|SG|$ is the cardinality of the set of reductions, $|att\_R|$ is the number of conditional attributes in a attribute reduction, $|att\_minR|$ is the number of conditional attributes in a minimal attribute reduction, and CPU(s) is the CPU time (by second) of the process.

Take the "ZOO" data set as an example, which has 17 attributes, denoted as $\{A, B, \cdots, P, Q\}$. We know that there are 11 attributes in the minimal reduction according to the results of Algorithm 4 [15]. Let $|att\_minR|$ be the number of conditional attributes in a minimal attribute reduction, that is $|att\_minR| = 11$.

Algorithm RSACO [6] finds three of the minimal reductions, they are $SG=$ $\{\{A, E, F, G, H, J, K, M, N, O, P\}; \{A, C, E, F, G, I, K, M, N, O, P\}; \{A, D, E, F, G, H, K, M, N, O, P\}\}$.

Otherwise, Algorithm R-ACO proposed in this paper finds six minimal reductions, they are $SG=\{\{A, C, E, F, G, H, K, M, N, O, P\}; \{A, C, E, F, G, I, K, M, N, O, P\}; \{A, E, F, G, H, J, K, M, N, O, P\}; \{A, E, F, G, H, K, M, N, O, P, Q\}; \{A, C, E, F, G, J, K, M, N, O, P\}; \{A, C, E, F, G, K, M, N, O, P, Q\}\}$.

It is obviously that the two algorithms can find the minimal reductions, but Algorithm R-ACO can find more minimal reductions than Algorithm RSACO. Furthermore, Algorithm R-ACO spends less time in 0.003 seconds than Algorithm RSACO in 6.570 seconds.

From Table 5, we can observe that Algorithm R-ACO is more efficient than Algorithm RSACO, and Algorithm R-ACO can find more minimal reductions. Furthermore, we can see that Algorithm R-ACO developed in this paper is a feasible solution to the MARP and the approach can acquire the minimal attribute reductions in most cases.

8   Conclusion

More minimal attribute reductions are expected to help clients make decisions in some application fields. Some researchers solve the minimal reductions problem based on the ant colony optimization metaheuristic. The most popular model is based on a complete graph whose nodes represent conditional attributes, with the edges between them denoting the choice of the next conditional attribute. The search for the minimal attribute reduction is then an ant traversal through the graph where a minimum number of nodes are visited that satisfies the traversal stopping criterion. The heuristic desirability of traversing between attributes usually is based on the roughs set dependency measure or based on the information entropy.

In this paper, we transfer the minimal reduction problem to a constraint satisfaction problem firstly, the goal is to find an assignment which minimizes the cost in the graph, the novel R-Graph model, whose nodes represent the distinct entries of the discernibility matrix. Intuitively, the R-Graph enlarges the search space. However, there are so many redundant entries in a discernibility matrix that we can simplify the search space by a preprocessing step proposed in this paper. Therefore, the novel algorithm R-ACO spends less time than the contrastive algorithms, another reason is that there are no computing of the roughs set dependency measure or the information entropy. And the simulation results show that our approach can find more minimal attribute reductions in most cases.

Acknowledgments

The authors would like to thank the anonymous reviewers for their helpful comments and suggestions. We would like to thank Professor Y.Y. Yao for his support when writing this paper. In particular, he suggested the idea for modeling the question. The research is partially funded by the China NNSF grant 60773113, and the Chongqing of China grant 2009BB2082 and KJ080510.

References

1. Bazan, J., Nguyen, H.S., Nguyen, S.H., Synak, P., Wroblewski, J.: Rough set algorithms in classification problem. In: Polkowski, L., Tsumoto, S., Lin, T.Y. (eds.) Rough Set Methods and Applications, pp. 49–88. Physica-Verlag, Heidelberg (2000)
2. Dorigo, M., Blum, C.: Ant colony optimization theory: A survey. Theoretical Computer Science 344, 243–278 (2005)
3. Dorigo, M., Sttzle, T.: Ant Colony Optimization. MIT Press, Cambridge (2004)
4. Deng, T.Q., Yang, C.D., Zhang, Y.T., Wang, X.X.: An Improved Ant Colony Optimization Applied to Attributes Reduction. In: Cao, B., Zhang, C., Li, T. (eds.) Fuzzy Information and Engineering, vol. 54, pp. 1–6. Springer, Heidelberg (2009)
5. Jensen, R., Shen, Q.: Finding rough set reducts with ant colony optimization. In: Proc. 2003 UK Workshop on Computational Intelligence, Bristol, UK, pp. 15–22 (2003)

6. Jiang, Y.C., Liu, Y.Z.: An Attribute Reduction Method Based on Ant Colony Optimization. In: WCICA: Intelligent Control and Automation, Dalian, China, pp. 3542–3546 (2006)
7. Ke, L.J., Feng, Z.R., Ren, Z.G.: An efficient ant colony optimization approach to attribute reduction in rough set theory. Pattern Recognition Letters 29(9), 1351–1357 (2008)
8. Liang, L., Xu, G.: Reduction of Rough Set Attribute Based on Immune Clone Selection. Journal of Xi,an Jiaotong University 39(11), 1231–1235 (2005)
9. Pawlak, Z.: Rough Sets. International Journal of Computer and Information Sciences 11, 341–356 (1982)
10. Skowron, A., Rauszer, C.: The discernibility matrices and functions in information systems. Fundamenta Informaticae 15(2), 331–362 (1991)
11. Solnon, C.: Ants can solve constraint satisfaction problems. IEEE Transactions on Evolutionary Computation 6(4), 347–357 (2002)
12. Slezak, D., Ziarko, W.: Attribute reduction in the Bayesian version of variable precision rough set model. Electronic Notes in Theoretical Computer Science 82(4), 263–273 (2003)
13. Tsang, E.P.K.: Foundations of Constraint Satisfaction. Academic Press, London (1993)
14. UCIrvine Machine Learning Repository, http://archive.ics.uci.edu/ml/
15. Wang, G.Y., Wu, Y., Fisher, P.S.: Rule Generation Based on Rough Set Theory. In: Dasarathy, B.V. (ed.) Proceedings of SPIE, Data Mining and Knowledge Discovery: Theory, Tools, and Technology II, vol. 4057, pp. 181–189 (2000)
16. Wang, G.Y., Yu, H., Yang, D.C.: Decision table reduction based on conditional information entropy. Chinese Journal of Computers (25), 759–766 (2002)
17. Wang, G.Y.: Calculation Methods for Core Attributes of Decision Table. Chinese Journal of Computers (26), 611–615 (2003)
18. Wang, J., Wang, J.: Reduction algorithms based on discernibility matrix: the ordered attributes method. Journal of Computer Science and Technology 16(6), 489–504 (2001)
19. Wang, X., Yang, J., Teng, X., Xia, W., Jensen, R.: Feature selection based on rough sets and particle swarm optimization. Pattern Recognition Letters 28, 459–471 (2007)
20. Wong, S.K.M., Ziarko, W.: On optimal decision rules in decision tables. Bulletin of Polish Academy of Sciences (33), 693–696 (1985)
21. Wu, W.Z., Zhang, M., Li, H.Z., Mi, J.S.: Knowledge reduction in random information systems via Dempster-Shafer theory of evidence. Information Sciences (174), 143–164 (2005)
22. Yao, Y.Y., Zhao, Y.: Discernibility Matrix Simplification for Constructing Attribute Reducts. Information Science 179(7), 867–882 (2009)
23. Zhang, W.X., Fang, Q.G., Wu, W.Z.: A general approach to attribute reduction in rough set theory. Science in China Series F: Information Sciences 50(2), 188–197 (2007)
24. Zhao, K., Wang, J.: A reduction algorithm meeting users' requirements. Journal of Computer Science and Technology 17, 578–593 (2002)

A Note on Attribute Reduction in the Decision-Theoretic Rough Set Model

Yan Zhao, S.K.M. Wong, and Yiyu Yao

Department of Computer Science, University of Regina
Regina, Saskatchewan, Canada S4S 0A2
yanzhao@cs.uregina.ca, skmwong@rogers.com, yyao@cs.uregina.ca

Abstract. This paper studies the definitions of attribute reduction in the decision-theoretic rough set model, which focuses on the probabilistic regions that induce different types of decision rules and support different types of decision making successively. We consider two groups of studies on attribute reduction. Attribute reduction can be interpreted based on either decision preservation or region preservation. According to the fact that probabilistic regions are non-monotonic with respect to set inclusion of attributes, attribute reduction for region preservation is different from the classical interpretation of reducts for decision preservation. Specifically, the problem of attribute reduction for decision preservation is a decision problem, while for region preservation is an optimization problem.

1 Introduction

Attribute reduction is an important problem of rough set theory. For classification tasks, we consider two possible interpretations of the concept of a reduct. The first interpretation views a reduct as a minimal subset of attributes that has the same classification power as the entire set of condition attributes [11]. The second interpretation views a reduct as a minimal subset of attributes that produces positive and boundary decision rules with precision over certain tolerance levels [19,22]. Studies on attribute reduction can therefore be divided into two groups.

The first group concentrates on the description of the classification power, i.e., the decision class or classes to which an equivalence class belongs. An object indiscernible within an equivalence class is classified by one decision class in consistent decision tables, and may be classified by more than one decision class in inconsistent decision tables. To describe the classification power, for each equivalence class Skowron [13] suggests a *generalized decision* consisting of the set of decision classes to which the equivalence class belongs. Similarly, Slezak [16] proposes the notion of a *majority decision* that uses a binary vector to indicate the decision classes to which each equivalence class belongs. In general, a *membership distribution function* over decision classes may be used to indicate the degree to which an equivalence class belongs [15]. Zhang *et al.* [8,23] propose the *maximum distribution criterion* based on the membership distribution function, which pays more attention to the decision classes that take the maximum distribution over the set.

J.F. Peters et al. (Eds.): Transactions on Rough Sets XIII, LNCS 6499, pp. 260–275, 2011.
© Springer-Verlag Berlin Heidelberg 2011

A reduct can be defined as a minimal subset of attributes that has the same classification power for all objects in the universe in terms of the types of decision classes, such as generalized decision and majority decision, or the proportion of the decision classes, such as decision distribution and maximum distribution.

The second group concentrates on the evaluation of the classification power, i.e., the positive, boundary and/or negative regions of decision classes to which an equivalence class belongs. In the Pawlak rough set model [10], each equivalence class may belong to one of the two regions, positive or boundary. The positive region is the union of equivalence classes that induce certain classification rules. The boundary region is the union of equivalence classes that induce uncertain classification rules. The negative region is in fact the empty set. The positive and boundary regions induce two different types of decision rules, called the positive rules and boundary rules, respectively [19,22]. While a positive rule leads to a definite decision, a boundary rule leads to a "wait-and-see" decision. In the decision-theoretic rough set model [19,20,21], a probabilistic generalization of the Pawlak rough sets, probabilistic regions are defined by two threshold values that, in turn, are determined systematically from a loss function by using the Bayesian decision procedure. In this case, the probabilistic negative region may not be the empty set. It represents the fact that we do not want to make any positive or boundary decision for some equivalence classes. In other words, since the confidence is too low to support any decision making, one cannot indicate to which decision class the object or the equivalence class belongs [19,22].

The decision-theoretic rough set model has a special interest in attribute reduction that is based on these two types of probabilistic rules. Reduct construction may be viewed as a search for a minimal subset of attributes that produces positive and boundary decision rules satisfying certain tolerance levels of precision.

2   The Decision-Theoretic Rough Set Model

In many data analysis applications, objects are only perceived, observed, or measured by using a finite number of attributes, and are represented as an information table [10].

Definition 1. *An information table is the following tuple:*

$$S = (U, At, \{V_a \mid a \in At\}, \{I_a \mid a \in At\}),$$

where U is a finite nonempty set of objects, At is a finite nonempty set of attributes, V_a is a nonempty set of values of $a \in At$, and $I_a : U \to V_a$ is an information function that maps an object in U to exactly one value in V_a.

For classification problems, we consider an information table of the form $S = (U, At = \mathbf{C} \cup \{D\}, \{V_a\}, \{I_a\})$, where $\mathbf{C}$ is a set of condition attributes describing the objects, and D is a decision attribute that indicates the classes of objects.

Let $\pi_D = \{D_1, D_2, \ldots, D_m\}$ be a partition of the universe U defined by the decision attribute D. Each equivalence class $D_i \in \pi_D$ is called a decision class. Given another partition $\pi_A = \{A_1, A_2, \ldots, A_n\}$ of U defined by a condition attribute set $A \subseteq \mathbf{C}$, each equivalence class A_j also is defined as $[x]_A = \{y \in U \mid \forall a \in A(I_a(x) = I_a(y))\}$. If A is understood, we simply use π for the partition and $[x]$ for the equivalence class containing x.

The *precision* of an equivalence class $[x]_A \in \pi_A$ for predicting a decision class $D_i \in \pi_D$ is defined as:

$$p(D_i|[x]_A) = \frac{|[x]_A \cap D_i|}{|[x]_A|},$$

where $|\cdot|$ denotes the cardinality of a set. The precision is the ratio of the number of objects in $[x]_A$ that are correctly classified into the decision class D_i and the number of objects in $[x]_A$. The decision-theoretic rough set model utilizes ideas from Bayesian decision theory and computes two thresholds based on the notion of expected loss (conditional risk). For a detailed description, please refer to papers [19,20,21,22].

2.1   Probabilistic Approximations and Regions

In the decision-theoretic rough set model, we introduce tolerance thresholds for defining probabilistic lower and upper approximations, and probabilistic positive, boundary and negative regions of a decision class, as well as of the partition formed by all decision classes. The decision-theoretic model systematically calculates the thresholds by a set of loss functions based on the Bayesian decision procedure. The physical meaning of the loss functions can be interpreted based on more practical notions of costs and risks.

By using the thresholds, for a decision class $D_i \in \pi_D$, the probabilistic lower and upper approximations with respect to a partition π_A can be defined based on two thresholds $0 \leq \beta < \alpha \leq 1$ as:

$$\underline{apr}_{(\alpha,\beta)}(D_i|\pi_A) = \{x \in U \mid p(D_i \mid [x]_A) \geq \alpha\},$$

$$= \bigcup_{p(D_i|[x]_A) \geq \alpha} [x]_A; \tag{1}$$

$$\overline{apr}_{(\alpha,\beta)}(D_i|\pi_A) = \{x \in U \mid p(D_i \mid [x]_A) > \beta\},$$

$$= \bigcup_{p(D_i|[x]_A) > \beta} [x]_A \tag{2}$$

The probabilistic positive, boundary and negative regions of $D_i \in \pi_D$ with respect to π_A are defined by:

$$\mathrm{POS}_{(\alpha,\beta)}(D_i|\pi_A) = \underline{apr}_{(\alpha,\beta)}(D_i|\pi_A), \tag{3}$$

$$\mathrm{BND}_{(\alpha,\beta)}(D_i|\pi_A) = \overline{apr}_{(\alpha,\beta)}(D_i|\pi_A) - \underline{apr}_{(\alpha,\beta)}(D_i|\pi_A), \tag{4}$$

$$\mathrm{NEG}_{(\alpha,\beta)}(D_i|\pi_A) = U - \mathrm{POS}_{(\alpha,\beta)}(D_i|\pi_A) \cup \mathrm{BND}_{(\alpha,\beta)}(D_i|\pi_A)$$

$$= U - \overline{apr}_{(\alpha,\beta)}(D_i|\pi_A)$$

$$= (\overline{apr}_{(\alpha,\beta)}(D_i|\pi_A))^c. \tag{5}$$

We can extend the concept of probabilistic approximations and regions of a single decision D_i to the entire partition π_D. The lower and upper approximations of the partition π_D with respect to π_A are the families of the lower and upper approximations of all the equivalence classes of π_D. For simplicity, we assume that the same loss functions are used for all decisions. That is,

$$\underline{apr}_{(\alpha,\beta)}(\pi_D|\pi_A) = (\underline{apr}_{(\alpha,\beta)}(D_1|\pi_A), \underline{apr}_{(\alpha,\beta)}(D_2|\pi_A)), \ldots, \underline{apr}_{(\alpha,\beta)}(D_m|\pi_A));$$

$$\overline{apr}_{(\alpha,\beta)}(\pi_D|\pi_A) = (\overline{apr}_{(\alpha,\beta)}(D_1|\pi_A), \overline{apr}_{(\alpha,\beta)}(D_2|\pi_A)), \ldots, \overline{apr}_{(\alpha,\beta)}(D_m|\pi_A)).$$

(6)

We can define the three regions of the partition π_D for the decision-theoretic rough set model as [19]:

$$\text{POS}_{(\alpha,\beta)}(\pi_D|\pi_A) = \bigcup_{1 \leq i \leq m} \text{POS}_{(\alpha,\beta)}(D_i|\pi_A),$$

$$= \{x \in U \mid \exists D_i \in \pi_D \ (p(D_i|[x]_A) \geq \alpha)\}; \quad (7)$$

$$\text{BND}_{(\alpha,\beta)}(\pi_D|\pi_A) = \bigcup_{1 \leq i \leq m} \text{BND}_{(\alpha,\beta)}(D_i|\pi_A),$$

$$= \{x \in U \mid \exists D_i \in \pi_D \ (\beta < p(D_i|[x]_A) < \alpha)\}; \quad (8)$$

$$\text{NEG}_{(\alpha,\beta)}(\pi_D|\pi_A) = U - \text{POS}_{(\alpha,\beta)}(\pi_D|\pi_A) \cup \text{BND}_{(\alpha,\beta)}(\pi_D|\pi_A),$$

$$= \{x \in U \mid \exists D_i \in \pi_D \ (p(D_i|[x]_A) \leq \beta)\}. \quad (9)$$

The union of the probabilistic positive and boundary regions is called a probabilistic non-negative region.

The Pawlak model, as a special case, can be derived by setting a loss function that produces $\alpha = 1$ and $\beta = 0$ [12]. The Pawlak regions of partition π_D are simplified as $\text{POS}(\pi_D|\pi_A)$, $\text{BND}(\pi_D|\pi_A)$ and $\text{NEG}(\pi_D|\pi_A)$, respectively, without indicating the thresholds. We can also derive the 0.50 probabilistic model [12], the symmetric variable precision rough set model [24], and the asymmetric variable precision rough set model [6]. More specifically, we may have the following probabilistic rough set models [19]:

- If $\alpha > 0.50$, $\text{POS}_{(\alpha,\beta)}(\pi_D|\pi_A)$ contains pairwise disjoint sets.
- If $\beta > 0.50$, $\text{POS}_{(\alpha,\beta)}(\pi_D|\pi_A)$, $\text{BND}_{(\alpha,\beta)}(\pi_D|\pi_A)$ and $\text{NEG}_{(\alpha,\beta)}(\pi_D|\pi_A)$ contain pairwise disjoint sets.
- If $\beta = 0$, $\text{NEG}_{(\alpha,\beta)}(\pi_D|\pi_A) = \emptyset$.

We can verify the following properties of the three regions in decision-theoretic models:

(1) The three regions of a certain decision class $D_i \in \pi_D$ are pairwise disjoint. The union of the three regions of the decision class is a covering of U, i.e., $\text{POS}_{(\alpha,\beta)}(D_i|\pi_A) \cup \text{BND}_{(\alpha,\beta)}(D_i|\pi_A) \cup \text{NEG}_{(\alpha,\beta)}(D_i|\pi_A) = U$.

(2) The three regions of the partition π_D are not necessarily pairwise disjoint. For example, it may happen that the intersection of the probabilistic positive and boundary regions are not empty, i.e., $\text{POS}_{(\alpha,\beta)}(\pi_D|\pi_A) \cap \text{BND}_{(\alpha,\beta)}(\pi_D|\pi_A) \neq \emptyset$. Furthermore, the probabilistic negative region of the partition π_D,

$\mathrm{NEG}_{(\alpha,\beta)}(\pi_D|\pi_A)$, is not necessarily empty. Nevertheless, the union of the three regions of the partition π_D is a covering of U, i.e., $\mathrm{POS}_{(\alpha,\beta)}(\pi_D|\pi_A) \cup \mathrm{BND}_{(\alpha,\beta)}(\pi_D|\pi_A) \cup \mathrm{NEG}_{(\alpha,\beta)}(\pi_D|\pi_A) = U$.

(3) The family of probabilistic positive regions of all decision classes, $\{\mathrm{POS}_{(\alpha,\beta)}(D_i|\pi_A) \mid 1 \le i \le m\}$, does not necessarily contain pairwise disjoint sets. It may happen that $\mathrm{POS}_{(\alpha,\beta)}(D_i|\pi_A) \cap \mathrm{POS}_{(\alpha,\beta)}(D_j|\pi_A) \ne \emptyset$, for some $i \ne j$.

(4) The family of probabilistic boundary regions of all decision classes, $\{\mathrm{BND}_{(\alpha,\beta)}(D_i|\pi_A) \mid 1 \le i \le m\}$, does not necessarily contain pairwise disjoint sets. It may happen that $\mathrm{BND}_{(\alpha,\beta)}(D_i|\pi_A) \cap \mathrm{BND}_{(\alpha,\beta)}(D_j|\pi_A) \ne \emptyset$, for some $i \ne j$.

When generalizing results from the Pawlak rough set model to the decision-theoretic rough set models, it is necessary to consider the implications of those properties.

Example 1. Consider a simple information table $S = (U, At = \mathbf{C} \cup \{D\}, \{V_a\}, \{I_a\})$ shown in Table 1. The condition attribute set $\mathbf{C}$ partitions the universe into six equivalence classes: $[o_1]_\mathbf{C}$, $[o_2]_\mathbf{C}$, $[o_3]_\mathbf{C}$, $[o_4]_\mathbf{C}$, $[o_5]_\mathbf{C}$ and $[o_7]_\mathbf{C}$.

Table 1. An information table

	c_1	c_2	c_3	c_4	c_5	c_6	D
			$\mathbf{C}$				D
o_1	1	1	1	1	1	1	M
o_2	1	0	1	0	1	1	M
o_3	0	1	1	1	0	0	Q
o_4	1	1	1	0	0	1	Q
o_5	0	0	1	1	0	1	Q
o_6	1	0	1	0	1	1	F
o_7	0	0	0	1	1	0	F
o_8	1	0	1	0	1	1	F
o_9	0	0	1	1	0	1	F

Suppose $\alpha = 0.75$ and $\beta = 0.60$. We can obtain the following probabilistic regions defined by $\pi_\mathbf{C}$.

$\mathrm{POS}_{(\alpha,\beta)}(\{M\}|\pi_\mathbf{C}) = \{o_1\}$,
$\mathrm{BND}_{(\alpha,\beta)}(\{M\}|\pi_\mathbf{C}) = \emptyset$,
$\mathrm{NEG}_{(\alpha,\beta)}(\{M\}|\pi_\mathbf{C}) = \{o_2, o_3, o_4, o_5, o_6, o_7, o_8, o_9\}$;

$\mathrm{POS}_{(\alpha,\beta)}(\{Q\}|\pi_\mathbf{C}) = \{o_3, o_4\}$,
$\mathrm{BND}_{(\alpha,\beta)}(\{Q\}|\pi_\mathbf{C}) = \emptyset$,
$\mathrm{NEG}_{(\alpha,\beta)}(\{Q\}|\pi_\mathbf{C}) = \{o_1, o_2, o_5, o_6, o_7, o_8, o_9\}$;

$\mathrm{POS}_{(\alpha,\beta)}(\{F\}|\pi_\mathbf{C}) = \{o_7\}$,
$\mathrm{BND}_{(\alpha,\beta)}(\{F\}|\pi_\mathbf{C}) = \{o_2, o_6, o_8\}$,
$\mathrm{NEG}_{(\alpha,\beta)}(\{F\}|\pi_\mathbf{C}) = \{o_1, o_3, o_4, o_5, o_9\}$;

Table 2. A reformation of Table 1 indicating the probabilistic regions associated with each equivalence class $[x]_C$

	C						$\text{POS}_{(\alpha,\beta)}$	$\text{BND}_{(\alpha,\beta)}$	$\text{NEG}_{(\alpha,\beta)}$
	c_1	c_2	c_3	c_4	c_5	c_6			
$[o_1]_C$	1	1	1	1	1	1	1	0	0
$[o_2]_C$	1	0	1	0	1	1	0	1	0
$[o_3]_C$	0	1	1	1	0	0	1	0	0
$[o_4]_C$	1	1	1	0	0	1	1	0	0
$[o_5]_C$	0	0	1	1	0	1	0	0	1
$[o_6]_C$	1	0	1	0	1	1	0	1	0
$[o_7]_C$	0	0	0	1	1	0	1	0	0
$[o_8]_C$	1	0	1	0	1	1	0	1	0
$[o_9]_C$	0	0	1	1	0	1	0	0	1

$$\text{POS}_{(\alpha,\beta)}(\pi_D|\pi_C) = \{o_1, o_3, o_4, o_7\},$$
$$\text{BND}_{(\alpha,\beta)}(\pi_D|\pi_C) = \{o_2, o_6, o_8\},$$
$$\text{NEG}_{(\alpha,\beta)}(\pi_D|\pi_C) = \{o_5, o_9\}.$$

A reformation of Table 1, shown as Table 2, indicates the belonging relationship of all equivalence classes $[x]_C$ to the a probabilistic region $\text{POS}_{(\alpha,\beta)}(\pi_D|\pi_C)$, $\text{BND}_{(\alpha,\beta)}(\pi_D|\pi_C)$ or $\text{NEG}_{(\alpha,\beta)}(\pi_D|\pi_C)$, where 1 means belong and 0 means not belong.

2.2   Decision Making

Skowron proposes a generalized decision δ as the set of all decision classes an equivalence class belongs [13]. For an equivalence class $[x]_A \in \pi_A$ the generalized decision is denoted as:

$$\delta([x]_A) = \{I_D(x) \mid x \in [x]_A\}$$
$$= \{D_i \in \pi_D \mid p(D_i|[x]_A) > 0\}.$$

By introducing precision thresholds, we can separate Skowron's generalized decision into three parts. The part of *positive decisions* is the set of decision classes with the precision higher than or equal to α. A positive decision may lead to a definite and immediate action. The part of *boundary decisions* is the set of decision classes with the precision lower than α but higher than β. A boundary decision may lead to a "wait-and-see" action. A decision with the precision lower than or equal to β is not strong enough to support any further action. The union of positive decisions and boundary decisions can be called the set of *general decisions* that support actual decision making. Let $D_{\text{POS}_{(\alpha,\beta)}}$, $D_{\text{BND}_{(\alpha,\beta)}}$ and $D_{\text{GEN}_{(\alpha,\beta)}}$ denote the positive, boundary and general decision sets, respectively. For an equivalence class $[x]_A \in \pi_A$,

$$D_{\text{POS}_{(\alpha,\beta)}}([x]_A) = \{D_i \in \pi_D \mid p(D_i|[x]_A) \geq \alpha\},$$
$$D_{\text{BND}_{(\alpha,\beta)}}([x]_A) = \{D_i \in \pi_D \mid \beta < p(D_i|[x]_A) < \alpha\},$$
$$D_{\text{GEN}_{(\alpha,\beta)}}([x]_A) = D_{\text{POS}_{(\alpha,\beta)}}([x]_A) \cup D_{\text{BND}_{(\alpha,\beta)}}([x]_A). \tag{10}$$

In the rest of this paper, we only focus on positive and general decision making, and the corresponding probabilistic positive and non-negative regions. For other probabilistic rough set regions, one can refer to Inuiguchi's study [4].

Example 2. Consider Table 1 and thresholds $\alpha = 0.75$ and $\beta = 0.60$. We can reformat the table by including δ, $D_{\text{POS}_{(\alpha,\beta)}}$, $D_{\text{BND}_{(\alpha,\beta)}}$ and $D_{\text{GEN}_{(\alpha,\beta)}}$ for all equivalence classes defined by **C**.

Table 3. A reformation of Table 1 indicating the decisions associated with each equivalence class $[x]_{\mathbf{C}}$

	C						δ	$D_{\text{POS}_{(\alpha,\beta)}}$	$D_{\text{BND}_{(\alpha,\beta)}}$	$D_{\text{GEN}_{(\alpha,\beta)}}$
	c_1	c_2	c_3	c_4	c_5	c_6				
$[o_1]_{\mathbf{C}}$	1	1	1	1	1	1	$\{M\}$	$\{M\}$	$\emptyset$	$\{M\}$
$[o_2]_{\mathbf{C}}$	1	0	1	0	1	1	$\{M,F\}$	$\emptyset$	$\{F\}$	$\{F\}$
$[o_3]_{\mathbf{C}}$	0	1	1	1	0	0	$\{Q\}$	$\{Q\}$	$\emptyset$	$\{Q\}$
$[o_4]_{\mathbf{C}}$	1	1	1	0	0	1	$\{Q\}$	$\{Q\}$	$\emptyset$	$\{Q\}$
$[o_5]_{\mathbf{C}}$	0	0	1	1	0	1	$\{Q,F\}$	$\emptyset$	$\emptyset$	$\emptyset$
$[o_6]_{\mathbf{C}}$	1	0	1	0	1	1	$\{M,F\}$	$\emptyset$	$\{F\}$	$\{F\}$
$[o_7]_{\mathbf{C}}$	0	0	0	1	1	0	$\{F\}$	$\{F\}$	$\emptyset$	$\{F\}$
$[o_8]_{\mathbf{C}}$	1	0	1	0	1	1	$\{M,F\}$	$\emptyset$	$\{F\}$	$\{F\}$
$[o_9]_{\mathbf{C}}$	0	0	1	1	0	1	$\{Q,F\}$	$\emptyset$	$\emptyset$	$\emptyset$

It should be noted, the generalized decision elaborates the set of decision classes tie to any object of the equivalence class, and is not related to the precision thresholds, while the positive, boundary and general decisions evaluate the decision class tie to the equivalence class, and is sensitive to the precision thresholds. We can compare the differences between the generalized decision and the general decision by the following example. For equivalence class $[o_2]_{\mathbf{C}}$, the generalized decision is $\{M, F\}$, while the general decision is the subset $\{F\}$. The decision class F is a boundary decision of $[o_2]_{\mathbf{C}}$ with respect to the β threshold. In other words, $[o_2]_{\mathbf{C}}$ can be partially classified as F, and weakly support the decision making associated with F. On the other hand, the decision class M is too weak to be used to classify $[o_2]_{\mathbf{C}}$, and thus $[o_2]_{\mathbf{C}}$ does not support the decision making associated with M.

2.3   Monotocity Property of the Regions

By considering the two thresholds separately, we obtain the following observations. For a partition π_A, the decrease of the precision threshold α can result an increase of the probabilistic positive region $\text{POS}_{(\alpha,\beta)}(\pi_D|\pi_A)$. Thus, we can make positive decisions for more objects. The decrease of the precision threshold β can result an increase of the probabilistic non-negative region $\neg\text{NEG}_{(\alpha,\beta)}(\pi_D|\pi_A)$, thus we can make general decisions for more objects.

Consider any two subsets of attributes $A, B \subseteq \mathbf{C}$ with $A \subseteq B$. For any $x \in U$, we have $[x]_B \subseteq [x]_A$. In the Pawlak model, if $[x]_A \in \text{POS}(\pi_D|\pi_A)$, then its subset $[x]_B$ also is in the positive region, i.e., $[x]_B \in \text{POS}(\pi_D|\pi_B)$. At the same time, if $[x]_A \in \text{BND}(\pi_D|\pi_A)$, its subset $[x]_B$ may be in the positive region or the boundary region. If $[x]_A \in \neg\text{NEG}(\pi_D|\pi_A)$, its subset $[x]_B$ may also belong to the positive region or the boundary region. We immediately obtain the monotonic property of the Pawlak positive and non-negative regions with respect to set inclusion of attributes:

$$A \subseteq B \Longrightarrow \text{POS}(\pi_D|\pi_A) \subseteq \text{POS}(\pi_D|\pi_B);$$
$$A \subseteq B \Longrightarrow \neg\text{NEG}(\pi_D|\pi_A) \subseteq \neg\text{NEG}(\pi_D|\pi_B).$$

That is, a larger subset of $\mathbf{C}$ induces a larger positive region and a larger non-negative region. The entire condition attribute set $\mathbf{C}$ induces the largest positive and non-negative regions.

The *quality of classification*, or the *degree of dependency of D*, is defined as [11]:

$$\gamma(\pi_D|\pi_A) = \frac{|\text{POS}(\pi_D|\pi_A)|}{|U|}, \tag{11}$$

which is equal to the generality of the positive region. Based on the monotocity of the Pawlak positive region, we can obtain the monotocity of the γ measure. That is, $A \subseteq B \Longrightarrow \gamma(\pi_D|\pi_A) \leq \gamma(\pi_D|\pi_B)$.

In the decision-theoretic model, for a subset $[x]_B$ of an equivalence class $[x]_A$, no matter to which region $[x]_A$ belongs, we do not know to which region $[x]_B$ belongs. Therefore, we cannot obtain the monotocity of the probabilistic regions with respect to set inclusion of attributes. The probabilistic positive and non-negative regions are monotonically increasing with respect to the decreasing of the α and β thresholds, respectively, but are non-monotonic with respect to the set inclusion of attributes. Intuitively, the largest condition attribute set $\mathbf{C}$ may not be able to induce the largest probabilistic positive and non-negative regions.

In the decision-theoretic model, the quantitative γ measure can be extended to indicate the quality of a probabilistic classification. A straightforward transformation of the γ measure is denoted as follows [24]:

$$\gamma_{(\alpha,\beta)}(\pi_D|\pi_A) = \frac{|\text{POS}_{(\alpha,\beta)}(\pi_D|\pi_A)|}{|U|}. \tag{12}$$

Since the probabilistic positive region is non-monotonic, the $\gamma_{(\alpha,\beta)}$ measure is also non-monotonic with respect to the set inclusion of attributes. In other words, the largest condition attribute set $\mathbf{C}$ may not be able to induce the largest value regarding the γ measure.

3   Definitions and Interpretations of Attribute Reduction

Based on the previous discussion on probabilistic regions and probabilistic decision making, we can have the following two definitions of attribute reduction for the decision-theoretic rough set model.

A reduct $R \subseteq \mathbf{C}$ for positive decision preservation can be defined by requiring that the positive decisions of all objects are unchanged.

Definition 2. *Given an information table* $S = (U, At = \mathbf{C} \cup \{D\}, \{V_a \mid a \in At\}, \{I_a \mid a \in At\})$, *an attribute set* $R \subseteq \mathbf{C}$ *is a reduct of* $\mathbf{C}$ *with respect to the positive decisions of all objects if it satisfies the following two conditions:*

(i) $\forall x \in U(D_{\mathrm{POS}_{(\alpha,\beta)}}([x]_R) = D_{\mathrm{POS}_{(\alpha,\beta)}}([x]_{\mathbf{C}}))$;

(ii) *for any* $R' \subset R$ *the condition (i) does not hold.*

The definition for general decision preservation can be similarly defined by having the condition (i) stated as: $\forall x \in U(D_{\mathrm{GEN}_{(\alpha,\beta)}}([x]_R) = D_{\mathrm{GEN}_{(\alpha,\beta)}}([x]_{\mathbf{C}}))$. The problem of constructing a decision-preserved reduct is a decision problem, i.e., it is a question with a yes-or-no answer.

A reduct $R \subseteq \mathbf{C}$ for positive region preservation can be defined by requiring that the induced positive region is not degraded.

Definition 3. *Given an information table* $S = (U, At = \mathbf{C} \cup \{D\}, \{V_a \mid a \in At\}, \{I_a \mid a \in At\})$, *an attribute set* $R \subseteq \mathbf{C}$ *is a reduct of* $\mathbf{C}$ *with respect to the probabilistic positive region of* π_D *if* $R = \arg\max_{A \subseteq \mathbf{C}} \{\mathrm{POS}_{(\alpha,\beta)}(\pi_D | \pi_A)\}$.

The definition indicates that a reduct R induces the global maximum probabilistic positive region regarding all subsets of $\mathbf{C}$. In some cases, this definition can be loosely stated as,

(i) $\mathrm{POS}_{(\alpha,\beta)}(\pi_D | \pi_R) \supseteq \mathrm{POS}_{(\alpha,\beta)}(\pi_D | \pi_{\mathbf{C}})$;

(ii) *for any* $R' \subset R$, $\mathrm{POS}_{(\alpha,\beta)}(\pi_D | \pi_{R'}) \subset \mathrm{POS}_{(\alpha,\beta)}(\pi_D | \pi_{\mathbf{C}})$.

In this loose definition, a reduct R induces the local maximum probabilistic positive region regarding all its own proper subsets. Strictly, the global maximum is not equal to the local maximum.

The definition for non-negative region preservation can be similarly defined as $R = \arg\max_{A \subseteq \mathbf{C}} \{\neg\mathrm{NEG}_{(\alpha,\beta)}(\pi_D | \pi_A)\}$. For simplicity, the qualitative measure can be replaced by the quantitative measure. For example, the set-theoretic measure of a region can be replaced by the cardinality of the region [10,18], or the entropy of the region [9,15,18]. The problem of constructing a region-preserved reduct is a optimization problem, i.e., it is concerned with finding the best answer to the given function.

3.1   An Interpretation of Region Preservation in the Pawlak Model

Pawlak defines a reduct based on the crisp positive region of π_D[10]. A reduct is an attribute set satisfying the following two conditions.

Definition 4. *[10] Given an information table* $S = (U, At = \mathbf{C} \cup \{D\}, \{V_a \mid a \in At\}, \{I_a \mid a \in At\})$, *an attribute set* $R \subseteq \mathbf{C}$ *is a reduct of* $\mathbf{C}$ *with respect to the positive region of* π_D *if it satisfies the following two conditions:*

(i) $\mathrm{POS}(\pi_D | \pi_R) = \mathrm{POS}(\pi_D | \pi_{\mathbf{C}})$;

(ii) *for any attribute* $a \in R$, $\mathrm{POS}(\pi_D | \pi_{R-\{a\}}) \neq \mathrm{POS}(\pi_D | \pi_R)$.

Based on the fact that the Pawlak positive region is monotonic with respect to set inclusion of attributes, the attribute set $\mathbf{C}$ must produce the largest positive region. A reduct R produces a positive region as big as $\mathbf{C}$ does, and all proper subsets of R cannot produce a bigger positive region than R does. Thus, only all proper subsets $R - \{a\}$ for all $a \in R$ need to be checked.

Many authors [1,3,10,18] use an equivalent quantitative definition of a Pawlak reduct, i.e., $\gamma(\pi_D|\pi_R) = \gamma(\pi_D|\pi_\mathbf{C})$. In other words, R and $\mathbf{C}$ induce the same quantitative measurement of the Pawlak positive region.

In the Pawlak model, for a reduct $R \subseteq \mathbf{C}$ we have $\mathrm{POS}(\pi_D|\pi_R) \cap \mathrm{BND}(\pi_D|\pi_R)$ $= \emptyset$, and $\mathrm{POS}(\pi_D|\pi_R) \cup \mathrm{BND}(\pi_D|\pi_R) = U$. The condition $\mathrm{POS}(\pi_D|\pi_R) = \mathrm{POS}(\pi_D|\pi_\mathbf{C})$ is equivalent to $\mathrm{BND}(\pi_D|\pi_R) = \mathrm{BND}(\pi_D|\pi_\mathbf{C})$. The requirement of the same boundary region is implied by the definition of a Pawlak reduct. It is sufficient to consider only the positive region in the Pawlak model.

3.2 Difficulties with the Interpretations of Region Preservation in the Decision-Theoretic Model

Parallel to Pawlak's definition, an attribute reduct in a decision-theoretic model can be defined by requiring that the probabilistic positive region of π_D is undegraded. Such a definition has been proposed by Kryszkiewicz as a β-reduct [7] for probabilistic rough set models in general, and by Inuiguchi as a β-low approximation reduct [4,5] for the variable precision rough set model in specific. A typical definition is defined as follows.

Definition 5. [7] *Given an information table* $S = (U, At = \mathbf{C} \cup \{D\}, \{V_a \mid a \in At\}, \{I_a \mid a \in At\})$, *an attribute set* $R \subseteq \mathbf{C}$ *is a reduct of* $\mathbf{C}$ *with respect to the probabilistic positive region of* π_D *if it satisfies the following two conditions:*

(i) $\mathrm{POS}_{(\alpha,\beta)}(\pi_D|\pi_R) = \mathrm{POS}_{(\alpha,\beta)}(\pi_D|\pi_\mathbf{C})$;

(ii)   *for any attribute* $a \in R$, $\mathrm{POS}_{(\alpha,\beta)}(\pi_D|\pi_{R-\{a\}}) \neq \mathrm{POS}_{(\alpha,\beta)}(\pi_D|\pi_\mathbf{C})$.

In probabilistic models, many proposals have been made to extend the Pawlak attribute reduction by using the extended and generalized measure $\gamma_{(\alpha,\beta)}$. Accordingly, the condition (i) of the definition can be re-expressed as $\gamma_{(\alpha,\beta)}(\pi_D|\pi_R) = \gamma_{(\alpha,\beta)}(\pi_D|\pi_\mathbf{C})$. Although the definition, especially the definition based on the extended $\gamma_{(\alpha,\beta)}$ measure, is adopted by many researchers [2,3,7,17,24], the definition itself is inappropriate for attribute reduction in probabilistic models. We can make the following three observations.

Problem 1: In probabilistic models, the probabilistic positive region is non-monotonic regarding set inclusion of attributes. The equality relation in condition (i) is not enough for verifying a reduct, and may miss some reducts. At the same time, the condition (ii) should consider all subsets of a reduct R, not only the subsets $R - \{a\}$ for all $a \in R$.

Example 3. Suppose $\alpha = 0.75$ and $\beta = 0.60$ for Table 1. Compare the probabilistic positive regions defined by $\mathbf{C}$ and all subsets of $\{c_1, c_2, c_5\}$ listed in Table 1.

Table 4. Probabilistic positive and non-negative regions defined by some attribute sets

$A \subseteq \mathbf{C}$	$\mathrm{POS}_{(\alpha,\beta)}(\pi_D|\pi_A)$	$\neg\mathrm{NEG}_{(\alpha,\beta)}(\pi_D|\pi_A)$
$\mathbf{C}$	$\{o_1, o_3, o_4, o_7\}$	$\{o_1, o_2, o_3, o_4, o_6, o_7, o_8\}$
$\{c_1, c_2, c_5\}$	$\{o_1, o_3, o_4, o_7\}$	$\{o_1, o_2, o_3, o_4, o_6, o_7, o_8\}$
$\{c_1, c_2\}$	$\{o_3\}$	$\{o_2, o_3, o_5, o_6, o_7, o_8, o_9\}$
$\{c_1, c_5\}$	$\{o_4, o_7\}$	$\{o_3, o_4, o_5, o_7, o_9\}$
$\{c_2, c_5\}$	$\{o_1, o_2, o_3, o_4, o_6, o_7, o_8\}$	$\{o_1, o_2, o_3, o_4, o_6, o_7, o_8\}$
$\{c_1\}$	$\emptyset$	$\emptyset$
$\{c_2\}$	$\emptyset$	U
$\{c_5\}$	$\{o_3, o_4, o_5, o_9\}$	$\{o_3, o_4, o_5, o_9\}$

It is clear that $\mathrm{POS}_{(\alpha,\beta)}(\pi_D|\pi_\mathbf{C}) = \mathrm{POS}_{(\alpha,\beta)}(\pi_D|\pi_{\{c_1,c_2,c_5\}})$, and none of the subset of $\{c_1, c_2, c_5\}$ keeps the same probabilistic positive region. Though, according to the non-monotocity of the probabilistic positive region, we can verify that $\mathrm{POS}_{(\alpha,\beta)}(\pi_D|\pi_{\{c_2,c_5\}})$ is a superset of $\mathrm{POS}_{(\alpha,\beta)}(\pi_D|\pi_\mathbf{C})$, and thus support positive decision for more objects. We can verify that $\{c_2, c_5\}$ is a reduct regarding the positive region preservation, and $\{c_1, c_2, c_5\}$ is not.

Problem 2: In probabilistic models, both the probabilistic positive region and the probabilistic boundary region, i.e., the probabilistic non-negative region, need to be considered for general decision making. The definition only reflects the probabilistic positive region and does not evaluate the probabilistic boundary region. Inuiguchi's definition for a β-upper approximation reduct also considers the general decision making [4]. However, the equality relation used may be inappropriate.

Example 4. We use the same Table 1 to demonstrate this problem. Suppose $\alpha = 0.75$ and $\beta = 0.60$. Compare the probabilistic non-negative regions defined by $\mathbf{C}$ and all subsets of $\{c_1, c_2, c_5\}$ listed in Table 3. The probabilistic non-negative regions are equal regarding the attribute sets $\{c_1, c_2, c_5\}$, $\{c_2, c_5\}$ and $\mathbf{C}$. Furthermore, $\neg\mathrm{NEG}_{(\alpha,\beta)}(\pi_D|\pi_{\{c_2\}}) = U$ is a superset of $\neg\mathrm{NEG}_{(\alpha,\beta)}(\pi_D|\pi_\mathbf{C})$, and thus supports general decision for more objects. Therefore, $\{c_2\}$ is a reduct regarding the non-negative region preservation, and none of its superset is.

Problem 3: Based on the condition $\gamma_{(\alpha,\beta)}(\pi_D|\pi_R) = \gamma_{(\alpha,\beta)}(\pi_D|\pi_\mathbf{C})$, we can obtain $|\mathrm{POS}_{(\alpha,\beta)}(\pi_D|\pi_R)| = |\mathrm{POS}_{(\alpha,\beta)}(\pi_D|\pi_\mathbf{C})|$, but not $\mathrm{POS}_{(\alpha,\beta)}(\pi_D|\pi_R) = \mathrm{POS}_{(\alpha,\beta)}(\pi_D|\pi_\mathbf{C})$. This means that the quantitative equivalence of the probabilistic positive regions does not imply the qualitative equivalence of the probabilistic positive regions.

Example 5. Quantitatively, $|\mathrm{POS}_{(\alpha,\beta)}(\pi_D|\pi_{\{c_5\}})| = |\mathrm{POS}_{(\alpha,\beta)}(\pi_D|\pi_\mathbf{C})|$ indicates $\gamma_{(\alpha,\beta)}(\pi_D|\pi_{\{c_5\}}) = \gamma_{(\alpha,\beta)}(\pi_D|\pi_\mathbf{C})$. Qualitatively, they indicate two different sets of objects. The positive decision will be made for the two different sets of objects. For example, suppose $\alpha = 0.75$ and $\beta = 0.60$,

$$\mathrm{POS}_{(\alpha,\beta)}(\pi_D|\pi_\mathbf{C}) = \{o_1, o_3, o_4, o_7\}, \text{ and}$$
$$\mathrm{POS}_{(\alpha,\beta)}(\pi_D|\pi_{\{c_5\}}) = \{o_3, o_4, o_5, o_9\}.$$

Thus, $\text{POS}_{(\alpha,\beta)}(\pi_D|\pi_{\mathbf{C}}) \neq \text{POS}_{(\alpha,\beta)}(\pi_D|\pi_{\{c_5\}})$, and $\gamma_{(\alpha,\beta)}(\pi_D|\pi_{\mathbf{C}}) = \gamma_{(\alpha,\beta)}(\pi_D|\pi_{\{c_5\}})$. Similarly, the quantitative equivalence of two regions $\neg\text{NEG}_{(\alpha,\beta)}(\pi_D|\pi_{\{c_1,c_2\}})$ and $\neg\text{NEG}_{(\alpha,\beta)}(\pi_D|\pi_{\mathbf{C}})$ does not imply the qualitative equivalence of them. They lead to general decision for two different sets of objects.

3.3 Constructing Decision-Preserved Reducts in the Decision-Theoretic Model

Constructing a reduct for decision preservation can apply any traditional methods, for example, the methods based on the discernibility matrix [14]. Both rows and columns of the matrix correspond to the equivalence classes defined by $\mathbf{C}$. An element of the matrix is the set of all attributes that distinguish the corresponding pair of equivalence classes. Namely, the matrix element consists of all attributes on which the corresponding two equivalence classes have distinct values and, thus, distinct decision making. A discernibility matrix is symmetric. The elements of a positive decision-based discernibility matrix $M_{D_{\text{POS}}}$ and a general decision-based discernibility matrix $M_{D_{\text{GEN}}}$ are defined as follows. For any two equivalence classes $[x]_{\mathbf{C}}$ and $[y]_{\mathbf{C}}$,

$$M_{D_{\text{POS}}}([x]_{\mathbf{C}},[y]_{\mathbf{C}}) = \{a \in \mathbf{C} \mid I_a(x) \neq I_a(y) \wedge D_{\text{POS}_{(\alpha,\beta)}}([x]_{\mathbf{C}}) \neq D_{\text{POS}_{(\alpha,\beta)}}([y]_{\mathbf{C}})\};$$

$$M_{D_{\text{GEN}}}([x]_{\mathbf{C}},[y]_{\mathbf{C}}) = \{a \in \mathbf{C} \mid I_a(x) \neq I_a(y) \wedge D_{\text{GEN}_{(\alpha,\beta)}}([x]_{\mathbf{C}}) \neq D_{\text{GEN}_{(\alpha,\beta)}}([y]_{\mathbf{C}})\}.$$

Skowron and Rauszer showed that the set of attribute reducts are in fact the set of prime implicants of the reduced disjunctive form of the discernibility function [14]. Thus, a positive decision reduct is a prime implicant of the reduced disjunctive form of the discernibility function

$$f(M_{D_{\text{POS}}}) = \bigwedge\{\bigvee(M_{D_{\text{POS}}}([x]_{\mathbf{C}},[y]_{\mathbf{C}}))|\forall x, y \in U \ (M_{D_{\text{POS}}}([x]_{\mathbf{C}},[y]_{\mathbf{C}}) \neq \emptyset)\}. \tag{13}$$

The expression $\bigvee(M_{D_{\text{POS}}}([x]_{\mathbf{C}},[y]_{\mathbf{C}}))$ is the disjunction of all attributes in $M_{D_{\text{POS}}}([x]_{\mathbf{C}},[y]_{\mathbf{C}})$, indicating that the pair of equivalence classes $[x]_{\mathbf{C}}$ and $[y]_{\mathbf{C}}$ can be distinguished by any attribute in M. The expression $\bigwedge\{\bigvee(M_{D_{\text{POS}}}([x]_{\mathbf{C}},[y]_{\mathbf{C}}))\}$ is the conjunction of all $\bigvee(M_{D_{\text{POS}}}([x]_{\mathbf{C}},[y]_{\mathbf{C}}))$, indicating that the family of discernible pairs of equivalence classes can be distinguished by a set of attributes satisfying $\bigwedge\{\bigvee(M_{D_{\text{POS}}}([x]_{\mathbf{C}},[y]_{\mathbf{C}}))\}$. In order to derive the reduced disjunctive form, the discernibility function $f(M_{D_{\text{POS}}})$ is transformed by using the absorption and distribution laws. Accordingly, finding the set of reducts can be modelled based on the manipulation of a Boolean function.

Analogically, a general decision reduct is a prime implicant of the reduced disjunctive form of the discernibility function

$$f(M_{D_{\text{GEN}}}) = \bigwedge\{\bigvee(M_{D_{\text{GEN}}}([x]_{\mathbf{C}},[y]_{\mathbf{C}}))|\forall x, y \in U \ (M_{D_{\text{GEN}}}([x]_{\mathbf{C}},[y]_{\mathbf{C}}) \neq \emptyset)\}. \tag{14}$$

Y. Zhao, S.K.M. Wong, and Y. Yao

Table 5. A discernibility matrix defined by Equation 13

	$[o_1]_\mathbf{C}$	$[o_2]_\mathbf{C}$	$[o_3]_\mathbf{C}$	$[o_4]_\mathbf{C}$	$[o_5]_\mathbf{C}$	$[o_6]_\mathbf{C}$	$[o_7]_\mathbf{C}$	$[o_8]_\mathbf{C}$	$[o_9]_\mathbf{C}$
$[o_1]_\mathbf{C}$	$\emptyset$								
$[o_2]_\mathbf{C}$	{2,4}	$\emptyset$							
$[o_3]_\mathbf{C}$	{1,5,6}	{1,2,4,5,6}	$\emptyset$						
$[o_4]_\mathbf{C}$	{4,5}	{2,5}	$\emptyset$	$\emptyset$					
$[o_5]_\mathbf{C}$	{1,2,5}	$\emptyset$	{2,6}	{1,2,4}	$\emptyset$	$\emptyset$			
$[o_6]_\mathbf{C}$	{2,4}	$\emptyset$	{1,2,4,5,6}	{2,5}	$\emptyset$	$\emptyset$			
$[o_7]_\mathbf{C}$	{1,2,3,6}	{1,3,4,6}	{2,3,5}	C	{3,5,6}	{1,3,4,6}	$\emptyset$		
$[o_8]_\mathbf{C}$	{2,4}	$\emptyset$	{1,2,4,5,6}	{2,5}	$\emptyset$	$\emptyset$	{1,3,4,6}	$\emptyset$	
$[o_9]_\mathbf{C}$	{1,2,5}	$\emptyset$	{2,6}	{1,2,4}	$\emptyset$	$\emptyset$	{3,5,6}	$\emptyset$	$\emptyset$

Example 6. Based on Table 3, we can construct a discernibility matrix defined by Equation 13. According to this discernibility matrix, we can find and verify the following reducts for positive decision preservation: $\{c_1, c_2, c_5\}$, $\{c_2, c_3, c_5\}$, $\{c_2, c_4, c_5\}$, $\{c_2, c_4, c_6\}$, $\{c_2, c_5, c_6\}$, $\{c_4, c_5, c_6\}$, and $\{c_1, c_2, c_3, c_4\}$.

3.4 Constructing Region-Preserved Reducts in the Decision-Theoretic Model

Based on the non-monotocity of the regions, the construction for region-based reduct is not trivial. One needs to exhaustively search all subsets of **C** in order to find the global optimal attribute set that induces the largest probabilistic positive region or the largest probabilistic non-negative region.

Comparing to the global search, a more practical means is to find a *local* optimal solution with respect to an attribute set, which keeps the same probabilistic positive region as **C** does. That is, for the local optimization, if the probabilistic positive regions are equivalent regarding two attribute sets A and **C** for all objects in the universe, then there exists a subset of A which is a local optimal positive region reduct regarding A. By this means, we can somehow solve an optimization problem based on the solution of a decision problem. Specifically, if we can find one, we can then check all its subsets for a local optimal positive region reduct, which is an optimization problem. Similarly, we can construct a local optimal non-negative region reduct.

Constructing an attribute set keeps the same probabilistic positive region also can apply the discernibility matrix methods. Both rows and columns of the matrix correspond to the equivalence classes defined by **C**. An element of the matrix is the set of all attributes that distinguish the corresponding pair of equivalence classes. Namely, the matrix element consists of all attributes on which the corresponding two equivalence classes belong to distinct regions. The elements of a positive region-based discernibility matrix M_{POS} and a general region-based discernibility matrix M_{GEN} are defined as follows. For any two equivalence classes $[x]_\mathbf{C}$ and $[y]_\mathbf{C}$,

$$M_{\text{POS}}([x]_\mathbf{C}, [y]_\mathbf{C}) = \{a \in \mathbf{C} \mid I_a(x) \neq I_a(y) \wedge D_{\text{POS}_{(\alpha,\beta)}}([x]_\mathbf{C}) \neq D_{\text{POS}_{(\alpha,\beta)}}([y]_\mathbf{C})\};$$

$$M_{\text{GEN}}([x]_\mathbf{C}, [y]_\mathbf{C}) = \{a \in \mathbf{C} \mid I_a(x) \neq I_a(y) \wedge D_{\text{GEN}_{(\alpha,\beta)}}([x]_\mathbf{C}) \neq D_{\text{GEN}_{(\alpha,\beta)}}([y]_\mathbf{C})\}.$$

A prime implicant of the reduced disjunctive form of the discernibility function is an attribute set that keeps the same probabilistic region as $\mathbf{C}$ does. That is,

$$f(M_{\text{POS}}) = \bigwedge \{\bigvee(M_{\text{POS}}([x]_{\mathbf{C}}, [y]_{\mathbf{C}}))| \forall x, y \in U \ (M_{\text{POS}}([x]_{\mathbf{C}}, [y]_{\mathbf{C}}) \neq \emptyset)\}. \tag{15}$$

$$f(M_{\text{GEN}}) = \bigwedge \{\bigvee(M_{\text{GEN}}([x]_{\mathbf{C}}, [y]_{\mathbf{C}}))| \forall x, y \in U \ (M_{\text{GEN}}([x]_{\mathbf{C}}, [y]_{\mathbf{C}}) \neq \emptyset)\}. \tag{16}$$

Example 7. For our running example, we can construct a discernibility matrix defined by Equation 16. Attribute sets $\{c_2, c_3\}$, $\{c_2, c_6\}$, $\{c_1, c_2, c_5\}$, $\{c_2, c_4, c_5\}$ and $\{c_4, c_5, c_6\}$ can keep the probabilistic positive region as $\mathbf{C}$ does. From $\{c_2, c_3\}$ we can verify that regarding the first four attribute sets, $\{c_2\}$ is a local optimal positive region reduct; regarding attribute set $\{c_4, c_5, c_6\}$, we verify the local optimal positive region reduct is the set itself.

Table 6. A discernibility matrix defined by Equation 16

	$[o_1]_{\mathbf{C}}$	$[o_2]_{\mathbf{C}}$	$[o_3]_{\mathbf{C}}$	$[o_4]_{\mathbf{C}}$	$[o_5]_{\mathbf{C}}$	$[o_6]_{\mathbf{C}}$	$[o_7]_{\mathbf{C}}$	$[o_8]_{\mathbf{C}}$	$[o_9]_{\mathbf{C}}$
$[o_1]_{\mathbf{C}}$	$\emptyset$								
$[o_2]_{\mathbf{C}}$	{2,4}	$\emptyset$							
$[o_3]_{\mathbf{C}}$	$\emptyset$	{1,2,4,5,6}	$\emptyset$						
$[o_4]_{\mathbf{C}}$	$\emptyset$	{2,5}	$\emptyset$	$\emptyset$					
$[o_5]_{\mathbf{C}}$	{1,2,5}	$\emptyset$	{2,6}	{1,2,4}	$\emptyset$	$\emptyset$			
$[o_6]_{\mathbf{C}}$	{2,4}	$\emptyset$	{1,2,4,5,6}	{2,5}	$\emptyset$	$\emptyset$			
$[o_7]_{\mathbf{C}}$	$\emptyset$	{1,3,4,6}	$\emptyset$	$\emptyset$	{3,5,6}	{1,3,4,6}	$\emptyset$		
$[o_8]_{\mathbf{C}}$	{2,4}	$\emptyset$	{1,2,4,5,6}	{2,5}	$\emptyset$	$\emptyset$	{1,3,4,6}	$\emptyset$	
$[o_9]_{\mathbf{C}}$	{1,2,5}	$\emptyset$	{2,6}	{1,2,4}	$\emptyset$	$\emptyset$	{3,5,6}	$\emptyset$	$\emptyset$

Conclusively, the discernibility matrix approach can solve a decision problem, i.e., to find the attribute set that keeps exactly the property the entire condition attribute set $\mathbf{C}$ has. Though, for an optimization problem, this approach is not enough. We need either to search the global optimization exhaustively, or to search the local optimization based on the results of the decision problem. Finding a decision-preservation reduct is a decision problem. The decision-preservation reducts can be constructed based on a decision-based matrix. Finding a region-preservation reduct is an optimization problem. The region-preservation reducts can be locally constructed based on a region-based matrix.

4   Conclusion

Definitions of attribute reduction in the decision-theoretic rough set model are examined in this paper, regarding both decision preservation and region preservation. While attribute construction for decision preservation can be explored by utilizing the monotonicity of attribute sets, attribute construction for region preservation cannot be done in a similar manner. Decision-based reducts can be constructed by the traditional approaches such as the ones on the discernibility matrix, while region-based reducts require exhaustive search methods for reaching the global optimization. A local optimization approach is suggested based on the discernibility matrix. Heuristics and algorithms need to be further studied.

References

1. Beaubouef, T., Petry, F.E., Arora, G.: Information-theoretic measures of uncertainty for rough sets and rough relational databases. Information Sciences 109, 185–195 (1998)
2. Beynon, M.: Reducts within the variable precision rough sets model: a further investigation. European Journal of Operational Research 134, 592–605 (2001)
3. Hu, Q., Yu, D., Xie, Z.: Information-preserving hybrid data reduction based on fuzzy-rough techniques. Pattern Recognition Letters 27, 414–423 (2006)
4. Inuiguchi, M.: Attribute reduction in variable precision rough set model. International Journal of Uncertainty, Fuzziness and Knowledge-Based Systems 14, 461–479 (2006)
5. Inuiguchi, M.: Structure-based attribute reduction in variable precision rough set models. Journal of Advanced Computational Intelligence and Intelligent Informatics 10, 657–665 (2006)
6. Katzberg, J.D., Ziarko, W.: Variable precision rough sets with asymmetric bounds. In: Ziarko, W. (ed.) Rough Sets, Fuzzy Sets and Knowledge Discovery, pp. 167–177. Springer, Heidelberg (1994)
7. Kryszkiewicz, M.: Maintenance of reducts in the variable precision rough sets model, ICS Research Report 31/94, Warsaw University of Technology (1994)
8. Mi, J.S., Wu, W.Z., Zhang, W.X.: Approaches to knowledge reduction based on variable precision rough set model. Information Sciences 159, 255–272 (2004)
9. Miao, D.Q., Hu, G.R.: A heuristic algorithm for reduction of knowledge. Chinese Journal of Computer Research and Development 36, 681–684 (1999)
10. Pawlak, Z.: Rough sets. International Journal of Computer and Information Sciences 11, 341–356 (1982)
11. Pawlak, Z.: Rough Sets: Theoretical Aspects of Reasoning About Data. Kluwer Academic Publishers, Boston (1991)
12. Pawlak, Z., Wong, S.K.M., Ziarko, W.: Rough sets: probabilistic versus deterministic approach. International Journal of Man-Machine Studies 29, 81–95 (1988)
13. Skowron, A.: Boolean reasoning for decision rules generation. In: Komorowski, J., Raś, Z.W. (eds.) ISMIS 1993. LNCS, vol. 689, pp. 295–305. Springer, Heidelberg (1993)
14. Skowron, A., Rauszer, C.: The discernibility matrices and functions in information systems. In: Slowiński, R. (ed.) Intelligent Decision Support, Handbook of Applications and Advances of the Rough Sets Theory. Kluwer, Dordrecht (1992)
15. Slezak, D.: Approximate reducts in decision tables. In: Proceedings of Information Processing and Management of Uncertainty, pp. 1159–1164 (1996)
16. Slezak, D.: Normalized decision functions and measures for inconsistent decision tables analysis. Fundamenta Informaticae 44, 291–319 (2000)
17. Swiniarski, R.W.: Rough sets methods in feature reduction and classification. International Journal of Applied Mathematics and Computer Science 11, 565–582 (2001)

18. Wang, G.Y., Zhao, J., Wu, J.: A comparative study of algebra viewpoint and information viewpoint in attribute reduction. Foundamenta Informaticae 68, 1–13 (2005)
19. Yao, Y.Y.: Decision-theoretic rough set models. In: Yao, J., Lingras, P., Wu, W.-Z., Szczuka, M.S., Cercone, N.J., Ślęzak, D. (eds.) RSKT 2007. LNCS (LNAI), vol. 4481, pp. 1–12. Springer, Heidelberg (2007)
20. Yao, Y.Y., Wong, S.K.M.: A decision theoretic framework for approximating concepts. International Journal of Man-machine Studies 37, 793–809 (1992)
21. Yao, Y.Y., Wong, S.K.M., Lingras, P.: A decision-theoretic rough set model. In: Ras, Z.W., Zemankova, M., Emrich, M.L. (eds.) ISMIS 1991. LNCS, vol. 542, pp. 17–24. Springer, Heidelberg (1991)
22. Yao, Y.Y., Zhao, Y.: Attribute reduction in decision-theoretic rough set models. Information Sciences 178, 3356–3373 (2008)
23. Zhang, W.X., Mi, J.S., Wu, W.Z.: Knowledge reduction in inconsistent information systems. Chinese Journal of Computers 1, 12–18 (2003)
24. Ziarko, W.: Variable precision rough set model. Journal of Computer and System Sciences 46, 39–59 (1993)

Author Index